ISNM 107:
International Series of Numerical Mathematics
Internationale Schriftenreihe zur Numerischen Mathematik
Série Internationale d'Analyse Numérique
Vol. 107

Edited by
K.-H. Hoffmann, München; H. D. Mittelmann, Tempe;
J. Todd, Pasadena

Springer Basel AG

Optimization, Optimal Control and Partial Differential Equations

First Franco-Romanian Conference, Iasi, September 7–11, 1992

Edited by

V. Barbu
J. F. Bonnans
D. Tiba

Springer Basel AG

Editors

Prof. Viorel Barbu
Faculty of Mathematics
University of Iasi
Bld Copou 11
6600 Iasi
Romania

Dr. J. Frédéric Bonnans
INRIA-Rocquencourt
Domaine de Voluceau
BP 105
78153 Le Chesnay Cedex
France

Dr. Dan Tiba
Institute of Mathematics
Romanian Academy of Science
Bld Pacii 212
79622 Bucharest
Romania

A CIP catalogue record for this book is available from the Library of Congress,
Washington D.C., USA

Deutsche Bibliothek Cataloging-in-Publication Data

Optimization, optimal control and partial differential equations:
first Franco-Romanian conference, Iasi, September 7–11,
1992 / ed. by V. Barbu ... – Basel ; Boston ; Berlin : Birkhäuser,
1992
  (International series of numerical mathematics ; Vol. 107)
  ISBN 978-3-0348-9704-4   ISBN 978-3-0348-8625-3 (eBook)
  DOI 10.1007/978-3-0348-8625-3
NE: Barbu, Viorel [Hrsg.]; GT

© 1992 Springer Basel AG
Originally published by Birkhäuser Verlag Basel in 1992
Softcover reprint of the hardcover 1st edition 1992

ISBN 978-3-0348-9704-4

Contents

VI

Organization

Organizers

Institut National de Recherche en Informatique et en Automatique,
    Rocquencourt, France
Institute of Mathematics, Bucharest, Romania

Sponsorship

Société de Mathématiques Appliquées et Industrielles, France
University of Iasi, Romania

Funding Support

MRT   (Ministère de la Recherche et de la Technologie)
UNESCO   (United Nations Educational, Scientific and Cultural Organization)

Programme Committee

J.P. Aubin	University of Paris IX, France
V. Barbu	University of Iasi (**co-Chairperson**), Romania
A. Bensoussan	INRIA and University of Paris IX, France
J.F. Bonnans	INRIA - Rocquencourt, France
D. Cioranescu	University of Paris VI, France
G. Gussi	Institute of Mathematics, Bucharest, Romania
A. Halanay	University of Bucharest, Romania
J.P. Hiriart-Urruty	University of Toulouse III, France
J.L. Lions	CNES and Collège de France (**co-Chairperson**)
D. Tiba	Institute of Mathematics, Bucharest, Romania

Organization Committee

D. Tiba	Institute of Mathematics, Bucharest (**Chairperson**)
J.F. Bonnans	INRIA - Rocquencourt
D. Cioranescu	University of Paris VI

Organization

C. Genest	INRIA - Rocquencourt
C. Thenault	INRIA - Rocquencourt

INRIA - Rocquencourt
Domaine de Voluceau - BP 105
78153 Le Chesnay Cedex - France

I. M.   **Institutul de Matematica, Academia Româna**
Bd Pacii 220, Bucuresti 79622 - Romania

PREFACE

This book collects research papers presented in the **First Franco-Romanian Conference on Optimization, Optimal Control and Partial Differential Equations** held at Iasi on 7-11 september 1992.

The aim and the underlying idea of this conference was to take advantage of the new social developments in East Europe and in particular in Romania to stimulate the scientific contacts and cooperation between French and Romanian mathematicians and teams working in the field of optimization and partial differential equations.

This volume covers a large spectrum of problems and result developments in this field in which most of the participants have brought notable contributions. The following topics are discussed in the contributions presented in this volume.

1 - **Variational methods in mechanics and physical models**

Here we mention the contributions of D. Cioranescu, P. Donato and H.I. Ene (fluid flows in dielectric porous media), R. Stavre (the impact of a jet with two fluids on a porous wall), C. Lefter and D. Motreanu (nonlinear eigenvalue problems with discontinuities), I. Rus (maximum principles for elliptic systems), and on asymptotic properties of solutions of evolution equations (R. Latcu and M. Megan, R. Luca and R. Morozanu, R. Faure).

2 - The controllability of infinite dimensional and distributed parameter systems with the contribution of P. Grisvard (singularities and exact controllability for hyperbolic systems), G. Geymonat, P. Loreti and V. Valente (exact controllability of a shallow shell model), C. Benard, M. Rosset-Louerat, X.F. Wang (identification of position of melting front). N. Burq and G. Lebeau (local approach to control of plate equation), C. Varsan (bounded solutions for controlled hyperbolic systems), J.F. Pommaret (controllability and turbulence),

3 - The H_∞ control problem with the contributions of V. Barbu on H_∞ control problem for boundary control systems of hyperbolic type and A. Bensoussan and P. Bernhard on robust control and differential games .

4 - The dynamic programming method and optimal control is present in the works of N. Caroff and H. Frankowska (Hamilton-Jacobi equation in optimal control), S. Mirica (verification theorem of dynamic programming type in optimal control), H. Frankowska and M. Quincampoix (Isaac's equation for value functions of differential games), A. Khoukhi and Y. Hamam (optimal control for robotic manipulators), C. Lobry (application of control theory to environmental problems), J. Morozan (optimal stochastic control).

5 - Optimal control of nonlinear partial differential equations
We mention here the works of F. Bonnans and E. Casas on boundary optimal control of nonlinear elliptic systems with state constraints, D. Tiba (controllability properties for elliptic systems, the fictitious domain method and optimal shape design problems), R.

Tahraoui (optimal control of elliptic equations and applications), V. Maksimov (inverse problems for variational inequalities), J.A. Bello, E. Fernandez-Cara and J.C. Simon (differentiation with respect to domain in a Navier-stokes flow).

6 - **Mathematical programming and nonsmooth optimization** with the contributions of F. Precupanu (scalar minmax properties in vectorial optimization), P. Loridan and J. Morgan (regularization for two-level optimization problem), Y. Sonntag (continuity of optimal value function with respect to the set of constraints), logconcavity and integral inequalities (R. Michel and M. Volle), P. Laborde (numerical solution of free boundary problems).

Finally we thank the members of the Program Commitee, Claudie Thénault and Christine Genest for their active participation to the organization of this conference, SMAI and the university of Iasi who accepted to sponser the meeting, and the Ministère de la Recherche et Technologie and UNESCO for their financial support.

V. Barbu, J.F. Bonnans, D. Tiba

Variational methods in mechanics
and physical models

International Series of Numerical Mathematics, Vol. 107, © 1992 Birkhäuser Verlag Basel

FLUID FLOWS IN DIELECTRIC POROUS MEDIA

Doina CIORANESCU, Patrizia DONATO and Horia I. ENE

1. INTRODUCTION AND FORMULATION OF THE PROBLEM.

Let Ω be an open set in $\mathbb{R}^n$ $(n \geq 2)$ and $Y = [0, l_1[\times .. \times [0, l_n[$ the representative cell. Denote by T an open subset of Y with smooth boundary ∂T, such that $\overline{T} \subset Y$. Denote by $\tau(\varepsilon\overline{T})$ the set of all translated images of $\varepsilon\overline{T}$ of the form $\varepsilon(kl + \overline{T})$, $k \in Z^n$, $kl = (k_1 l_1, .., k_n l_n)$.

We make the following assumption:

The holes $\tau(\varepsilon\overline{T})$ do not intersect the boundary $\partial\Omega$.

Denote now by T_ε the set of the holes contained in Ω and set

$$\Omega_\varepsilon = \Omega \backslash \overline{T_\varepsilon}.$$

By this construction Ω_ε is periodically perforated by holes of size of the same order as the period (see Figure 1).

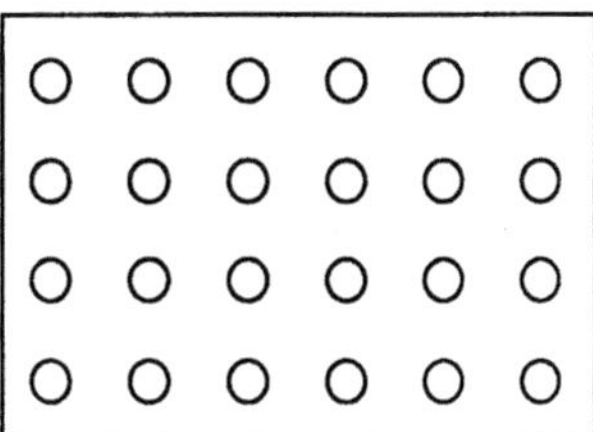

Figure 1

In the sequel we use the notations
- $Y^* = Y \backslash \overline{T}$,
- $|\omega|$ = the Lebesgue mesure of ω (n-dimensional if ω is a n-dimensional open set, $(n-1)$-dimensional if ω is a curve),
- χ_ω = the characteristic function of ω,
- $\tilde{v}$ = the zero extension to the whole Ω for any fonction v defined on Ω_ε,
- $\mathcal{M}_\omega(\phi) = \dfrac{1}{|\omega|} \displaystyle\int_\omega \phi(x)dx$, the mean value of ϕ on ω.

Consider in Ω_ε the Stokes system with non homogeneous slip boundary condition on ∂T_ε

$$(1.1) \qquad \begin{cases} -\Delta u^\varepsilon + \nabla p^\varepsilon = f & \text{in } \Omega_\varepsilon \\[2mm] \operatorname{div} u^\varepsilon = 0 & \text{in } \Omega_\varepsilon \\[2mm] u^\varepsilon = 0 & \text{on } \partial\Omega \\[2mm] -p^\varepsilon \cdot n + \dfrac{\partial u^\varepsilon}{\partial n} + \alpha\varepsilon^\gamma u^\varepsilon = g^\varepsilon & \text{on } \partial T_\varepsilon, \end{cases}$$

where $u^\varepsilon = (u_1^\varepsilon, .., u_n^\varepsilon)$ stands for the velocity field, p^ε for the pression, f is the field of exterior body forces, α and γ are positive constants , n is the exterior unit normal to Ω_ε.

The boundary condition on ∂T_ε means that the stress vector gives rice firstly, to a braking phenomenon due to the presence of the term $\alpha\varepsilon^\gamma u^\varepsilon$ and secondly, to a proportionality with the exterior surface forces due to the presence of g^ε.

Assume that the data f and g^ε satisfy

$$(1.2) \qquad \begin{cases} i) \ \ f \in (L^2(\Omega))^n \\[2mm] ii) \ \ g^\varepsilon = g^0(x) + g(\dfrac{x}{\varepsilon}), \quad \text{with } g^0 \in (H^1(\Omega))^n \text{ and} \\[2mm] \qquad g \in (L^2(\partial T))^n \ \ Y\text{-periodic such that } \mathcal{M}_{\partial T}(g_i) \neq 0. \end{cases}$$

Let introduce the space

$$V_\varepsilon = \{v/v \in (H^1(\Omega_\varepsilon))^n, \ v = 0 \text{ on } \partial\Omega\}.$$

The variational formulation of system (1.1) is then the following:

$$\begin{cases} \text{Find } u^\varepsilon \in V_\varepsilon, \ p^\varepsilon \in L^2(\Omega_\varepsilon) \text{ such that} \\[3mm] \displaystyle\int_{\Omega_\varepsilon} \nabla u^\varepsilon \nabla\varphi \, dx + \alpha\varepsilon^\gamma \int_{\partial T_\varepsilon} u^\varepsilon\varphi \, d\sigma - \int_{\Omega_\varepsilon} p^\varepsilon \operatorname{div}\varphi \, dx = \\[4mm] \qquad\qquad = \displaystyle\int_{\Omega_\varepsilon} f\,\varphi \, dx + \int_{\partial T_\varepsilon} g^\varepsilon\varphi \, d\sigma, \quad \forall\varphi \in V_\varepsilon, \\[4mm] \displaystyle\int_{\Omega_\varepsilon} u^\varepsilon \nabla\varphi \, dx = 0, \quad \forall\varphi \in V_\varepsilon. \end{cases}$$

Classical results give the existence of a unique solution of this problem. We are interested to give the asymptotic behaviour of $(u^\varepsilon, p^\varepsilon)$ when $\varepsilon \to 0$. We are here in the classical homogenization framework.

In [7] H.I. Ene and E. Sanchez-Palencia studied the Stokes flow in a periodic porous medium with Dirichlet conditions on the boundary of the holes. The limit law describing the homogenized medium is a Darcy's law. In [3] D. Cioranescu and P. Donato consider the Laplace equation with non homogeneous Fourier conditions on the boundary of the holes containing a term of type $\alpha \varepsilon^\gamma u^\varepsilon$. Following the values of γ, several a priori estimates are obtained which lead to different limit laws.

In the case we present here (i.e. Stokes system with non homogeneous Fourier boundary conditions) we obtain at the limit, following the values of γ, a Darcy's law ($\gamma < 1$), a Brinkmann equation ($\gamma = 1$) or the Stokes equation ($\gamma > 1$). This phenomenon was already observed by C. Conca [6] when studying the Stokes equation with homogeneous Fourier bondary conditions. It was also noticed by G. Allaire who considers the Stokes equation in a perforated domain with holes of size r_ε with $r_\varepsilon \ll \varepsilon$. The boundary conditions on the holes are either of Dirichlet type [1] or of slip type [2]. In this situation it is the geometry of the domain, more precisely the size r_ε which determines the type of the limit law.

In this paper we give the main results we obtained for system (1.1) and their physical interpretation. We refer the reader to [4] for complete proofs and for furher results and comments.

2. RESULTS.

I. Case $\gamma < 1$ (Darcy's law).

From system (1.1) we have the following estimates:

$$\|\varepsilon^\gamma u^\varepsilon\|_{(L^2(\Omega_\varepsilon))^n} \leq c$$
$$\|\varepsilon^{\frac{1+\gamma}{2}} \nabla u^\varepsilon\|_{(L^2(\Omega_\varepsilon))^n} \leq c$$

where c is a constant independent of ε.

Hence, up to a subsequence

$$(2.1) \qquad \varepsilon^\gamma \widetilde{u^\varepsilon} \rightharpoonup u \quad \text{in } (L^2(\Omega))^n \text{ weakly.}$$

Following along the lines of [3], let us introduce the linear form μ_h^ε defined by

$$< \mu_h^\varepsilon, \varphi > = \varepsilon \int_{\partial T_\varepsilon} h(\frac{x}{\varepsilon})\varphi \, d\sigma, \quad \forall \varphi \in H^1(\Omega_\varepsilon)$$

where $h \in L^2(\partial T)$.

From lemma 3.1 of [3] we can easily prove the following proposition:

PROPOSITION 2.1. *Let $\{v^\varepsilon\} \subset H^1(\Omega_\varepsilon)$ be a sequence satisfying*

$$\varepsilon^\gamma \widetilde{v^\varepsilon} \rightharpoonup u \quad \text{in } L^2(\Omega) \text{ weakly}$$

and suppose there exists $\delta \in]0,1[$ such that

$$\|\varepsilon^\delta \nabla v^\varepsilon\|_{L^2(\Omega_\varepsilon)} \leq c$$

with c a constant independent of ε. Then

$$< \mu_h^\varepsilon, v^\varepsilon > \; \rightarrow \; < \mu_h, v >$$

where $\mu_h = \dfrac{1}{|Y|} \displaystyle\int_{\partial T} h(y)\, ds$.

Applying lemma 5.1 of [6] one has an extension P^ε of the pressure p^ε such that

$$\|P^\varepsilon\|_{L^2(\Omega)} \leq c\, \varepsilon^{-1}.$$

Consequently, up to a subsequence

$$(2.2) \qquad\qquad\qquad \varepsilon P^\varepsilon \rightharpoonup P \quad \text{in } L^2(\Omega) \text{ weakly.}$$

We can now state the homogenization result:

THEOREM 2.2. *The limit fonction u given by (2.1) satisfies*

$$u = \frac{1}{\alpha\mu_1}(\mu_1 g^0 + \mu_g - \nabla P)$$

with $\mu_1 = \dfrac{|\partial T|}{|Y|}$ and $\mu_g = \dfrac{1}{|Y|}\displaystyle\int_{\partial T} g(y)ds$.

II. Case $\gamma = 1$ (Brinkmann's equation).

When $\gamma = 1$ we derive from (1.1) the estimate

$$\|\varepsilon u^\varepsilon\|_{(H^1(\Omega_\varepsilon))^n} \leq c$$

with c independent of ε. Then there exist (see D. Cioranescu-J. Saint Jean Paulin [5]) extension operators $Q^\varepsilon \in \mathcal{L}(H^1(\Omega_\varepsilon); H^1(\Omega))$ such that

$$(2.3) \qquad\qquad\qquad Q^\varepsilon(\varepsilon u^\varepsilon) \rightharpoonup u \quad \text{in } (H_0^1(\Omega))^n \text{ weakly.}$$

Moreover, convergence (2.2) still holds. Let χ_Λ, q_Λ be solution of the following Stokes system given on the reference cell Y:

$$(2.4) \quad \begin{cases} -\Delta\chi_\Lambda + \nabla q_\Lambda = 0 & \text{in } Y^* \\ \operatorname{div}\chi_\Lambda = 0 & \text{in } Y^* \\ -\dfrac{\partial(\chi_\Lambda - \Lambda y)}{\partial n} + q_\Lambda \cdot n = 0 & \text{on } \partial T, \\ \chi_\Lambda \ \ Y\text{-periodic} \end{cases}$$

for any matrix $\Lambda \in \mathbb{R}^{n^2}$.

Define

$$w_\Lambda = \Lambda y - \chi_\Lambda.$$

Let $Q \in \mathcal{L}(H^1(Y^*); H^1(Y^*))$ be an extension operator defined over Y as constructed in [5] and set

$$w_\Lambda^\varepsilon(x) = \varepsilon Q(w_\Lambda - \Lambda y)(\frac{x}{\varepsilon}) + \Lambda x, \quad x \in \Omega.$$

Then the following convergences hold:

$$w_\Lambda^\varepsilon \rightharpoonup \Lambda x \quad \text{in } (H^1_{loc}(\mathbb{R}^n))^n \text{ weakly}$$
$$\nabla w_\Lambda^\varepsilon \rightharpoonup \Lambda \quad \text{in } (L^2(\Omega))^{n^2} \text{ weakly}.$$

Analogously, if we define

$$q_\Lambda^\varepsilon(x) = \widetilde{q_\Lambda}(\frac{x}{\varepsilon}), \quad x \in \Omega,$$

we have

$$q_\Lambda^\varepsilon \rightharpoonup \mathcal{P}_\Lambda \quad \text{in } L^2(\Omega) \text{ weakly}$$

where $\mathcal{P}_\Lambda$ is the mean value of $\widetilde{q_\Lambda}$ over Y. Introduce finally

$$\eta_\Lambda^\varepsilon = \widetilde{\nabla w_\Lambda^\varepsilon}$$

which, thanks to (2.4), satisfies the equation

$$-\operatorname{div}\eta_\Lambda^\varepsilon + \nabla q_\Lambda^\varepsilon = 0 \quad \text{in}\Omega.$$

Due to the periodicity, one has the convergence

$$\eta_\Lambda^\varepsilon \rightharpoonup \mathcal{A}_\Lambda \quad \text{in } (L^2(\Omega))^{n^2} \text{ weakly}.$$

Moreover, $\mathcal{A}_\Lambda$ and $\mathcal{P}_\Lambda$ being linear in Λ one has

$$(2.5) \qquad \begin{cases} \mathcal{A}_\Lambda = a_{ijkh}\lambda_{kh} \\ \mathcal{P}_\Lambda = \mathcal{P}_{kh}\lambda_{kh}. \end{cases}$$

We can give a more precise form to the coefficients a_{ijkh}. To do that, remark that system (2.4) defines n^2 solutions $\{\chi^{kh}, q^{kh}\}k, h = 1, ..., n$ of

$$(2.6) \qquad \begin{cases} -\Delta\chi^{kh} + \nabla q^{kh} = 0 \quad \text{in } Y^* \\ \operatorname{div}\chi^{kh} = 0 \quad \text{in } Y^* \\ -\dfrac{\partial(\chi^{kh} - \Pi^{kh})}{\partial n} + q^{kh} \cdot n = 0 \quad \text{on } \partial T \\ \chi^{kh} \ Y\text{-periodic} \end{cases}$$

where $\Pi^{kh} = (\Pi_i^{kh})_i$ with $\Pi_i^{kh} = \delta_{ki}y_h$.

Then,(2.5) and (2.6) yield the formulae

$$a_{ijkh} = \int_{Y^*} \frac{\partial}{\partial y_l}\left(\chi^{kh} - \Pi^{kh}\right)\frac{\partial}{\partial y_l}\left(\chi^{ij} - \Pi^{ij}\right) dx.$$

Let now give the homogenization result.

THEOREM 2.3. *The fonction u defined by (2.3) satisfies the equation*

$$\begin{cases} -\dfrac{\partial}{\partial x_j}\left[(a_{ijkh} - \mathcal{P}_{ij}\delta_{kh})\dfrac{\partial u_k}{\partial x_h}\right] + \mu_\alpha u_i = \mu_1 g_i^0 + \mu_{g_i} \quad \text{in } \Omega \\ u = 0 \quad \text{on } \partial\Omega. \end{cases}$$

III. Case $\gamma > 1$ (Equation de Stokes).

In this case convergences (2.2) and (2.3) still hold and we can prove the following result:

THEOREME 2.4. *The fonction u defined by (2.3) is solution of the equation*

$$\begin{cases} -\dfrac{\partial}{\partial x_j}\left[(a_{ijkh} - \mathcal{P}_{ij}\delta_{kh})\dfrac{\partial u_k}{\partial x_h}\right] = \mu_1 g_i^0 + \mu_{g_i} \quad \text{in } \Omega \\ u = 0 \quad \text{on } \partial\Omega. \end{cases}$$

IV. Variant of system (1.1) (Darcy's law).

We can also consider the Stokes equation with a slightly different slip condition

$$\begin{cases} -\Delta u^\varepsilon + \nabla p^\varepsilon = f & \text{in } \Omega_\varepsilon \\ \operatorname{div} u^\varepsilon = 0 & \text{in } \Omega_\varepsilon \\ u^\varepsilon = 0 & \text{sur } \partial\Omega \\ u^\varepsilon \cdot n = 0 & \text{on } \partial T_\varepsilon \\ \dfrac{\partial u^\varepsilon}{\partial n} \cdot \tau + \alpha\varepsilon^\gamma u^\varepsilon \cdot \tau = g^\varepsilon \cdot \tau & \text{on } \partial T_\varepsilon, \end{cases}$$

where τ is the unit tangent to Ω_ε. In this case, for any value of γ, we obtain always at the limit a Darcy's type law (see [4] for details).

3. RELATED MECHANICAL MODELS.

Problem (1.1) describes the flow of an incompressible viscous fluid through a porous medium under the action of an exterior electric field. One generally knows (cf. J. R. Melcher [8]), that the electric surface charges act on the boundary between the solid and fluid part of the medium. These charges give rice to a double layer which permits the slip of the fluid.

The boundary condition on ∂T_ε can be rewritten under the form

$$\sigma_{ij}^\varepsilon \cdot n_j + \alpha\varepsilon^\gamma u_i^\varepsilon = g_i^\varepsilon.$$

This means that the stress vector $\sigma_{ij}^\varepsilon \cdot n_j$ induces a slowing effect on the motion of the fluid, expressed by the coefficient $\alpha\varepsilon^\gamma$. Moreover, if there are exterior forces like, for instance, an electric field, then the non homogeneity of the boundary condition on the holes is expressed in terms of surface charges contained in g^ε. We know that in a periodic heterogeneous medium the electric field can be obtained by standard homogenization and, consequently, g^ε has the form (1.2)ii (cf. E. Sanchez-Palencia [9]). In all the cases studied here, at the limit appear additional terms issued from g^ε. Let us point out that these terms, usually introduced by the physicists as a result of observations, are obtained rigorously by the homogenization method.

In the case $\gamma < 1$ we obtain a Darcy's law with the additinal terms μ_g and μ_1. The braking of the motion beeing important, one has a slow flow. It is interesting to note that the term $\dfrac{1}{\alpha\mu_1}$ shows that the viscosity of the fluid increases.

The case $\gamma = 1$ is a critical one. We have at the limit a Brinkmann type law: the slowing effect is not too important.

In the third case, when $\gamma > 1$, we get at the limit the Stokes equation with supplementary terms. The fluid behaves like a free fluid. The slip beeing quite important the behaviour of the fluid is not affected by the presence of solid inclusions.

We have also to mention that in all the cases, at the limit, the fluid is not any more incompressible.

To conclude, a last remark deals with the differences between the boundary condition in system (1.1) and the boundary conditions used by C. Conca [6] and G. Allaire [2]. They studied purely mechanical slip conditions without exterior contributions whereas we are in the presence of an applied electric field which is at the origin of the slip of the fluid.

REFERENCES

[1] G. ALLAIRE, Homogenization of Navier-Stokes equations in open sets perforated with tiny holes. Arch. Rat. Mech. Anal.,6 (1989), 497-537.

[2] G. ALLAIRE, Homogenization of the Navier-Stokes equations with a slip boundary condition. Comm. Pure Appl. Math., XLIV, 6 (1991), 605-642.

[3] D. CIORANESCU and P. DONATO, Homogénéisation du problème du Neumann non homogène dans des ouverts perforés, Asymptotic Analysis, 1 (1988), 115-138.

[4] D. CIORANESCU, P. DONATO and H. I. ENE, to appear.

[5] D. CIORANESCU and J. SAINT JEAN PAULIN,Homogenization in open sets with holes, J. Math. Anal. Appl.,71 (1979), 590-607.

[6] C. CONCA, On the application of the homogenization theory to a class of problems arising in fluid mechanics, Journal Math. pures et Appl., 64 (1985), 31-75

[7] H. I. ENE and E. SANCHEZ - PALENCIA, Equation et phènomènes de surface pour l'écoulement dans un modèle de milieu poreux, Journal Mécanique, 14 (1975), 73-108.

[8] J. R. MELCHER, *Continuum Electromechanics*, MIT Press, Cambridge, Massachusetts and London, England (1981).

[9] E. SANCHEZ - PALENCIA, *Non homogeneous Media and Vibration Theory*, Lecture Notes in Physics 127, Springer Verlag (1980).

Doina CIORANESCU
    Laboratoire Analyse Numérique
    Université Pierre et Marie Curie
    Tour 55-65, 5ème étage
    4 place Jussieu
    75252 Paris Cedex 05, France.

Patrizia DONATO
    Istituto di Matematica
    Facoltà di Scienze M.F.N.
    Università di Salerno
    84081 Baronissi (Salerno), Italy.

Horia I.ENE
    Institutul de Matematica al Academiei
    Str. Academiei 14
    P. O. Box 1-764
    70700 Bucuresti, Romania.

International Series of Numerical Mathematics, Vol. 107, © 1992 Birkhäuser Verlag Basel

The impact of a jet with two fluids on a porous wall

Ruxandra Stavre

Abstract. We study the two-dimensional flow of two incompressible, irrotational, inviscid jets incident on a porous wall. Existence and uniqueness theorems and some properties of the flow regions are established.

1. Introduction.

In this paper we consider the incompressible and irrotational flow of two ideal fluids moving towards a porous wall. The impact of one fluid on a porous wall was studied in Stavre (1991), Stavre (1990). In Jenkins and Barton (1988) and King (1990) some numerical results concerning the same problem are established. The jet problem without a permeable wall was studied in Alt et al. (1982), Alt et al. (1983) (for one fluid) and in Alt et al. (1984a), Alt et al. (1984b) (for two fluids).

The present problem has two free boundaries: the boundary between the two fluids and the boundary between the outer fluid and the air. We consider a jet of two inviscid fluids with densities d_1 and d_2, $d_1 > d_2$, exiting from a small opening of a nozzle and moving towards a porous wall at which the normal velocity is prescribed (and is less than the normal velocity on the mouth of the nozzle). We present the physical problem and the variational formulation of the problem. Then, existence and uniqueness results are proved. We also give some results about the monotonicity of the solution and we study the flow regions.

2. Mathematical formulation

We suppose that the nozzle and the porous wall have an horizontal axis of symmetry, Ox. We denote by F_i, $i=1,2$ the two flow regions and by u_i, $i=1,2$ the stream functions. We study a steady, irrotational, incompressible flow, without the presence of body forces. The first fluid with constant density d_1 is the inner fluid and the second one with constant density d_2 is the outer fluid. The normal velocity V_0 on the mouth of the nozzle and the normal velocity V_f on the porous wall are known, and $V_0 > V_f > 0$.

Using the symmetry of the motion, we can study the problem in $\{y>0\}$. We denote by "a" the distance between the nozzle and the porous wall and by $AB_1 \cup B_1B_2$ the mouth of the nozzle, $B_1(a,b_1)$, $B_2(a,b_2)$, $b_1 < b_2$. If we denote by C_i the intersections of the free boundaries l_i with the porous wall, we obtain $y_{C_i} = \dfrac{V_0}{V_f} b_i$

$i = 1,2$, by using the continuity of the stream functions on the boundary of the flow regions in $\{y > 0\}$ (see Fig.1)

On the free boundary l_1 (the boundary which separates the two fluids) the Bernoulli's law becomes:

(2.1) $d_1|\nabla u_1|^2 - d_2|\nabla u_2|^2 = (d_1 - d_2)V_0^2$,

and on the free boundary l_2:

(2.2) $|\nabla u_2| = V_0$.

Since on the mouth of the nozzle the normal velocity is V_0 and on the porous wall, V_f, by denoting:

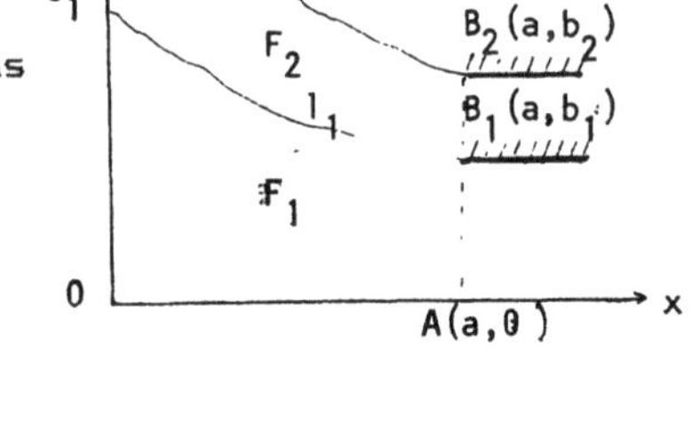

(2.3) $u = \begin{cases} d_1^{1/2} u_1 \text{ in } F_1, \\[2mm] d_2^{1/2} u_2 \text{ in } F_2, \end{cases}$

and by taking $u_1 = u_2 = 0$ on the free stream line l_1 we obtain the following formulation of the physical problem:

$$\Delta u = 0 \text{ in } F_i, \quad u > 0 \text{ in } F_1, \quad d_2^{1/2} V_0(b_1 - b_2) < u < 0 \text{ in } F_2,$$

$$u(x,0) = d_1^{1/2} V_0 b_1 \text{ on } OA$$

$$u(a,y) = d_1^{1/2} V_0 (b_1 - y) \text{ on } AB_1, \quad u(a,y) = d_2^{1/2} V_0 (b_1 - y) \text{ on } B_1B_2,$$

(2.4.) $u(0,y)=d_1^{1/2}(V_0 b_1 - V_f y)$ on OC_1, $u(0,y)=d_2^{1/2}(V_0 b_1 - V_f y)$ on $C_1 C_2$

$u=0$, $[|\nabla u|^2]=(d_1-d_2)V_0^2$ on l_1,

$u=d_2^{1/2}V_0(b_1-b_2)$, $|\nabla u|=V_0$ on l_2,

where, by $[|\nabla u|^2]$ we denoted the jump of $|\nabla u|^2$ across the free boundary.

For obtaining the variational formulation of the mechanical problem we define:

(2.5) $D=(0,a)\times(0,\infty)$,

(2.6) $f:\partial D \longrightarrow \mathbb{R}$,

$$
f=\begin{cases}
d_1^{1/2}V_0 b_1 & \text{on} \quad 0<x<a,\ y=0, \\[2ex]
d_1^{1/2}V_0(b_1-y) & \text{on} \quad x=a,\ 0<y<b_1, \\[2ex]
d_2^{1/2}V_0(b_1-y) & \text{on} \quad x=a,\ b_1<y<b_2, \\[2ex]
d_2^{1/2}V_0(b_1-b_2) & \text{on} \quad x=a,\ y>b_2, \\[2ex]
d_1^{1/2}(V_0 b_1 - V_f y) & \text{on} \quad x=0,\ 0<y<\dfrac{V_0}{V_f}b_1, \\[2ex]
d_2^{1/2}(V_0 b_1 - V_f y) & \text{on} \quad x=0,\ \dfrac{V_0}{V_f}b_1<y<\dfrac{V_0}{V_f}b_2, \\[2ex]
d_2^{1/2}V_0(b_1-b_2) & \text{on} \quad x=0,\ y>\dfrac{V_0}{V_f}b_2,
\end{cases}
$$

(2.7) $K=\{v\in H^1(D\cap B_R),\ R>0 \ / \ v=f$ on ∂D, $v\geq d_2^{1/2}V_0(b_1-b_2)$ a.e. in $D\}$.

$$
(2.8)\quad H(u)=\begin{cases}
d_1 & , \quad \text{if } u>0, \\
d_2 & , \quad \text{if } d_2^{1/2}V_0(b_1-b_2)<u\leq 0, \\
0 & , \quad \text{if } u\leq d_2^{1/2}V_0(b_1-b_2),
\end{cases}
$$

(2.9) $J(u)=\displaystyle\int_D |\nabla u|^2+V_0^2 H(u)$.

and we consider the minimum problem:

(2.10) Find u in K such that $J(u)\leq J(v)$ for all v in K.

For a solution u of (2.10) with $u_y \leq 0$ a.e. in D we can obtain as in Alt et al. (1984a): $\partial\{u>0\}\cap D = \partial\{u<0\}\cap D = l_1$ and l_1 is continuously differentiable (see Alt et al. 1984c).

Moreover $\partial\{u>d_2^{\frac{1}{2}}V_0(b_1-b_2)\}\cap D = l_2$ is analytic; this follows from Alt and Caffarelli (1981). Thus we can prove as in Alt and Caffarelli (1981) that a solution u of (2.10) satisfies the physical problem (2.4), u is in $C^{0,1}(D)$, and u is harmonic in the open sets $\{u>0\}$ and $\{d_2^{\frac{1}{2}}V_0(b_1-b_2)<u<0\}$.

Lemma 2.1.   There exists v in K such that $J(v)$ is finite.

Proof. If we take v in $H^1(D\cap B_R)$, $R>0$, $v=f$ on ∂D, $v\geq d_2^{\frac{1}{2}}V_0(b_1-b_2)$ a.e. in D and $v=d_2^{\frac{1}{2}}V_0(b_1-b_2)$ for $y\geq \dfrac{V_0}{V_f}b_2$ then $J(v)$ is finite.

Lemma 2.2.   If u is a solution of (2.10) then $u\leq d_1^{\frac{1}{2}}V_0 b_1$ a.e. in D.

Proof. If for any c in $(0,1)$ we define $v=u-c(u-d_1^{\frac{1}{2}}V_0 b_1)^+$ which is an element of K, from (2.10) we get:

$$(2.11)\quad 2c\int_D \nabla u\cdot\nabla(u-d_1^{\frac{1}{2}}V_0 b_1)^+ - c^2\int_D |\nabla(u-d_1^{\frac{1}{2}}V_0 b_1)^+|^2 \leq$$
$$V_0^2\int_D H(u-c(u-d_1^{\frac{1}{2}}V_0 b_1)^+)-H(u).$$

Since $u-c(u-d_1^{\frac{1}{2}}V_0 b_1)^+\leq u$ and H is defined by (2.8) we obtain:

$H(u-c(u-d_1^{\frac{1}{2}}V_0 b_1)^+)\leq H(u)$ and hence, by making c tend to zero it

follows: $\displaystyle\int_D \nabla u\cdot\nabla(u-d_1^{\frac{1}{2}}V_0 b_1)^+\leq 0$ or $\displaystyle\int_D |\nabla(u-d_1^{\frac{1}{2}}V_0 b_1)^+|^2\leq 0$ which

completes the proof of the lemma.

3. Existence and uniqueness

For n in $\mathbb{N}$ we define:

$$(3.1)\quad D_n=(0,a)\times(0, \dfrac{V_0}{V_f}b_2+n),$$

(3.2) $K_n = \{v \in H^1(D_n) \; / \; v=f \text{ on } \partial D_n \cap \partial D, \; v=d_2^{1/2}V_0(b_1-b_2) \text{ on}$

$$y = \frac{V_0}{V_f} b_2 + n, \quad v \geq d_2^{1/2}V_0(b_1-b_2) \text{ a.e. in } D_n\},$$

(3.3) $J_n(u_n) = \int_{D_n} |\nabla u_n|^2 + V_0^2 H(u_n),$

and we consider the problem:

(3.4) Find u_n in K_n such· that $J_n(u_n) \leq J_n(v_n)$ for all v_n in K_n. We shall prove the existence of a solution of (2.10) as a weak limit of $(u_n)_{n \geq 0}$ in $H^1(D \cap B_R)$ for any $R>0$. Since the proof is similar to the one presented in Stavre (1991) we shall give only some details.

Theorem 3.1. The problem (2.10) has at least one solution.

Proof. One can easily prove that for a fixed n, $J_n(u)$ tends to infinity when $\|u\|_{H^1(D_n)}$ tends to infinity. Since $(H(v_n^k))_{k \geq 0}$ is bounded in $L^\infty(D_n)$ we have on a subsequence $H(v_n^k) \xrightarrow[k \to \infty]{} \tilde{H}$ weakly star in $L^\infty(D_n)$, and $\tilde{H} \geq H(v_n)$ a.e. in D_n, where v_n is the weak limit in $H^1(D_n)$ of $(v_n^k)_{k \geq 0}$. Hence J_n is coercive and lower semicontinuous with respect to the weak topology in $H^1(D_n)$ on a weakly closed set K_n and, by using a Weierstrass theorem, the existence of a solution u_n of (3.4) is obtained. Moreover $(u_n)_{n \geq 0}$ is bounded in $H^1(D \cap B_R)$ for any $R>0$. If we denote by u the weak limit of $(u_n)_{n \geq 0}$ in $H^1(D \cap B_R)$ for any $R>0$ (u does not depend on R) we have to prove next that u satisfies (2.10). For $v \in K$ we define $v_n = \min(v - d_2^{1/2}V_0(b_1-b_2), \; d_1^{1/2}(V_0 b_2 - V_f y + V_f n)^+) + d_2^{1/2}V_0(b_1-b_2)$ for any n in $\mathbb{N}$ and v_n in K_n. Hence:

(3.5) $J_n(u_n) \leq J_n(\min(v - d_2^{1/2}V_0(b_1-b_2), d_1^{1/2}(V_0 b_2 - V_f y + V_f n)^+) + d_2^{1/2}V_0(b_1-b_2)).$

If $R>0$ is fixed, we have:

(3.6) $J_n(u_n) \geq \int_{D \cap B_R} |\nabla u_n|^2 + V_0^2 H(u_n).$

By using (3.5) and (3.6), we obtain:

(3.7) $\int\limits_{D\cap B_R} |\nabla u|^2 + V_0^2 H(u) \leq$

$\leq \lim\limits_{\overline{n\to\infty}} J_n(\min(v-d_2^{1/2}V_0(b_1-b_2),\ d_1^{1/2}(V_0b_2-V_f y+V_f n)^+)+d_2^{1/2}V_0(b_1-b_2)).$

If $\int\limits_D |\nabla v|^2 = \infty$ or $\int\limits_D H(v) = \infty$ we get from (3.7):

(3.8) $\int\limits_{D\cap B_R} |\nabla u|^2 + V_0^2 H(u) \leq J(v)$ for any v in K.

We suppose next that $J(v) < \infty$. Since $\min(v-d_2^{1/2}V_0(b_1-b_2),$

$d_1^{1/2}(V_0b_2-V_f y+V_f n)^+) \leq v-d_2^{1/2}V_0(b_1-b_2)$, it follows from (3.7):

(3.9) $\int\limits_{D\cap B_R} |\nabla u|^2 + V_0^2 H(u) \leq$

$\leq \lim\limits_{\overline{n\to\infty}} \int\limits_{D_n} |\nabla \min(v-d_2^{1/2}V_0(b_1-b_2),\ d_1^{1/2}(V_0b_2-V_f y+V_f n)^+)|^2 + \int\limits_D V_0^2 H(v).$

In order to obtain again (3.8), we have to prove that:

(3.10) $\lim\limits_{\overline{n\to\infty}} \int\limits_{D_n} |\nabla \min(v-d_2^{1/2}V_0(b_1-b_2),\ d_1^{1/2}(V_0b_2-V_f y+V_f n)^+)|^2 \leq$

$\leq \int\limits_D |\nabla v|^2.$

The inferior limit is equal to:

$\lim\limits_{\overline{n\to\infty}} \left(\int\limits_{D_n\cap\{v-d_2^{1/2}V_0(b_1-b_2)\leq d_1^{1/2}(V_0b_2-V_f y+V_f n)^+\}} |\nabla v|^2 \right.$

$\left. + \int\limits_{D_n\cap\{v-d_2^{1/2}V_0(b_1-b_2)>d_1^{1/2}(V_0b_2-V_f y+V_f n)^+\}} d_1 |\nabla(V_0b_2-V_f y+V_f n)^+|^2 \right) \leq$

$\leq \int\limits_D |\nabla v|^2 + d_1 V_f^2 \lim\limits_{\overline{n\to\infty}} (\text{mes}(D\cap\{v-d_2^{1/2}V_0(b_1-b_2)>d_1^{1/2}(V_0b_2-V_f y+V_f n)^+\})).$

If we denote by $A_n = D\cap\{v-d_2^{1/2}V_0(b_1-b_2)>d_1^{1/2}(V_0b_2-V_f y+V_f n)^+\}$

we get a decreasing sequence $(A_n)_{n \geq 0}$ with:

$$\text{mes } A_0 = \text{mes}(D \cap \{v - d_2^{1/2} V_0 (b_1 - b_2) > d_1^{1/2} (V_0 b_2 - V_f y)^+ \leq$$

$$\leq \text{mes}(D \cap \{v > d_2^{1/2} V_0 (b_1 - b_2)\}) \leq c \int_{D_\infty} H(v) < \infty.$$

It follows that $\varliminf_{n \to \infty} (\text{mes } A_n) = \text{mes}(\bigcap_{n=0} A_n) = 0$ and hence, the inequality (3.10) holds.

From (3.8) when R tends to infinity we obtain that u is a solution of (2.10).

In the next theorem we establish a uniqueness result.

Theorem 3.2. The solution of (2.10) is unique.

Proof. Let u_1 and u_2 be two solutions of (2.10). We shall prove first that $u_1 = u_2$ in $\{u_1 > 0\} \cap \{u_2 > 0\}$, which is an open, non empty set. Let C be an open, connected component of $\{u_1 > 0\} \cap \{u_2 > 0\}$ such that $\partial C \cap (OA \cup AB_1 \cup OC_1) = \emptyset$. Hence $\partial C \cap \partial D = \emptyset$. We define: $v = \min(u_1, u_2) - \aleph(C) \min(u_1, u_2)$, which is an element of K (since $\partial C \subset D$ and $\min(u_1, u_2) = 0$ on ∂C). Hence, from (2.10) we get for $i = 1, 2$: $J(u_i) \leq J(v)$.

On the other hand: $J(u_1) + J(u_2) = J(\min(u_1, u_2)) + J(\max(u_1, u_2))$. This implies $J(u_1) = J(u_2) = J(\min(u_1, u_2)) = J(\max(u_1, u_2))$. It follows that:

$$(3.11) \quad \int_C |\nabla \min(u_1, u_2)|^2 + V_0^2 H(\min(u_1, u_2)) \leq d_2 V_0^2 \text{ mes}(C).$$

Since $u_i > 0$ in C, $i = 1, 2$, from (3.11) we obtain:

$$(3.12) \quad \int_C |\nabla \min(u_1, u_2)|^2 \leq V_0^2 (d_2 - d_1) \text{mes}(C),$$

which is not possible if $\text{mes}(C) > 0$ $(d_2 < d_1)$. Hence, there exists no connected components C of $\{u_1 > 0\} \cap \{u_2 > 0\}$ with $\partial C \cap (OA \cup AB_1 \cup OC_1) = \emptyset$. The solutions being positive in a neighbourhood of $OA \cup AB_1 \cup OC_1$, it follows that $\{u_1 > 0\} \cap \{u_2 > 0\}$ is a connected set. Further on, following the ideas of Alt et al. (1982), we can prove as in Stavre (1991) that we cannot have $u_1 > u_2$ (or $u_1 < u_2$) in $\{u_1 > 0\} \cap \{u_2 > 0\}$ which means that there exists $(x_0, y_0) \in \{u_1 > 0\} \cap \{u_2 > 0\}$ with $u_1(x_0, y_0) = u_2(x_0, y_0)$. It follows that in any neighbourhood $V = V(x_0, y_0)$ we have $u_1 \geq u_2$ (or $u_1 \leq u_2$) (for details see Stavre 1991), and hence, u_1, u_2 being analytic in $\{u_1>0\}\cap\{u_2>0\}$, we obtain $u_1=u_2$ in $\{u_1>0\}\cap\{u_2>0\}$. The next step is to prove that $u_1=u_2$ in $\{d_2^{1/2}V_0(b_1-b_2)<u_1<0\}\cap\{d_2^{1/2}V_0(b_1-b_2)<u_2<0\}$. Since in a neighbourhood of C_1C_2 and B_1B_2 we have $u_i\in(d_2^{1/2}V_0(b_1-b_2),0)$, $i=1,2$, it follows that the set is non-empty. If we consider as before an open, connected component C of $\{d_2^{1/2}V_0(b_1-b_2)<u_1<0\}\cap\{d_2^{1/2}V_0(b_1-b_2)<u_2<0\}$ with $\partial C\cap(C_1C_2\cup B_1B_2)=\emptyset$ and $v=\min(u_1-d_2^{1/2}V_0(b_1-b_2),\ u_2-d_2^{1/2}V_0(b_1-b_2))-\aleph(C)$ $\min(u_1-d_2^{1/2}V_0(b_1-b_2),\ u_2-d_2^{1/2}V_0(b_1-b_2))+d_2^{1/2}V_0(b_1-b_2)$ we get $v\in K$ and:

(3.13) $\quad J(u_1)=J(u_2)=J(\min(u_1,u_2))\le J(v)$.

(3.13) implies:

(3.14) $\quad \int_C |\nabla\min(u_1,u_2)|^2+V_0^2 d_2\,\mathrm{mes}(C)\le 0$,

and, hence, $\mathrm{mes}(C)=0$.

Moreover $\{d_2^{1/2}V_0(b_1-b_2)<u_1<0\}\cap\{d_2^{1/2}V_0(b_1-b_2)<u_2<0\}$ has at most two connected components. Let C be such a connected component which is in contact with B_1B_2 or C_1C_2. We shall prove that $u_1=u_2$ in C. We cannot have $u_1>u_2$ (or $u_1<u_2$) in C; hence there exists $(x_0,y_0)\in C$ such that $u_1(x_0,y_0)=u_2(x_0,y_0)$. The proof being similar with the one mentioned before, it follows that $u_1=u_2$ in C and hence $\dot u_1=u_2$ in $\{d_2^{1/2}V_0(b_1-b_2)<u_1<0\}\cap\{d_2^{1/2}V_0(b_1-b_2)<u_2<0\}$.

Thus the statement of the theorem holds.

4. Properties of the flow regions and of the solution.

In this section we shall obtain that the flow region $\{u>d_2^{1/2}V_0(b_1-b_2)\}$ is bounded. We also prove the monotonicity of u with respect to x and y.

Theorem 4.1. For any n in $\mathbb{N}$ we obtain $\{u_n>d_2^{1/2}V_0(b_1-b_2)\}\subset\tilde D_0$, where u_n is a solution of (3.4) and $\tilde D_0=(0,a)\times\left(0,\left(\dfrac{d_1}{d_2}\right)^{1/2}\dfrac{V_0}{V_f}b_2\right)$.

Proof. We define: $v=\min\left(d_1^{1/2}V_0\left(\dfrac{V_0}{V_f}b_2-\left(\dfrac{d_2}{d_1}\right)^{1/2}y\right)^+,d_1^{1/2}V_0b_2\right)+$

$+d_2^{1/2}V_0(b_1-b_2)$ and $v_n=\min(u_n,v)$. After some computations we obtain $v_n\in K_n$, which implies:

$$(4.1) \quad J_n(u_n) \leq J_n(\min(u_n,v)), \text{ or:}$$

$$(4.2) \quad \int_{D_n \cap \{u_n > v\}} \nabla(u_n-v) \cdot \nabla(u_n+v) + V_0^2 \int_{D_n} H(u_n) - H(\min(u_n,v)) \leq 0.$$

If $D_n \subset \tilde{D}_0$ it is clear that $\{u_n > d_2^{1/2} V_0 (b_1-b_2)\} \subset \tilde{D}_0$. The proof will be made only for n in $\mathbb{N}$ such that $D_n \setminus \tilde{D}_0 \neq \emptyset$. From (2.8) we get:

$$\int_{\tilde{D}_0} H(u_n) - H(\min(u_n,v)) \geq 0 \text{ and hence (4.2) becomes:}$$

$$(4.3) \quad \int_{D_n} |\nabla(u_n-v)^+|^2 + 2\int_{D_n} \nabla(u_n-v)^+ \cdot \nabla v + V_0^2 \int_{D_n \setminus \tilde{D}_0} H(u_n) - H(\min(u_n,v)) \leq 0.$$

For $y > \left(\dfrac{d_1}{d_2}\right)^{1/2} \dfrac{V_0}{V_f} b_2$ we have $v = d_2^{1/2} V_0 (b_1-b_2)$ and v is not constant only for y in $\left(\left(\dfrac{d_1}{d_2}\right)^{1/2}\left(\dfrac{V_0}{V_f}-1\right)b_2, \ \left(\dfrac{d_1}{d_2}\right)^{1/2}\dfrac{V_0}{V_f}b_2\right)$. Thus:

$$(4.4) \quad \int_{D_n} |\nabla(u_n-v)^+|^2 - 2d_2^{1/2} V_0 \int_{(0,a) \times \left(\left(\frac{d_1}{d_2}\right)^{1/2}\left(\frac{V_0}{V_f}-1\right)b_2, \ \left(\frac{d_1}{d_2}\right)^{1/2}\frac{V_0}{V_f}b_2\right)} [(u_n-v)^+]_y +$$

$$+ V_0^2 \int_{D_n \setminus \tilde{D}_0} H(u_n) \leq 0.$$

The second integral is equal to:

$$\int_{\left\{y=\left(\frac{d_1}{d_2}\right)^{1/2}\frac{V_0}{V_f} b_2\right\}} (u_n - d_2^{1/2} V_0 (b_1-b_2)) \, ds, \text{ since on } y = \left(\dfrac{d_1}{d_2}\right)^{1/2}\left(\dfrac{V_0}{V_f}-1\right)b_2 \text{ we have}$$

$$(u_n-v)^+ = (u_n - d_1^{1/2} V_0 b_2 - d_2^{1/2} V_0 (b_1-b_2))^+ = (u_n - d_1^{1/2} V_0 b_1 + (d_1^{1/2}-d_2^{1/2}) V_0 (b_1-b_2))^+ = 0$$

(we obtain, as in Lemma 2.2, $u_n \leq d_1^{1/2} V_0 b_1$). As for $y > \left(\dfrac{d_1}{d_2}\right)^{1/2} \dfrac{V_0}{V_f} b_2$,

$v = d_2^{1/2} V_0 (b_1 - b_2)$, (4.4) becomes:

$$(4.5) \quad \int_{\tilde{D}_0} |\nabla(u_n - v)^+|^2 + \int_{\tilde{D}_n \backslash \tilde{D}_0} |\nabla u_n|^2 + 2d_2^{1/2} V_0 \int_{\tilde{D}_n \backslash \tilde{D}_0} \left[u_n - d_2^{1/2} V_0 (b_1 - b_2) \right]_y^+$$
$$+ V_0^2 \int_{\tilde{D}_n \backslash \tilde{D}_0} H(u_n) \leq 0, \quad \text{or}$$

$$(4.6) \quad \int_{\tilde{D}_0} |\nabla(u_n - v)^+|^2 + \int_{(\tilde{D}_n \backslash \tilde{D}_0) \cap \{u_n > d_2^{1/2} V_0 (b_1 - b_2)\}} |\nabla u_n|^2 + 2d_2^{1/2} V_0 (u_n)_y + d_2 V_0^2 \leq 0,$$

and hence $\text{mes}((\tilde{D}_n \backslash \tilde{D}_0) \cap \{u_n > d_2^{1/2} V_0 (b_1 - b_2)\}) = 0$ which completes the proof.

Since the solution u of (2.10) can be obtained as a weak limit of $(u_n)_{n \geq 0}$, and $\{u_n > d_2^{1/2} V_0 (b_1 - b_2)\} \subset \tilde{D}_0$ for all n in $\mathbb{N}$, it follows that $\{u > d_2^{1/2} V_0 (b_1 - b_2)\} \subset \tilde{D}_0$.

We shall prove next that the solution of (2.10) is monotone decreasing with respect to y and x.

We denote by u^* the monotone decreasing rearrangement in direction y and by u^{**} the monotone decreasing rearrngement in x (see Kawohl 1985).

Theorem 4.2. The solution u of (2.10) has the properties:

(4.7) $u_y \leq 0$, $u_x \leq 0$ a.e. in D.

Proof. The values of f being decreasing in y on $\{(0,y)/y > 0\}$ and on $\{(a,y)/y > 0\}$ and $u \leq d_1^{1/2} V_0 b_1$ in D we get $u^* = f$ on ∂D and hence, $u^* \in K$. Moreover, $J(u^*) \leq J(u)$ (Kawohl 1985). By the unicity result, we obtain $u = u^*$.

For obtaining the monotonicity in x, we show first that:

(4.8) $u(x,y) \geq f_1(y)$, for all (x,y) in D,

where $f_1 : [0, \infty) \longrightarrow \mathbb{R}$,

$$
f_1(y) = \begin{cases} d_1^{1/2} V_0 (b_1 - y) & \text{if } 0 \le y < b_1, \\[2ex] d_2^{1/2} V_0 (b_1 - y) & \text{if } b_1 \le y \le b_2, \\[2ex] d_2^{1/2} V_0 (b_1 - b_2) & \text{if } y > b_2. \end{cases}
$$

For proving (4.8) we take $v = \max(u, f_1(y))$ which is a test function for (2.10); hence we obtain:

$$
\int_{\widetilde{D}_0} |\nabla \min(u - f_1(y), 0)|^2 + 2 \int_{\widetilde{D}_0} \nabla \min(u - f_1(y), 0) \cdot \nabla f_1(y) +
$$

$$
+ V_0^2 \int_{\widetilde{D}_0} H(u) - H(\max(u, f_1(y))) \le 0 \quad \text{or}
$$

$$
\int_{(0,a) \times (0,b_1) \cap \{u < d_1^{1/2} V_0 (b_1 - y)\}} |\nabla u|^2 + 2 d_1^{1/2} V_0 u_y + V_0^2 H(u) +
$$

$$
+ \int_{(0,a) \times (b_1,b_2) \cap \{d_2^{1/2} V_0 (b_1 - b_2) < u < d_2^{1/2} V_0 (b_1 - y)\}} |\nabla u|^2 + 2 d_2^{1/2} V_0 u_y + d_2 V_0^2 \quad -
$$

$$
- 2 d_1^{1/2} V_0 \int_{\{y = b_1\}} \min(u, 0) \, ds + 2 d_2^{1/2} V_0 \int_{\{y = b_1\}} \min(u, 0) \, ds \le 0.
$$

(we used $\min(u - d_2^{1/2} V_0 (b_1 - b_2), 0) = 0$ on $y = b_2$).

Since $\displaystyle \int_{\{y = b_1\}} \min(u, 0) \, ds = \int_{(0,a) \times (0,b_1)} [\min(u,0)]_y$ we get:

$$
\int_{(0,a) \times (0,b_1) \cap \{0 < u < d_1^{1/2} V_0 (b_1 - y)\}} (u_x)^2 + (u_y + d_1^{1/2} V_0)^2 \; + \int_{(0,a) \times (0,b_1) \cap \{d_2^{1/2} V_0 (b_1 - b_2) < u \le 0\}} (u_x)^2 + (u_y + d_2^{1/2} V_0)^2 \; +
$$

$$
+ \int_{(0,a) \times (b_1,b_2) \cap \{d_2^{1/2} V_0 (b_1 - b_2) < u < d_2^{1/2} V_0 (b_1 - y)\}} (u_x)^2 + (u_y + d_2^{1/2} V_0)^2 \le 0.
$$

This implies: $\mathrm{mes}((0,a) \times (0,b_1) \cap \{d_2^{1/2} V_0 (b_1 - b_2) < u < d_1^{1/2} V_0 (b_1 - y)\}) = 0$ and $\mathrm{mes}((0,a) \times (b_1,b_2) \cap \{d_2^{1/2} V_0 (b_1 - b_2) < u < d_2^{1/2} V_0 (b_1 - y)\}) = 0$. Hence:
$u \ge d_1^{1/2} V_0 (b_1 - y)$ in $(0,a) \times (0,b_1)$ and $u \ge d_2^{1/2} V_0 (b_1 - y)$ in $(0,a) \times (b_1,b_2)$, i.e. (4.8).

Next, we have to prove that:

(4.9) $u(x,y) \leq f_2(y)$, for all (x,y) in D,

where $f_2: [0,\infty) \longrightarrow \mathbb{R}$;

$$f_2(y) = \begin{cases} d_1^{1/2}(V_0 b_1 - V_f y) & \text{if} \quad 0 \leq y < \dfrac{V_0}{V_f} b_1 \\[2ex] d_2^{1/2}(V_0 b_1 - V_f y) & \text{if} \quad \dfrac{V_0}{V_f} b_1 \leq y \leq \dfrac{V_0}{V_f} b_2, \\[2ex] d_2^{1/2} V_0 (b_1 - b_2) & \text{if} \quad y > \dfrac{V_0}{V_f} b_2. \end{cases}$$

We obtain the inequality (4.9) by taking $v = \min(u, f_2(y))$ in (2.10), using some similar ideas as in proving (4.8).

Since u is constant on $\{(x,0)/0 < x < a\}$ we get from (4.8) and (4.9) that u^{**} is in K and hence u is monotone decreasing with respect to x.

As a consequence of (4.9) it follows that $u = d_2^{1/2} V_0 (b_1 - b_2)$ in $(0,a) \times$ $\times (\dfrac{V_0}{V_f} b_2, \infty)$ and hence the flow region $\{u > d_2^{1/2} V_0 (b_1 - b_2)\}$ is contained in a smaller rectangle than $\tilde{D}_0$, i.e. D_0. Moreover the rectangle $(0,a) \times (0, b_1)$ is contained in the flow region of the inner fluid.

References.

Stavre R. (1991), On a free boundary problem in fluid mechanics, Eur.J.Mech., B/Fluids, 10, no.1.

Stavre R. (1990), Ph.D.Thesis, Bucharest, (in romanian).

Jenkins D.R., Barton N.G. (1988), Computation of the free surface shape of an inviscid jet incident on a porous wall, IMA Journal of Appl.Math.,41.

King A.C.(1990), A note on the impact of a jet on a porous wall, IMA Journal of Appl.Math., 45.

Alt H.W. et al. (1982), Asymmetric jet flows, Comm.Pure.Appl.Math., 35.

Alt H.W. et al. (1983), Axially symmetric jet flows, Arch.Rat.Mech.Anal., 81.

Alt H.W. et al. (1984a), Jets with two fluids I: One free

boundary, Indiana Univ.Math.J., 33, no.2.

Alt H.W. et al. (1984b), Jets with two fluids II: Two free boundaries,Indiana Univ.Math.J., 33, no.3.

Alt H.W. et al. (1984c), Variational problems with two phases and their free boundaries, Trans.Amer.Math.Soc., 282, no.2.

Alt H.W., Caffarelli L.A. (1981), Existence and regularity for a minimum problem with free boundary, J.Reine Angew.Math., 325.

Kawohl B. (1985), Rearrangements and convexity of level sets in PDE, Lect.Notes Math., Springer-Verlag Berlin Heidelberg, 1150.

Author's address:

Dr.Ruxandra Stavre
Institute of Mathematics
of the Romanian Academy
P.O. Box 1-764
70700 Bucharest
Romania.

International Series of Numerical Mathematics, Vol. 107, © 1992 Birkhäuser Verlag Basel

Critical Point Methods
in Nonlinear Eigenvalue Problems with Discontinuities

Cătălin Lefter and Dumitru Motreanu

Abstract. This Paper contains existence and multiplicity results for two different eigenvalue problems with discontinuous nonlinearities. One treats situations that are not covered by the corresponding results of K.C.Chang and P.H.Rabinowitz. The approach is based on minimax methods in nonsmooth critical point theory.

1. Introduction

Let H be a Hilbert space that is continuously embedded in $L^{s+1}(\Omega)$ for a bounded domain Ω in R^n with some $s \geq 0$ and let $L:H \longrightarrow H$ be a bounded self-adjoint linear operator for which we denote

$$m = \inf\left\{(Lu,u)_H / \|u\|_H^2 \mid u \in H, u \neq 0\right\}. \tag{1.1}$$

Let $p : \Omega \times R \longrightarrow R$ be a measurable function such that for every $x \in \Omega$, $p(x,.) : R \longrightarrow R$ is locally bounded. Following Chang (1981) we associate to p the next functions

$$\underline{p}(x,t) = \lim_{\delta \to 0} \; \operatorname*{ess\,inf}_{|\tau-t|<\delta} p(x,\tau)$$

$$\overline{p}(x,t) = \lim_{\delta \to 0} \; \operatorname*{ess\,sup}_{|\tau-t|<\delta} p(x,\tau), \; (x,t) \in \Omega \times R. \tag{1.2}$$

This paper deals with the below two eigenvalue problems with discontinuous nonlinearities

(E1) $(Lu - \lambda u)(x) \in [\underline{p}(x,u(x)),\overline{p}(x,u(x))]$ for a.e. $x \in \Omega$

(E2) $Lu(x) \in [\lambda \underline{p}(x,u(x)), \lambda \overline{p}(x,u(x))]$ for a.e. $x \in \Omega$.

The statements (E1),(E2) express in fact a general concept of solution for the corresponding problem in the equality form (see Chang 1981). It reduces to the notion of solution considered by Massabo and Stuart (1978) in their case of a discrete set of discontinuity points for p.

Concerning problem (E1) we prove in Theorem 2.1 the existence of solutions under conditions which are weaker than those used in Rabinowitz (1986), p.14,25,30, just when p is a continuous function on $\overline{\Omega} \times R$. In particular, we do not assume

$$0 \in [\underline{p}(x,0),\overline{p}(x,0)] \quad \text{for a.e. } x \in \Omega , \tag{1.3}$$

so excepting the situations in (1.3), the obtained solution of (E1) is necessarily nontrivial. As shown by examples, Theorem 2.1 applies equally to elliptic and hyperbolic nonlinear eigenvalue problems. Other applications can be given in the Panagiotopoulos' theory of hemivariational inequalities (see Panagiotopoulos 1991).

For problem (E2) we establish several multiplicity results depending on the growth condition satisfied by p. Theorems 3.1, 3.2 and 3.3 improve Theorem 5.5 of Chang (1981) in that the existence of nontrivial solutions is proved for $\lambda \in R$ belonging to a half-line or a segment. Theorem 3.5 gives a sharper conclusion than in Theorem 5.6 of Chang (1981) by precising that (E2) possesses a prescribed number of distinct pairs of solutions for any sufficiently large $\lambda \in R$.

The basic technical tool used here is the nonsmooth version of Palais-Smale condition due to Chang (1981). We mention that recently Wang (1991) formulated a new version of it in terms of quasi-tangent vectors introduced by Motreanu and Pavel (1982). For other version and applications of smooth Palais-Smale condition we refer to Motreanu (1986).

The remaining of the paper is organized as follows. Section 2 concerns the existence of solutions of problem (E1). Section 3 treats the existence and the multiplicity of solutions of problem (E2).

2. Eigenvalue problem (E1)

Let H be a Hilbert space with the scalar product $(.,.)_H$ and the induced norm $\|.\|_H$ which is compactly and densely embedded in $L^2(\Omega)$ for some bounded domain Ω in R^n. Fix a positive real constant $C(\Omega)$ such that

$$\|u\|_{L^2(\Omega)} \leq C(\Omega)\|u\|_H \quad \text{for all } u \in H. \tag{2.1}$$

Throughout this Section $\Lambda: H \longrightarrow H^\times$ stands for the duality isomorphism

$$(\Lambda u)v = (u,v)_H \quad \text{for all } u,v \in H.$$

Consider a measurable function $p : \Omega \times R \longrightarrow R$ satisfying the growth condition

$$|p(x,t)| \leq a_1 + a_2|t| \quad \text{for all } (x,t) \in \Omega \times R, \tag{2.2}$$

where a_1 and a_2 are positive numbers independent of t. Denote by $P : \Omega \times R \longrightarrow R$ the primitive of p

$$P(x,t) = \int_0^t p(x,\tau)d\tau , \quad (x,t) \in \Omega \times R. \tag{2.3}$$

The generalized gradient (in the sense of Clarke) $\partial_t P(x,t)$ of P with respect to t is a closed interval in R

$$\partial_t P(x,t) = [q(x,t),r(x,t)] \subseteq [\underline{p}(x,t),\bar{p}(x,t)],$$
$$(x,t) \in \Omega \times R, \tag{2.4}$$

with $\underline{p},\bar{p}$ introduced in (1.2).

We suppose that the following measurability hypothesis of Chang (1981) is valid

$$q \text{ and } r \text{ are N-measurable functions on } \Omega \times R. \tag{2.5}$$

Let $L:H \longrightarrow H$ be a bounded self-adjoint linear operator.

The below theorem is our main existence result for problem (E1).

THEOREM 2.1. If $\lambda \in R$ is chosen so that

$$\lambda < m - a_2(C(\Omega))^2, \tag{2.6}$$

where m, a_2 and $C(\Omega)$ are the constants entering $(1.1),(2.2)$ and (2.1), there exists a solution $u \in H$ of eigenvalue problem (E1).

In order to prove Theorem 2.1 we need a preliminary result.

LEMMA 2.2. Consider the locally Lipschitz functional
$J : L^2(\Omega) \longrightarrow R$,

$$J(u) = \int_\Omega P(x,u(x))dx, \quad u \in L^2(\Omega). \tag{2.7}$$

Let (u_n) be a bounded sequence in H and let $w_n \in H^\times$ belong to the generalized gradient $\partial(J|_H)(u_n)$ such that

$$\Lambda(Lu_n - \lambda u_n) - w_n \longrightarrow 0 \quad \text{in } H^\times \text{ as } n \longrightarrow \infty \tag{2.8}$$

for some $\lambda \in R$ verifying

$$\lambda < m. \tag{2.9}$$

Then (u_n) contains a convergent subsequence in H.

PROOF. In view of (2.2) the definition of J in (2.7),(2.3) makes sense. The boundedness of (u_n) in H and the compactness of the inclusion $H \subset L^2(\Omega)$ imply the convergence of a subsequence of (u_n) in $L^2(\Omega)$. The density of H in $L^2(\Omega)$ insures

$$\partial(J|_H)(u) \subset \partial J(u) \quad \text{for all } u \in H$$

(cf. Theorem 2.2 in Chang 1981), so the corresponding subsequence (w_n) is bounded in $L^2(\Omega)$. Hence one can find a convergent subsequence of (w_n) in $H^\times$.

Taking in account (2.8) one deduces the convergence along a subsequence in H of $Lu_n - \lambda u_n$. Since one has

$$(m - \lambda)\|u_n - u_m\|_H \leq \|L(u_n - u_m) - \lambda(u_n - u_m)\|_H \quad \text{for all } n, m,$$

by (2.9) it follows the desired conclusion.

PROOF OF THEOREM 2.1. For some fixed $\lambda \in R$ satisfying (2.6)
let us denote

$$\gamma = m - a_2(C(\Omega))^2 - \lambda \; > 0. \tag{2.10}$$

We introduce the locally Lipschitz functional $I_\lambda : H \longrightarrow R$ by

$$I_\lambda (u) = \tfrac{1}{2}((Lu,u)_H - \lambda \|u\|_H^2) - J(u), \quad u \in H, \tag{2.11}$$

with J written in (2.7).
From (2.1),(2.2),(2.7),(2.10) and (2.11) one derives

$$I_\lambda (u) \geq \tfrac{1}{2}\gamma \|u\|_H^2 - a_1(\text{meas}(\Omega)^{\frac{1}{2}}C(\Omega)\|u\|_H \tag{2.12}$$

$$\text{for all } u \in H.$$

Inequalities (2.10) and (2.12) show that the functional I_λ on
H is bounded from below.

We check now the Palais-Smale condition, as stated by Chang
(1981), for the functional I_λ. According to this, let (u_n) be
a sequence in H such there is a constant M > 0 with

$$I_\lambda (u_n) \leq M \quad \text{for every n,} \tag{2.13}$$

and

$$\inf \left\{ \|w\|_{H^*} \mid w \in \partial I_\lambda (u_n) \right\} \longrightarrow 0 \quad \text{as } n \longrightarrow \infty. \tag{2.14}$$

The estimates (2.12) and (2.13) imply the boundedness of (u_n)
in H. By (2.11) and (2.14) one finds a sequence (w_n) in H^* such
that

$$w_n \in \partial (J|_H)(u_n) \quad \text{for all n}$$

and (2.8) holds. Then, by Lemma 2.2, we conclude the existence
of a convergent subsequence of (u_n). Consequently, I_λ satisfies the Palais-Smale condition as claimed. The preceding properties allow to apply to the functional $I_\lambda : H \longrightarrow R$ Theorem
3.5 in Chang (1981). It turns out that $\inf_{u \in H} I(u)$ is a critical
value of I_λ. Therefore there is some $u \in H$ satisfying

$$0 \in \partial I_\lambda (u). \tag{2.15}$$

Relation (2.15) shows that the function $u \in H$ solves (E1) .

EXAMPLES 2.4. (a) Consider the nonsmooth Dirichlet eigenvalue problem

$$- \operatorname{div}(k(x) \nabla u) = \lambda u + p(x,u), \quad x \in \Omega$$
$$u = 0, \qquad\qquad\qquad x \in \partial\Omega , \qquad\qquad (2.16)$$

where $k \in C^1(\bar{\Omega}, R)$ and $p : \Omega \times R \longrightarrow R$ verifies (2.2) and (2.5). Taking $H = W_0^{1,2}(\Omega)$ with $L : H \longrightarrow H$ constructed in a standard way, Theorem 2.1 provides a solution of (2.16) in the sense of (E1).

(b) Replace in the nonsmooth eigenvalue equation (2.16) the Dirichlet boundary condition with the mixed one

$$\frac{\partial u}{\partial n} + b(x)u = 0, \quad x \in \partial\Omega ,$$

where $b \in C(\partial\Omega, R)$. We choose $H = W^{1,2}(\Omega)$ and let $L : H \longrightarrow H$ be the suitable operator. Then Theorem 2.1 gives a solution of the above boundary value problem in the sense of (E1).

(c) Let us consider the nonsmooth hyperbolic eigenvalue problem on the cylinder $Q = \Omega \times (0,T)$ in R^{n+1}

$$u_{tt} - \operatorname{div}(k(x) \nabla_x u) = \lambda u + p(x,t,u), \quad (x,t) \in Q,$$
$$u\big|_{t=0} = \varphi \quad \text{and} \quad u_t\big|_{t=0} = \Psi , \quad x \in \Omega \qquad (2.17)$$
$$u\big|_\Gamma = 0 \quad \text{on } \Gamma = \partial\Omega \times (0,T),$$

where $k \in C^1(\bar{\Omega} \times R)$, $p : Q \times R \longrightarrow R$ verifies (2.2) and (2.5) with Q in place of Ω and $\varphi, \Psi \in L^2(\Omega)$. Take in Theorem 2.1 $H = W^{1,2}(Q)$ and $L : H \longrightarrow H$ the following operator

$$(Lu,v)_H = \int_Q (k(x) \nabla_x u \, \nabla_x v - u_t v_t)dx \, dt - \int_\Omega \Psi(x)v(x,0)dx$$
$$\text{for } u,v \in H.$$

Then any solution $u \in H$ of problem (E1) as given by Theorem 2.1 satisfying the initial condition $u\big|_{t=0} = \varphi$ on Ω represents a generalized solution of (2.17).

REMARK 2.5. Theorem 2.1 is an extension of Proposition 2.41 and Theorems 2.42,5.16 in Rabinowitz (1986) to nonsmooth boundary value problems in the sublinear case. As seen in Examples 2.4 the lack of positiveness assumption upon the operator L permits to treat by Theorem 2.1 not only elliptic equations, but also hyperbolic problems.

3. Eigenvalue problem (E2)

In this Section we suppose that H is a Hilbert space which is densely and compactly embedded in $L^{s+1}(\Omega)$, where Ω is a bounded domain in R^n and $s \geq 0$. Fix a positive constant $C(\Omega)$ such that

$$\|u\|_{L^{s+1}(\Omega)} \leq C(\Omega)\|u\|_H \quad \text{for all } u \in H. \tag{3.1}$$

Assume throughout that $L:H \longrightarrow H$ is a bounded self-adjoint linear operator for which there is a constant $\alpha > 0$ such that

$$(Lu,u)_H \geq \alpha\|u\|_H^2 \quad \text{for each } u \in H. \tag{3.2}$$

We suppose also that $p:\Omega \times R \rightarrow R$ is a measurable locally bounded function verifying condition (2.5).

The next three theorems study the existence of nontrivial solutions to problem (E2).

THEOREM 3.1. Assume that the following conditions hold

(i)   there exist positive constants a_1, a_2 and $s \in [0,1)$ such that

$$|p(x,t)| \leq a_1 + a_2|t|^s \quad \text{for all } (x,t)\in\Omega \times R; \tag{3.3}$$

(ii) $p(x,t) = o(|t|)$ for $t \longrightarrow 0$, uniformly with respect to $x \in \Omega$;

(iii)   there is a point $t \in R$ with $P(x,t) > 0$ for a.e. $x \in \Omega$, where P designates the primitive of p in (2.3);

(iv)   H can be continuously embedded in some $L^{s'+1}(\Omega)$ with $s' > 1$.

Then there exists a constant $\lambda_0 > 0$ such that the eigenvalue problem (E2) has at least two nontrivial solutions for every $\lambda \geq \lambda_0$.

PROOF. Corresponding to any $\lambda \in R$ let us define the locally Lipschitz functional $I_\lambda : H \longrightarrow R$ by

$$I_\lambda (u) = \tfrac{1}{2}(Lu,u)_H - \lambda J(u), \quad u \in H, \tag{3.4}$$

where $J : L^{s+1}(\Omega) \longrightarrow R$ denotes the mapping given by relation (2.7). From (3.3) and (3.4) it is seen there exist positive constants a and b with the property

$$|J(u)| \leq a\|u\|_H + b\|u\|_H^{s+1} \quad \text{for all } u \in H. \tag{3.5}$$

Since s+1 < 2, (3.5) yields the estimate

$$|\lambda J(u)| \leq \alpha \epsilon \|u\|_H^2 + M \quad \text{for all } u \in H, \tag{3.6}$$

where M > 0 and $\epsilon \in (0,1/2)$ are constants and $\alpha > 0$.

By (3.2) and (3.5) we are allowed to apply Theorem 4.3 in Chang (1981) for deducing that I_λ satisfies the Palais-Smale condition and is bounded from below. Then Theorem 3.5 in Chang (1981) implies that

$$b_\lambda = \inf_{u \in H} I_\lambda (u) \tag{3.7}$$

is a critical value of I_λ , so there exists $u_\lambda \in H$ satisfying (2.15) with u_λ in place of u.

Hypothesis (iii) and the density of H in $L^{s+1}(\Omega)$ ensure the existence of $u_0 \in H$ and $\lambda_0 > 0$ with $I_\lambda (u_0) < 0$ provided $\lambda \geq \lambda_0$. We derive that

$$b_\lambda < 0 \quad \text{for all } \lambda \geq \lambda_0. \tag{3.8}$$

It is clear from (3.3) that there exist new positive constants a_1',a_2' for which the next estimate is true

$$|p(x,t)| \leq a_1' + a_2' |t|^{s'} \quad \text{for all } (x,t) \in \Omega \times R. \tag{3.9}$$

Then following an idea in Rabinowitz (1986), p.10, for an

arbitrary $\varepsilon > 0$ there is, by (3.9) and assumption (ii), a positive number A such that

$$|P(x,t)| < \varepsilon t^2 + A|t|^{s'+1} \quad \text{for all } (x,t) \in \Omega \times R. \qquad (3.10)$$

Hypothesis (iv) and inequality (3.10) show that, for some positive constant k, independent of ε , the below estimate holds,

$$|J(u)| \leq k\|u\|_H^2(\varepsilon + A\|u\|_H^{s'-1}) \quad \text{for every } u \in H. \qquad (3.11)$$

Since $s'>1$, we deduce from (3.11) that $J(u) = o(\|u\|_H^2)$ as $u \longrightarrow 0$. Consequently, for each $\lambda \in R$ there are positive numbers ρ , β with

$$I_\lambda(u) \geq \beta \quad \text{for every } u \in H \text{ on the sphere } \|u\|_H = \rho .$$

This fact combined with (3.8) enables us to apply the nonsmooth version of Mountain Pass Theorem (see Chang 1981 and Rabinowitz 1986) in the case of our functional I_λ for $\lambda \geq \lambda_0$. We get $v_\lambda \in H$ satisfying (2.15) and $I_\lambda(v_\lambda) \geq \beta$ for $\lambda \geq \lambda_0$. Comparing with (3.7),(3.8) we see that v_λ is a nontrivial solution of (E2) which is different from u_λ .

THEOREM 3.2. If the conditions (2.2) and (2.5) hold, then there exists a solution of problem (E2) for every $\lambda \in R$ satisfying

$$|\lambda| < \frac{\alpha}{a_2(C(\Omega))^2} . \qquad (3.12)$$

where $\alpha, a_2, C(\Omega)$ are the constants appearing in (3.2),(2.2) and (3.1) (with s=1), respectively.

PROOF. One checks that Theorems 4.3 and 3.5 in Chang (1981) apply to I_λ in (3.4), so $\inf_{u \in H} I_\lambda(u)$ is a critical value of I_λ .

THEOREM 3.3. Assume that conditions (ii),(iii) in Theorem 3.1 are verified together with the further assumptions

(i')   there exist positive constants a_1 and a_2 such that the
       growth condition (3.3) hold with an exponent $s \geq 1$;

(v)   the primitive $P : \Omega \times R \longrightarrow R$ of p has a strictly subqua-
         dratic growth, that is, there exist positive constants
         b_1, b_2 and $0 \leq \beta < 2$ such that

$$|P(x,t)| \leq b_1 + b_2 |t|^\beta \quad \text{for all } (x,t) \in \Omega \times R.$$

Then there exists a constant $\lambda_0 > 0$ such that for every $\lambda \geq \lambda_0$
the eigenvalue problem (E2) assumes at least two nontrivial so-
lutions.

PROOF. Hypothesis (v) implies evidently estimate (3.6). Then
we may proceed as in the proof of Theorem 3.1.

Now we need the following nonsmooth version of Clark 's
theorem (see Rabinowitz 1986, p.53).

THEOREM 3.4. Let X be a reflexive Banach space and let
$f:X \longrightarrow R$ be a locally Lipschitz function which is even, bounded from below and satisfies the nonsmooth Palais-Smale condition in the sense of Chang (1981). Assume in addition that
$f(0) = 0$ and there exists a subset K of X homeomorphic to the
unit (j-1)-sphere S^{j-1} by an odd map such that $\sup_{u \in K} f(u) < 0$.
Then f has at least j distinct pairs of critical points.

Our final result points out a multiplicity property in
solving the discontinuous eigenvalue problem (E2).

THEOREM 3.5. Assume that the function $p : \Omega \times R \longrightarrow R$ has the
properties (i) and

(vi)   the primitive P of p satisfies

   $P(x,t) > 0$ for all $x \in \Omega$ and $t \in R \setminus \{0\}$;

(vii) $p(x,.):R \longrightarrow R$ is an odd map for all $x \in \Omega$.

Then for every natural number j provided the operator $L:H \longrightarrow H$
has at least j eigenvalues (counted with their multiplicities),
there is some constant $\overline{\lambda}_j$ such that if $\lambda \geq \overline{\lambda}_j$ there exist
at least j distinct pairs of nontrivial symmetric solutions of
eigenvalue problem (E2).

PROOF. Let $(v_i)_{1 \le i \le j}$ denote the sequence of the corresponding normalized eigenvectors of L. We set

$$K = \left\{ \sum_{i=1}^{j} \alpha_i v_i \ \Big| \ \sum_{i=1}^{j} \alpha_i^2 = 1, \quad \alpha_i \in R, \ i=1,\dots,j \right\}. \quad (3.13)$$

Clearly K is homeomorphic to the spehere S^{j-1} by an odd homeo-morphism.

For each $\lambda \in R$, the functional $I_\lambda : H \longrightarrow R$ introduced by (3.4) is even, bounded from below and satisfies the Palais-Smale condition. These assertions are consequences of the first part of the proof of Theorem 3.1. The compactness of K in (3.13) and hypothesis (vi) ensure

$$I_\lambda(u) \le \frac{1}{2} \max_{1 \le n \le j} \lambda_n - \lambda \inf_{u \in K} J(u) < 0 \quad \text{for all } u \in K \text{ and}$$
$$\lambda > 0 \text{ sufficiently large.}$$

All the hypotheses of Theorem 3.4 are verified for $f = I_\lambda$ with λ as above. The application of Theorem 3.4 completes the proof.

REMARK 3.6. Theorems 3.1, 3.2 and 3.3 provide the existence of nontrivial solutions of eigenvalue problem (E2) with $\lambda \in R$ running on an entire half-line (or segment in Theorem 3.2), and not only that there exist such values λ as in Chang (1981) (Theorems 5.5 and 5.6). Theorem 3.5 is an extension to the non-smooth case of Theorem 9.10 in Rabinowitz (1986). It is also relevant that we cover the situation treated in Massabo and Stuart (1978). If we drop the hypothesis (ii) in Theorem 3.1 one still obtains a solution u_λ of (E2) given by (3.7) for each $\lambda \in R$.

References

Chang K.C. (1981), Variational methods for non differentiable functionals and their applications to partial differential e-quations. J.Math.Anal.Appl. 80, 102-129.

Massabo I. and Stuart C.A. (1978), Elliptic eigenvalue problems with discontinuous nonlinearities. J.Math.Anal.Appl.
66, 261-281.

Motreanu D. and Pavel N.H. (1982), Quasi-tangent vectors in
flow invariance and optimization problems on Banach manifolds.
J.Math.Anal.Appl. 88, 116-132.

Motreanu D. (1986), Existence for minimization with nonconvex
constraints. J.Math.Anal.Appl. 117, 128-139.

Panagiotopoulos P.D. (1991), Coercive and semicoercive hemivariational inequalities. Nonlinear Anal. TMA 16, 209-231.

Rabinowitz P.H. (1986), Minimax Methods in Critical Point Theory
with Applications to Differential Equations. CBMS Regional Conf.
Ser.in Math., No.65, Amer.Math.Soc., Providence, R.I.

Wang T. (1991), A minimax principle without differentiability.
J.Math.Research and Exposition 11, 111-116.

Authors' address:

Cătălin Lefter and Dumitru Motreanu
Department of Mathematics
University of Iaşi
6600 Iaşi, Romania

International Series of Numerical Mathematics, Vol. 107, © 1992 Birkhäuser Verlag Basel

MAXIMUM PRINCIPLES FOR ELLIPTIC SYSTEMS

Ioan A. Rus

1. $\underline{\text{Introduction}}$. Let Ω be a bounded domain of R^n. Consider the following system

$$\sum_{i,j=1}^{n} A_{ij} \frac{\partial^2 u}{\partial x_i \partial x_j} + F(x, u, \text{grad } u) = 0 \qquad (1)$$

where $A_{ij}: \bar{\Omega} \to M_{mm}(R)$ and $F: \bar{\Omega} \times R^m \times R^{nm} \to R^m$. We begin with the following definition

$\underline{\text{Definition 1}}$. The system (1) is called elliptic at a point $x \in \Omega$, if

$$\det \left(\sum_{i,j=1}^{n} A_{ij}(x) \lambda_i \lambda_j \right) \neq 0,$$

for all $\lambda \in R^n - \{0\}$.

$\underline{\text{Definition 2}}$. The system (1) is called strongly elliptic at a point $x \in \Omega$, if

$$\sum_{i,j=1}^{n} \left[\tau^T A_{ij}(x) \tau \right] \lambda_i \lambda_j > 0$$

for all $\tau \in R^m - \{0\}$ and $\lambda \in R^n - \{0\}$.

$\underline{\text{Definition 3}}$. The system (1) satisfies Somigliana's conditions at a point $x \in \Omega$, if

$$\sum_{i,j=1}^{n} \tau_i^T A_{ij}(x) \tau_j > 0$$

for all $\tau_k \in R^m$, $k = \overline{1,n}$, and $\sum_{i=1}^{n} \| \tau_i \| \neq 0$.

A systematic method for constructing functions, defined on the solutions of some nonlinear weakly coupled elliptic systems, which satisfy a maximum principle was initiated by Cosner and Schaefer

(1987). In the present paper we give further examples of such functions.

2. Maximum principles for $\max\left[u_1,\ldots,u_m\right]$.

Consider the following differential operators:

$$L_p := \sum_{i,j=1}^{n} a_{ij}^P \frac{\partial^2}{\partial x_i \partial x_j} + \sum_{i=1}^{n} b_i^P \frac{\partial}{\partial x_i} \ , \quad p=\overline{1,m}$$

where a_{ij}^P, $b_i^P:\overline{\Omega} \to R$ are bounded, $i,j=\overline{1,n}$, $p=\overline{1,m}$.

We suppose that L_p, $p=\overline{1,m}$ are uniformly elliptic on Ω and the matrix $\left[a_{ij}\right]$ is positive definite. In what follow, in this paragraph, we denote

$$H(u) = \max\left[u_1,\ldots,u_m\right].$$

We have

<u>Theorem 1</u>. Let $u \in C\left[\overline{\Omega},R^m\right] \cap C^2\left[\overline{\Omega},R^m\right]$ be a solution of the following system of inequations

$$L_p u_p + F_p\left[x,u_1,\ldots,u_m,\text{grad } u_p\right] \geq 0, \quad p=\overline{1,m}.$$

We suppose that

(i) $\qquad F_1\left[x,t_1,t_2',\ldots,t_m',0\right] \leq F_1\left[x,t_1,t_2'',\ldots,t_m'',0\right],$

for all $x \in \Omega$, $t_1, t_i', t_i'' \in R$, $t_i' \leq t_i''$, $i=\overline{2,m}$;

(ii) $\qquad F_p\left[x,t_1',\ldots,t_{p-1}',t_p,t_{p+1}',\ldots,t_m',0\right] \leq$

$$\leq F_p\left[x,t_1'',\ldots,t_{p-1}'',t_p,t_{p+1}'',\ldots,t_m'',0\right],$$

for all $x \in \Omega$, $t_p, t_i', t_i'' \in R$, $t_i' \leq t_i''$, $i \neq p$;

(iii) $\qquad F_m\left[x,t_1',\ldots,t_{m-1}',t_m,0\right] \leq F_m\left[x,t_2'',\ldots,t_{m-1}'',t_m,0\right]$

for all $x \in \Omega$, $t_i', t_i'', t_m \in R$, $i=\overline{1,m-1}$;

(iv) $\qquad F_p\left[x,r,\ldots,r,0\right] < 0,$

for $p=\overline{1,m}$ and all $x \in \Omega$ and $r > 0$.

If $H(u)(x_o) > 0$ is a relative maximum of $H(u)$ on $\overline{\Omega}$, then $x_o \in \partial\Omega$.

<u>Proof</u>. We suppose that $x_o \in \Omega$.

Let $H\left[u(x_o)\right] = \max\left[u_1(x_o),\ldots,u_m(x_o)\right] = u_k(x_o) = M > 0$. Then there exists a neighborhood $V(x_o)$ of x_o such that $u_i(x) \leq M$, $i=\overline{1,m}$, for all $x \in V(x_o)$ and $u_k(x_o)$ is a relative maximum of u_k.

We have

$$L_k\left[u_k(x_o)\right] + f_k\left[x_o, u_1(x_o), \ldots, u_m(x_o), \mathrm{grad}\ u_k(x_o)\right] = \sum_{i,j=1}^{n} a_{ij}^k\ \frac{\partial^2 u_k(x_o)}{\partial x_i\ \partial x_j} +$$

$$+ f_k\left[x_o, u_1(x_o), \ldots, u_m(x_o), \mathrm{grad}\ u_k(x_o)\right] \leq f_k\left[x_o, M, \ldots, M, 0\right] < 0$$

<u>Definition 4</u>. A function $F: \overline{\Omega} \times R^m \times R^n \to R^m$ which satisfies the conditions (i)+(ii)+(iii) from the Theorem 1 is said to be quasimonoton increasing.

By similar arguments we have

<u>Theorem 2</u>. Let $u \in C(\overline{\Omega}, R^m) \cap C^2(\Omega, R^m)$ be a solution of the following systems of inequations

$$L_p(u_p)(x) + F_p\left[x, u(x), \mathrm{grad}\ u_p(x)\right] > 0, \quad p=\overline{1,m},$$

for all $x \in \Omega$.

We suppose that

(i) F is a quasimonoton increasing functions;

(ii) $F_p\left[x, r, \ldots, r, 0\right] \leq 0$, for $p=\overline{1,m}$ and all $x \in \Omega$ and $r \in R_+$.

If $H(u)(x_o)$ 0 is a relative maximum of $H(u)$ on $\overline{\Omega}$, then $x_o \in \partial x$.

<u>Theorem 3</u>$\left[\text{see Protter (1982), Rus (1987)}\right]$.

Let $u \in C\left[\overline{\Omega}, R^m\right] \cap C^2\left[\Omega, R^m\right]$ be a solution of the following system of inequations

$$L_p u_p + \sum_{i=1}^{m} C_{pi} u_i \geq 0$$

We suppose that

(i) $C_{pi} \geq 0$ for $i \neq p$, $p=\overline{1,m}$;

(ii) $\sum_{i=1}^{n} C_{pi} \leq 0$, $p=\overline{1,m}$.

Then

$$\max_{\overline{\Omega}} H(u) = \max_{\partial\Omega} H(u).$$

<u>Remark 1</u>. In this way the Protter's vector maximum principle $\left[\text{see Protter 1982, Rus 1987, Schaefer 1988}\right]$, appears as a maximum principle for $H(u) = \max\left[u_1, \ldots, u_m\right]$.

3. Maximum principle for $\langle u, u \rangle$.

Let Ω be a bounded domain of R^n. Consider the following second order system

$$L(u) := \sum_{i,j=1}^{n} A_{ij} \frac{\partial^2 u}{\partial x_i \partial x_j} + \sum_{i=1}^{n} A_i \frac{\partial u}{\partial x_i} + A_o u = 0, \qquad (2)$$

where $A_{ij}, A_i, A_o : \overline{\Omega} \to M_{mm}(R)$.

The following general maximum principle is given in Rus(1968):

<u>Theorem 4</u>. Suppose that:

(i)   the system (1) is strongly elliptic,

(ii) $\displaystyle\sum_{i=1}^{m} e_i L_i(e) < 0$ for all $e \in C^2\left[\Omega, R^m\right]$, such that $\displaystyle\sum_{i=1}^{m} e_i^2 = 1$.

If $u \in C^2\left[\Omega, R^m\right] \cap C\left[\overline{\Omega}, R^m\right]$ is a solution of (2), then $\displaystyle\sum_{i=1}^{m} u_i^2$ attains the maximum values on $\overline{\Omega}$, in $\partial\Omega$.

<u>Remark 2</u>. The Theorem 4 generalies some results given by Ciorănescu, Giraud, Stys, Wasowski and Sabitov $\left[\text{see Pratter 1982,}\right.$ Rus 1968, Rus 1969, Sabitov 1991, Schaefer 1988$\left.\right]$.

<u>Remark 3</u>. If L satisfies the condition (ii) in the Theorem 4, then the operator $L + b_1 E \dfrac{\partial}{\partial x_1} + \ldots + b_n E \dfrac{\partial}{\partial x_n}$ satisfies this condition for all $B: \overline{\Omega} \to R^n$, $B = \left[b_1, \ldots, b_n\right]$.

From this remark the following problem arises:

<u>Problem 1</u>. $\left[\text{see Rus 1968}\right]$ Let $A \in M_{nn}(R)$.

Determine the minimum value of the function

$$\varphi: R \to R, \quad \mapsto \|A - xE\|$$

where $\|\cdot\|$ is the spectral norm.

Let $\langle\cdot,\cdot\rangle$ be a generalized inner product on $C\left[\bar{\Omega},R^m\right]$ $\left[\text{see Rus}\right.$ $\left.1968\right]$. The Theorem 4 represents a maximum principle for

$$H(u)=\langle u,u\rangle, \text{ where } \langle u,u\rangle=\sum_{i=1}^{m} u_i v_i.$$ In what follow we give another example of $H(u)$ as $\langle u,u\rangle$.

Let $\Omega=\Omega_1\times\Omega_2\subset R^n$ be a bounded domain, where $\Omega_1\subset R^k$, $\Omega_2\subset R^{n-k}$, $1\leq k<n$, are domains with smooth boundary. Consider the following system

$$L(u):=\frac{\partial^2 u}{\partial x_1^2}+\ldots+\frac{\partial^2 u}{\partial x_k^2}+\sum_{i,j=k+1}^{n}A_{ij}\frac{\partial^2 u}{\partial x_i\partial x_j}+\sum_{i=1}^{n}b_i\frac{\partial u}{\partial x_i}+Cu=0 \qquad (3)$$

We have

<u>Theorem 5</u>. Let $u\in C^2\left[\bar{\Omega},R^m\right]$ be a solution of (3) and

$$H(u)\left[x_1,\ldots,x_k\right]=\int_{\Omega_2}\left[u_1^2(x)+\ldots+u_m^2(x)\right]dx_{k+1}\ldots dx_n$$

We suppose that:

(i) $\qquad u\big|_{\Omega_1\times\partial\Omega_2}=0;$

(ii) $\qquad A_{ij},C\in M_{mm}(R)$ and $b_i\in R$, $i,j=\overline{1,n}$.

(iii) $\qquad \sum_{i=1}^{k}\tau_i^T\tau_i+\sum_{i,j=k+1}^{n}\tau_j^T A_{ij}\tau_i-\sum_{i=1}^{n}b_i\tau^T\tau_i-\tau^T C\tau\geq 0$

for all $\tau,\tau_i\in R^m$.

Then

$$\max_{\bar{\Omega}_1}H(u)=\max_{\partial\Omega_1}H(u).$$

<u>Proof</u>. $\Delta_k H(u)=2\int_{\Omega_2}\left[\sum_{j=1}^{m}\sum_{i=1}^{k}\left(\frac{\partial u_j}{\partial x_i}\right)^2+\sum_{i=1}^{k}\sum_{j=1}^{m}u_j\frac{\partial^2 u_j}{\partial x_i^2}\right]dx_{k+1}\ldots dx_m.$

But

$$\int_{\Omega_2}\left[\sum_{j=1}^{m}\sum_{i=1}^{k}u_j\frac{\partial^2 u_j}{\partial x_i^2}dx_{k+1}\ldots dx_n\right]=\int_{\Omega_2}u^T\left[-\sum_{i,j=k+1}^{n}A_{ij}\frac{\partial^2 u}{\partial x_i\partial x_j}-\sum_{i=1}^{n}b_i\frac{\partial u}{\partial x_i}-\right.$$

$$-Cu\bigg]dx_{k+1}\cdots dx_m = \int_{\Omega_2}\left[\sum_{i,j=k+1}^{n}\frac{\partial u^T}{\partial x_j}A_{ij}\frac{\partial u}{\partial x_i}-\sum_{i=1}^{n}b_i u^T\frac{\partial u}{\partial x_i}-u^T Cu\right]dx_{k+1}\cdots dx_m.$$

Thus, from (iii) we have that $\Delta_k H(u)\geq 0$. Now the theorem follows from the Hopf's maximum principle.

<u>Remark 4</u>. If $\Omega_1=[0,1]$, we have a result by Lavrentiev $\Big[$see Rus 1968$\Big]$.

<u>Remark 5</u>. If $b_i=0$, $c=0$ and the system (3) satisfies Somigliana's condition, then the condition (iii) in the Theorem 5 is satisfied.

4. Maximum principle for $\phi(u)$.

Consider the following differential operator

$$L(u):=\sum_{i,j=1}^{n}a_{ij}\frac{\partial^2}{\partial x_i\partial x_j}+\sum_{i=1}^{n}b_i\frac{\partial}{\partial x_i}+c \qquad (4)$$

We assume that L satisfies the maximum principle $\Big[$see Cosner and Schaefer 1987, and Peetre and Rus 1967$\Big]$.

Consider the following system

$$Lu_k+f_k(x,u)=0,\quad k=\overline{1,m},\quad x\in\Omega. \qquad (5)$$

Let $\phi\in C^2\Big[\mathbb{R}^m\Big]$. We have

<u>Theorem 6</u>. Let u be a solution of (5). We suppose that

(i)   the Hessian of ϕ is positive semidefinite;

(ii) $-\displaystyle\sum_{k=1}^{m}\frac{\partial\phi(y)}{\partial y_k}f_k(x,y)+c(x)\left[\phi(y)-\sum_{k=1}^{m}y_k\frac{\partial\phi(y)}{\partial y_k}\right]\geq 0$

for all $y\in\mathbb{R}^m$.

Then $\phi(u)$ satisfies the maximum principle.

<u>Remark 6</u>. For the case $f(x,u)=f(u)$ and $c=0,$ see Cosner and Schaefer 1987.

5. <u>Remarks</u>.

5.1. For other maximum principles for classical solutions see: Bertinger (1978), Cosner and Schaefer (1987), Hong (1983), Leung (1989), Maz'ya and Kresin (1984), Protter (1982), Rus (1968), Rus (1988), Sabitov (1991), Schaefer (1988), Schröder (1980),...

5.2. For maximum principles for generalized solutions see: Benilan (1989), Bertinger (1978), Gilbarg and Trudinger (1983), Kawohl (1983), Krylov (1987), Mandras (1976), Protter (1982),...
5.3. For some applications of maximum principles see: Amman (1976), Bénilan (1989), Clémend and Sweers (1987), Leung (1989), Schaefer (1988), Schröder (1980), Tsai (1980), Vidossich (1979).

REFERENCES

Amann H. (1976), Fixed point equations and nonlinear eigenvalue problems in ordered Banach spaces, SIAM Review, Vol.18, 620-709.

Bénilan P. et al. (1989), Recent advances in nonlinear elliptic and parabolic problems, Longman.

Bertiger W. (1978), Maximum principles, gradient estimates, and weak solutions for second order partial differential equations, Trans. A.M.S., Vol.238, 213-227.

Bertiger W. and Cosner C. (1979), Systems of second order equations with nonnegative characteristic form, Comm. in Partial Differential Equations, Vol.4, 701-737.

Clément P. and Sweers (1987), Getting a solution between sub- and supersolutions without monotone iteration, Rend. Istit. Mat. Univ. Triste, Vol.19, 189-194.

Cosner C. and Schaefer (1987), On the development of functionals which satisfy a maximum principle, Applicable Analysis, Vol.26, 45-60.

Gilbarg D. and Trudinger N.S. (1983), Elliptic partial differential equations of second order, Springer-Verlag, Berlin.

Hong C.W. (1983), Necessary and sufficient conditions for a class of generalized maximum principle, Acta Math. Sinica, Vol.26, No.3, 307-321.

Kawohl B. (1983), On a maximum principles and Liouville theorems for quasilinear elliptic equations and systems, Comm. Math. Univ. Caroline, Vol.4, No.4, 647-655.

Krylov N.V. (1987), Nonlinear elliptic and parabolic equations of the second order, Reidel Publ. Company, Boston.

Leung A.W. (1989), System of nonlinear partial differential equations, Kluwer Acad. Publ., Boston.

Mandras F. (1976), Principio di massimo debale per sotoosoluzioni di sistemi lineari elliptici debalmente accoppiati, Ball. U.M.I., Vol.13-A, 592-600.

Maz'ya V.G. and Kresin G.I. (1984), The maximum principle for second-order strongly elliptic and parabolic with constant coefficients (Russian), Vol.125, No.4, 458-480.

Peetre J. et Rus I.A. (1967), Sur la positivité de la fonction de Green, Math. Scand., Vol.21, 80-89.

Protter M.H. (1982), The maximum principle and the eigenvalue problems, Proceed. of the 1980 Beijing Symposium on Differential Geometry and Diff.Eq., 787-860, Scienses Press, Beijing.

Rus I.A. (1968), Sur les propriétés des normes des solutions d'un systém d'équations différentielles du second ordre, Studia Univ. Babeş-Bolyai, fas.1, 19-26.

Rus I.A. (1969), Un principe du maximum pour les solutions d'un système fortement elliptique, Glasnik Matematicki, Vol.4, 75-77.

Rus I.A. (1987), Some vector maximum principle for second order elliptic systems, Mathematica, Vol.29, 89-92.

Rus I.A. (1988), Maximum principle for strongly elliptic systems: a conjecture, Babeş-Bolyai Univ. Preprint Nr.8, 43-46.

Sabitov K.B. (1991), Maximum principles for some elliptic and hyperbolic systems of the second order (Russian), Vol.27, Nr.2, 272-278.

Schaefer P.W. (editor;1988), Maximum principles and eigenvalue problems in partial differential equations, Longman, New York.

Schröder J. (1980), Operator inequalities, Acad. Press, New York.

Tsai L.-Y. (1980), Existence of solutions of nonlinear elliptic systems, Bull.Inst.Math.Acad.Sinica, Vol.8, Nr.1, 111-127.

Vidossich G. (1979), Comparison, existence, uniqueness and successive approximations for the Dirichlet problem of elliptic equations, Univ. of Texas at Arlington, Tech.Report No.119.

Author's address:
Ioan A. Rus
University of Cluj-Napoca
Str. M. Kogălniceanu nr.1
3400 Cluj-Napoca
ROMANIA

International Series of Numerical Mathematics, Vol. 107, © 1992 Birkhäuser Verlag Basel

Exponential dichotomy of evolution operators in Banach spaces

Mihail Megan and Radu Latcu

Abstract. This paper introduces a concept of exponential dichotomy for a general class of nonlinear evolution operators. Necessary and sufficient conditions for exponential dichotomy are given. Obtained results are generalizations of well-known results of R. Datko (1973) and A. Ichikawa (1984) about exponential stability.

1. Introduction

In this paper we study asymptotical behaviours of nonlinear evolution operators defined on a Banach space, the main restriction on these operators being that their norms can increase not faster that an exponential function. A necessary and sufficient condition for the uniform exponential stability of a linear evolution operator in a Banach space has been obtained by Datko (1973). A nonlinear version of Datko's result is given by Ichikawa (1984). An extension of Datko's theorem to the case of uniform exponential dichotomy for linear evolution operators has been proved by Preda and Megan (1985). In this paper we define a concept of (nonuniform) exponential dichotomy for a class of nonlinear evolution operators introduced by Ichikawa (1984). We extend Datko's theorem for this general case. Thus the results of this paper may be regarded as generalizations of results of Datko and Ichikawa. They are applicable for a large class of systems described by Ichikawa (1984).

2. Terminology and preliminary results

Let X be a real or complex Banach space. The norm on X will be denoted by $\|.\|$. Let T be the set defined by

$$T = \{(t, t_0) : 0 \le t_0 \le t < \infty\}.$$

For every $(t, t_0) \in T$ we consider an X-valued application $\Phi(t, t_0)$ defined on X.

Definition 2.1. The family $\{\Phi(t,t_0)\}_{(t,t_0)\in T}$ is called an *evolution operator* (on X) if

e_1)
$$\Phi(t,s)\Phi(s,t_0) = \Phi(t,t_0)$$

for all $t \geq s \geq t_0 \geq 0$;

e_2) $\Phi(t,t) = I$ (the identity operator on X);

e_3) for each $t_0 \geq 0$ and $x_0 \in X$ the mapping $t \mapsto \Phi(t,t_0)x_0$ is continuous on $[t_0,\infty)$;

e_4) there is a nondecreasing continuous function $\phi : \mathbf{R}_+ = [0,\infty) \to \mathbf{R}_+^* = (0,\infty)$ such that

$$\|\Phi(t,t_0)x_0\| \leq \phi(t-t_0)\|x_0\| \text{ for all } (t,t_0) \in T \text{ and } x_0 \in X.$$

Throughout in this paper an evolution operator will always be denoted by Φ. We remark that $\Phi(t,t_0)$ is not necessary a linear operator.

Remark 2.1. If Φ is an evolution operator then (see Preda and Megan 1985) we can suppose in (e_4) that $\phi(t) = M\exp(\omega t)$, where $M \geq 1$ and $\omega > 0$ are independent of t. Let X_1 and X_2 be two subsets of X and let Φ be an evolution operator on X.

Definition 2.2. We say that the pair (X_1, X_2) induces an *exponential dichotomy* for the evolution operator Φ if there are N_1, N_2 and $\nu > 0$ such that

d_1) for every $x_1 \in X_1$ there is $t_1 \geq 0$ such that

$$\|\Phi(t,t_1)x_1\| \leq N_1 e^{-\nu(t-s)}\|\Phi(s,t_1)x_1\|$$

for all $t \geq s \geq t_1 + 1$;

d_2) for every $x_2 \in X_2$ there is $t_2 \geq 0$ such that

$$\|\Phi(t,t_2)x_2\| \geq N_2 e^{\cdot\nu(t-s)}\|\Phi(s,t_2)x_2\|$$

for all $t \geq s \geq t_2 + 1$.

Lemma 2.1. Let $x_1 \in X_1$, $t_1 \geq 0$ and $N_1, \nu, \delta > 0$ such that

$$\|\Phi(t,t_1)x_1\| \leq N_1 e^{-\nu(t-s)}\|\Phi(s,t_1)x_1\|$$

for all $t \geq s + \delta \geq s \geq t_1$. Then there is $\overline{N_1} > 0$ such that

$$\|\Phi(t,t_1)x_1\| \leq N_1 e^{-\nu(t-s)}\|\Phi(s,t_1)x_1\|$$

for all $t \geq s \geq t_1$.

Proof. If $s \geq t_1$ and $t \in [s, s + delta]$ then

$$\|\Phi(t,t_1)x_1\| \leq \phi(t-s)\|\Phi(s,t_1)x_1\| \leq \phi(\delta)e^{\nu\delta}e^{-\nu(t-s)}\|\Phi(s,t_1)x_1\| \leq$$
$$\overline{N_1}e^{-\nu(t-s)}\|\Phi(s,t_1)x_1\|$$

where $\overline{N_1} = M_1 + \phi(\delta)e^{\nu\delta}$. ∎

Lemma 2.2. Let $x_2 \in X_2$, $t_2 \geq 0$ and $N_2, \nu, \delta > 0$ such that

$$\|\Phi(t, t_2)x_2\| \geq N_2 e^{-\nu(t-s)}\|\Phi(s, t_2)x_2\| \text{ for all } t \geq s + \delta \geq s \geq t_2.$$

Then there exists $\overline{N_2} > 0$ such that

$$\|\Phi(t, t_2)x_2\| \geq N_2 e^{-\nu(t-s)}\|\Phi(s, t_2)x_2\| \text{ for all } t \geq s \geq t_2.$$

Proof. It is similar to the proof of the preceding Lemma. ∎

Lemma 2.3. Let $t_1 \geq 0$ and $x \in X$. The following statements are equivalent:
   i) there are $N, \nu > 0$ such that

$$\|\Phi(t, t_1)x\| \leq N e^{-\nu(t-s)}\|\Phi(s, t_1)x\|$$

   for all $t \geq s \geq t_1$;
   ii) there is a function $g : \mathbf{R}_+ \to \mathbf{R}_+^*$ such that $\inf\limits_{t \in \mathbf{R}_+^*} g(t) < 1$ and

$$\|\Phi(t, t_1)x\| \leq g(t - s)\|\Phi(s, t_1)x\|$$

   for all $t \geq s \geq t_1$.
   Proof. The implication (i) $\Longrightarrow$ (ii) is obvious. (ii)$\Longrightarrow$ (i) Let δ be a positive real number such that $g(\delta) < 1$. Then for $t \geq s \geq t_0$ there are a natural number n and a real number $r \in [0, \delta)$ such that $t = s + n\delta + r$. Hence

$$\|\Phi(t, t_1)x\| \leq \phi(t - s - n\delta)\|\Phi(s + n\delta, t_1)x\| \leq \phi(\delta)g(\delta)^n,$$
$$\|\Phi(s, t_1)x\| = \phi(\delta)e^{-\nu n\delta}\|\Phi(s, t_1)x\| \leq N e^{-\nu(t-s)}\|\Phi(s, t_1)x\|,$$

for all $t \geq s \geq t_1$, where $\nu = \dfrac{\ln g(\delta)}{\delta} > 0$ and $N = \phi(\delta)e^{\nu\delta} - \dfrac{\phi(\delta)}{g(\delta)} > 0$. ∎

Lemma 2.4. Let $t_2 \geq 0$ and $x \in X$. The following statements are equivalent:
   i) there are $N, \nu > 0$ such that

$$\|\Phi(t, t_2)x\| \leq N e^{-\nu(t-s)}\|\Phi(s, t_2)x\| \text{ for all } t \geq s \geq t_2;$$

   ii) there is a continuous function $h : \mathbf{R}_+ \to \mathbf{R}_+^*$ such that $\sup_{t \in \mathbf{R}_+} h(t) > 1$ and

$$\|\Phi(t, t_2)x\| \leq h(t - s)\|\Phi(s, t_2)x\| \text{ for all } t \geq s \geq t_2.$$

   Proof. (i)$\Longrightarrow$(ii) is obvious.

(ii)$\Longrightarrow$(i). Let $\delta > 0$ such that $h(\delta) > 1$. Then $m = \inf\{h(x) : x \in [0, \delta]\} > 0$. Let n be a natural number and let $r \in [0, \delta)$ such that $t - s = n\delta + r$. Then

$$\|\Phi(t, t_2)x\| \geq h(t - s - n\delta)\|\Phi(s + n\delta, t_2)x\| \geq$$
$$h(r)h(\delta)^n\|\Phi(s, t_2)x\| \geq me^{\nu(t-s)}\|\Phi(s, t_2)x\| \geq$$
$$me^{\nu(t-s)}e^{-\nu r}\|\Phi(s, t_2)x\| \geq$$
$$Ne^{\epsilon(t-s)}\|\Phi(s, t_2)x\|$$

for all $t \geq s \geq t_1$, where $N = me^{-\nu\delta} - \dfrac{m}{h(\delta)} > 0$ and $\nu = \dfrac{\ln h(\delta)}{\delta} > 0$. $\blacksquare$

3. The main results

In this section we present two necessary and sufficient conditions for exponential dichotomy. These results may be regarded as generalizations of well-known results of Datko (1973) and Ichikawa (1984).

Theorem 3.1. The pair (X_1, X_2) induces an exponential dichotomy for the evolution operator Φ if and only if there are two continuous functions $g, h : \mathbf{R}_+ \to \mathbf{R}_+^*$ with

$$\inf_{t\in\mathbf{R}_+^*} g(t) < 1 \quad \text{and} \quad \sup_{t\in\mathbf{R}_+} h(t) > 1$$

such that
d_1') for every $x_1 \in X - 1$ there is $t_1 \geq 0$ such that

$$\|\Phi(t, t_1)x_1\| \leq g(t - s)\|\Phi(s, t_1)x_1\| \text{ for all } t \geq s \geq t_1 + 1$$

and
d_2') for each $x_2 \in X_2$ there exists $t_2 \geq 0$ such that

$$\|\Phi(t, t_2)x_2\| \geq h(t - s)\|\Phi(s, t_2)x_2\| \text{ for all } t \geq s \geq t_2 + 1.$$

Proof. Necessity is obvious. Sufficiency is an imediate consequence of Lemmas 2.3 and 2.4 $\blacksquare$

The main result of this paper is the following:

Theorem 3.2. The pair (X_1, X_2) induces an exponential dichotomy for the evolution operator Φ if and only if there is $M > 0$ such that
d_1'') for every $x_1 \in X_1$ there is $t_1 \geq 0$ such that

$$\int_t^\infty \|\Phi(s, t_1)x_1\|ds \leq M\|\Phi(t, t_1)x_1\|$$

for all $t \geq t_1 + 1$, and

d_2'') for each $x_2 \in X_2$ there exists $t_2 \geq 0$ such that

$$\int_{t_2}^{t} \|\Phi(s,t_2)x_2\|\|ds \leq M\|\Phi(t,t_2)x_2\|$$

for every $t \geq t_2 + 1$.

Proof. *Necessity* is imediate by direct verification.

Sufficiency. Suppose that Φ satisfies the conditions (d_1'') and (d_2''). If $t \geq s+1 \geq s \geq t_1 + 1$ then by (d_1'') we have that

$$\|\Phi(t,t_1)x_1\| \int_0^{t-s} \frac{du}{\phi(u)} = \|\Phi(t,t_1)x_1\| \int_0^{t} \frac{d\tau}{\phi(t-\tau)} \leq$$

$$\int_0^{t} \|\Phi(\tau,t_1)x_1\|d\tau \leq M\|\Phi(s,t_1)x_1\|$$

which implies

$$\|\Phi(t,t_1)x_1\| \leq M_1\|\Phi(s,t_1)x_1\|$$

, for all $t \geq s \geq t_1 + 1$, where $M_1 = \phi(1) + \dfrac{M}{c_1}$ and $c_1 = \int_0^1 \dfrac{du}{\phi(u)}$. By integration on $[s,t]$ we obtain

$$(t - s)\|\Phi(t,t_1)x_1\| = \int_s^{t} \|\Phi(t,t_1)x_1\|d\tau \leq$$

$$M_1 \int_s^{t} \|\Phi(\tau,t_1)x_1\|d\tau \leq MM_1\|\Phi(s,t_1)x_1\|$$

and hence

$$\|\Phi(t,t_1)x_1\| \leq \frac{M_1(1+M)}{1+t-s}\|\Phi(s,t_1)x_1\|$$

for all $t \geq s \geq t_1 + 1$, which shows that the condition (d_1') from Theorem 3.1 is satisfied. Similarly, from the condition (d_2'') it follows that

$$\|\Phi(s,t_2)x_2\| \int_0^1 \frac{du}{\phi(u)} \leq \|\Phi(s,t_2)x_2\| \int_{t_2}^{s} \frac{du}{\phi(s-u)} \leq$$

$$\int_{t_2}^{s} \|\Phi(u,t_2)x_2\|du \leq \int_{t_2}^{t} \|\Phi(u,t_2)x_2\|du \leq M\|\Phi(t,t_2)x_2\|$$

and hence

$$\|\Phi(s,t_2)x_2\| \leq \frac{M}{c_1}\|\Phi(t,t_2)x_2\|$$

for all $t \geq s \geq t_2 + 1$. By integration on $[s, t]$ we obtain

$$(t - s)\|\Phi(s, t_2)x_2\| \leq \frac{M}{c_1} \int_s^t \|\Phi(u, t_2)x_2\|du \leq$$

$$\frac{M}{c_1} \int_{t_2}^t \|\Phi(u, t_2)x_2\|du \leq \frac{M^2}{c_1}\|\Phi(t, t_2)x_2\|$$

and hence

$$\frac{c_1(1 + t - s)}{M + M^2}\|\Phi(s, t_2)x_2\| \leq \|\Phi(t, t_2)x_2\|$$

for all $t \geq s \geq t_2 + 1$. Therefore and the condition (d_2') from Theorem 3.1 is satisfied. Finally, by Theorem 3.1, it follows that (X_1, X_2) induces an exponential dichotomy for Φ. ∎

References

Coppel W.A. (1978), Dichotomies in Stability Theory, Lecture Notes in Mathematics, Nr. 629, Springer-Verlag, Berlin.

Datko R. (1973), Uniform Asymptotic Stability of Evolutionary Processes in a Banach Space, SIAM J. Math. Anal., pp. 428-445.

Ichikawa A. (1984), Equivalence of Lp Stability and Exponential Stability for a Class of Nonlinear Semigroups, Nonlinear Analysis Theory, Methods and Applications , vol. 8, nr. 9, pp. 805-815. Massera J.L.,

Schaffer J.J.(1966), Linear Differential Equations and Function Spaces, Acad. Press, New York and London Preda P.,

Megan M. (1985), Exponential Dichotomy of Evolutionary Processes in Banach Spaces, Czechosl. Math. Journal, vol. 35(110), pp.312-323.

Author's address:

Mihail Megan and Radu Latcu
University of Timisoara
Departament of Mathematics
Bul. V. Parvan Nr. 4
1900-Timisoara-Romania.

International Series of Numerical Mathematics, Vol. 107, © 1992 Birkhäuser Verlag Basel

Asymptotic properties of solutions
to evolution equations

Rodica Luca and Gheorghe Moroşanu

Abstract. In this paper we deal with the asymptotic behaviour of solutions to some nonlinear evolution equations associated to monotone operators. Some first order difference equations are also investigated from the same point of view.

0. Introduction

We shall present some results concerning asymptotic behaviour of solutions to some nonlinear evolution equations and to some difference equations associated with monotone operators.

In the first section we are concerned with a time dependent periodic equation. The equation we had in mind is

$$u(t) + \int_0^t Au(s)ds = f(t), \quad t > 0,$$

where A is a single-valued maximal monotone operator acting in a Hilbert space H and $f \in L^\infty_{loc}(\mathbf{R}_+; H)$. Using the change $w = u - f$, we obtain the following time–dependent evolution equation:

$$w' + A(w + f(t)) = 0. \tag{1}$$

In this section we give some results relating to existence and asymptotic behaviour of solutions to equation (1), as $t \longrightarrow \infty$, for f a periodic function. We mention that A satisfies some severe assumptions including $D(A) = H$. However, we cannot avoid these conditions because, even in the case in which A is a subdifferential, with $D(A) \neq H$, the equation (1) admits in general only "weak solutions", without uniqueness (see Barbu 1979 and Moroşanu 1982). The main result of this section can be compared with that obtained by Baillon and Haraux (1977).

In Section 2 we give a comparison criterion for the averages of solutions to nonlinear evolution equations. In addition, a discrete version of this criterion is formulated.

The third section is devoted to the asymptotic behaviour of the solution of a first order finite–difference equation.

1. A nonlinear periodic evolution equation

Let H be a real Hilbert space with the inner product and the associated norm denoted by $(.,.)$ and $\| \cdot \|$, respectively. Let A be a maximal monotone operator from H into itself, with $D(A) = H$. We suppose the familiarity of the reader with the theory of nonlinear monotone operators and evolution equations of monotone type developed in Hilbert spaces. Background material on this subject can be found in the books by Brézis (1973) and Moroşanu (1988).

Consider the equation:

$$\frac{dw}{dt}(t) + A(w(t) + f(t)) \ni 0, \quad t > 0. \tag{E}$$

Theorem 1 (existence and uniqueness) *If we assume that $f \in L^\infty_{loc}(\mathbf{R}_+; H)$, $A : D(A) = H \longrightarrow H$ is a maximal monotone operator, bounded on bounded subsets of H then there exists a unique strong solution $w \in W^{1,\infty}_{loc}(\mathbf{R}_+; H)$ of the equation (E) with the initial condition $w(0) = w_0 \in H$.*

Remark 1 It seems that this result cannot be deduced from the known theorems relating to time–dependent nonlinear differential equations. Anyway, the proof of this fact is standard.

Proof of Theorem 1 If A is a Lipschitz operator, then the result is obvious. Therefore, for any $\lambda > 0$, there exists a unique solution $w_\lambda \in W^{1,\infty}_{loc}(\mathbf{R}_+; H)$ of the problem:

$$\begin{cases} w'_\lambda(t) + A_\lambda(w_\lambda(t) + f(t)) = 0, & t > 0 \\ w_\lambda(0) = w_0, \end{cases} \tag{3}$$

where A_λ is is the Yosida approximate of A.

We multiply (3) by $w_\lambda(t)$ and using the inequality:

$$(A_\lambda(w_\lambda(t) + f(t)), w_\lambda(t)) \geq (A_\lambda(f(t)), w_\lambda(t))$$

we obtain:

$$\frac{1}{2}\frac{d}{dt}\|w_\lambda(t)\|^2_2 \leq \|A^0(f(t))\| \cdot \|w_\lambda(t)\|, \quad a.e. \ t > 0, \tag{4}$$

where A^0 is the minimal section of A. For $T > 0$ fixed, we integrate (4) over $[0, T]$ and we get:

$$\|w_\lambda(t)\| \leq C(T), \quad 0 \leq t \leq T. \tag{5}$$

By (3) and (5) we deduce

$$\begin{aligned} \|w'_\lambda(t)\| = \|A_\lambda(w_\lambda(t) + f(t))\| &\leq \\ &\leq \|A^0(w_\lambda(t) + f(t))\| \leq C_1(T), \quad a.e. \ t \in (0, T). \end{aligned} \tag{6}$$

Now, we multiply the equation

$$w'_\lambda(t) - w'_\mu(t) + A_\lambda(w_\lambda(t) + f(t)) - A_\mu(w_\mu(t) + f(t)) = 0, \quad \lambda, \mu > 0, \quad a.e.\ t \in (0, T),$$

by $w_\lambda(t) - w_\mu(t)$ to obtain that:

$$\frac{1}{2}\frac{d}{dt}\|w_\lambda(t) - w_\mu(t)\|^2 =$$
$$= -(A_\lambda(w_\lambda(t) + f(t)) - A_\mu(w_\mu(t) + f(t)), J_\lambda(w_\lambda(t) + f(t)) - J_\mu(w_\mu(t) + f(t)))-$$
$$- (A_\lambda(w_\lambda(t) + f(t)) - A_\mu(w_\mu(t) + f(t)), \lambda A_\lambda(w_\lambda(t) + f(t)) - \mu A_\mu(w_\mu(t) + f(t))),$$
$$0 \le t \le T,$$

$$(7)$$

where $J_\lambda = (I + \lambda A)^{-1}$. Taking into account (6) and (7) we can easily show that

$$\frac{d}{dt}\|w_\lambda(t) - w_\mu(t)\|^2 \le 4C_1^2(T)(\lambda + \mu), \quad \forall \lambda, \mu > 0, \quad a.e.\ t \in (0, T).$$

Therefore

$$\|w_\lambda(t) - w_\mu(t)\| \le 2C_1(T)t^{\frac{1}{2}}(\lambda + \mu)^{\frac{1}{2}}, \quad 0 \le t \le T. \tag{8}$$

But (8) says that there exists a $w \in C([0, T]; H)$ such that $w_\lambda \longrightarrow w$, as $\lambda \longrightarrow 0$, in $C([0, T]; H)$.

Using a standard argument (see, e.g., Moroşanu 1988, p.50) we obtain that $w \in W_{loc}^{1,\infty}(\mathbf{R}_+; H)$ is the unique solution of (E) with the initial condition $w(0) = w_0$. Thus, the proof is complete. $\qquad$ Q.E.D.

The main result of this section is

Theorem 2 *Assume that $\varphi : H \longrightarrow \mathbf{R}$ is a convex function, Gâteaux differentiable on H and $A = \nabla\varphi$ is bounded on bounded subsets of H. Let $f \in L^\infty(\mathbf{R}_+; H)$ be a T-periodic function $(T > 0)$. Then equation (E) has a solution bounded on $\mathbf{R}_+$ if and only if it has at least one T-periodic solution. In this case, all the solutions of (E) are bounded on $\mathbf{R}_+$ and for every solution $w(t)$, $t \ge 0$ there exists a T-periodic solution p of (E) such that*

$$w(t) - p(t) \longrightarrow 0, \quad as\ t \longrightarrow \infty\ weakly\ in\ H. \tag{9}$$

Proof Using the monotonicity of $A = \nabla\varphi$ we deduce that, if the equation (E) has a bounded solution, then any other solution of (E) is bounded on $\mathbf{R}_+$.

Now, we are going to prove that the boundedness of solutions on $\mathbf{R}_+$ implies the existence of a periodic solution. To this purpose we define $Q : H \longrightarrow H$, $Q(x) = w(T; x)$, where $w(t; x), t \ge 0$ is the solution starting from $x \in H$ (see Theorem 1). It is easy to see that Q is nonexpansive and for each $x \in H$ the sequence $\{Q^n(x)\}$ is bounded, because $Q^n(x) = w(nT; x)$. Then, according to a fixed point theorem due to Browder and Petryshyn (1966), Q has at least one fixed point. But, this means that (E) has at least one T–periodic solution.

For the second part of the theorem, we shall use Opial's lemma (see, e.g., Moroşanu 1988, p.5). Let $w(t), t \geq 0$ be a solution of (E), bounded on $\mathbf{R}_+$ and let $w_n : [0, T] \longrightarrow H$, $w_n(t) = w(t + nT)$, $n \in \mathbf{N}^*$. We denote by $\mathbf{F}$ the set of all T–periodic solutions of (E) and by $\mathbf{H}$ the space $L^2(0, T; H)$. For any $p \in \mathbf{F}$ we have:

$$\|w_n(t) - p(t)\| \leq \|w_n(0) - p(0)\| = \|w_{n-1}(T) - p(T)\| \leq$$
$$\leq \|w_{n-1}(t) - p(t)\| \leq \|w_{n-1}(0) - p(0)\|, 0 \leq t \leq T.$$

Therefore

$$\lim_{n \to \infty} \|w_n(t) - p(t)\| = l(p), \tag{10}$$

uniformly on $[0, T]$ and, consequently,

$$\|w_n - p\|_{\mathbf{H}} \longrightarrow \hat{l}(p) := T^{\frac{1}{2}} l(p), \forall p \in \mathbf{F},$$

i.e. the first condition of Opial's lemma is verified.

Let $q \in L^\infty(0, T; H)$ be a weak cluster point of $\{w_n\}$. I.e. there exists a subsequence $\{w_{n_k}\}$ of $\{w_n\}$ such that $w_{n_k} \longrightarrow q$ weakly-star in $L^\infty(0, T; H)$. Moreover, because A is bounded on bounded subsets of H the sequence $\frac{dw_{n_k}}{dt}$ is bounded and we can suppose that $\frac{dw_{n_k}}{dt} \longrightarrow \frac{dq}{dt}$, weakly–star in $L^\infty(0, T; H)$.

Denote by $\bar{A}$ the operator $\bar{A} : D(\bar{A}) \subset \mathbf{H} \longrightarrow \mathbf{H}, D(\bar{A}) = \{u \in \mathbf{H}, \exists v \in \mathbf{H}$ such that $v(\xi) = \nabla\varphi(u(\xi) + f(\xi))$, a.e. $\xi \in (0, T)\}$, $\bar{A}(u) = \nabla\varphi(u + f)$. Obviously $\bar{A} = \partial\Phi$, where $\Phi : \mathbf{H} \longrightarrow] -\infty, \infty]$ is defined by :

$$\Phi(v) = \begin{cases} \int_0^T \varphi(v(\xi) + f(\xi))d\xi & \text{,if } \varphi \circ (v + f) \in L^1(0, T); \\ +\infty & \text{,otherwise.} \end{cases}$$

Now, from (10) we have:

$$\left\langle \frac{dw_n}{dt} - \frac{dp}{dt}, w_n - p \right\rangle_{\mathbf{H}} = \frac{1}{2}(\|w_n(T) - p(T)\|^2 -$$
$$- \|w_n(0) - p(0)\|^2) \longrightarrow 0, \quad \text{as } n \longrightarrow \infty, \quad \forall p \in \mathbf{F}. \tag{11}$$

On the other hand,

$$\Phi(w_n) - \Phi(v) \leq - \left\langle \frac{dw_n}{dt} - \frac{dp}{dt}, w_n - p \right\rangle_{\mathbf{H}} -$$
$$- \left\langle \frac{dp}{dt}, w_n - p \right\rangle_{\mathbf{H}}, \forall v \in D(\Phi), p \in \mathbf{F}. \tag{12}$$

Now, using (11) and (12) we can see that:

$$\Phi(q) - \Phi(v) \le \left\langle -\frac{dp}{dt}, q - v \right\rangle_{\mathbf{H}}, \forall v \in D(\Phi).$$

It follows that:

$$-\frac{dp}{dt} \in \bar{A}(q), \quad \forall p \in \mathbf{F}. \tag{13}$$

A similar resoning leads us to

$$-\frac{dq}{dt} \in \bar{A}(p). \quad \forall p \in \mathbf{F}.$$

Because $\bar{A}$ is single–valued, we obtain that $dp/dt = dq/dt$, $\forall p \in \mathbf{F}$.

Therefore (see also (13)), $q \in \mathbf{F}$. We see that every periodic solution differs from another one by an additive constant. Since both conditions of Opial's lemma are verified, there exists some $p \in \mathbf{F}$ such that $w_n \longrightarrow p$, weakly in $\mathbf{H}$, as $n \longrightarrow \infty$. This means that $q = p$ and all we have proved for w_{n_k} is still valid for w_n. Using the equality

$$tw_n(t) = \int_0^t [sw'_n(s) + w_n(s)]ds,$$

we deduce that $w_n(t) \longrightarrow p(t)$, weakly in H, as $n \longrightarrow \infty$, $\forall\, t \in [0,T]$. By the Arzela–Ascoli Criterion the sequences $\{(w_n(t), x)\}_n$ $(x \in H)$ are relatively compact. Therefore, we have (eventually on subsequences) $w_n(t) \longrightarrow p(t)$, weakly in H, as $n \longrightarrow \infty$, uniformly with respect to $t \in [0,T]$. Now, for $t \in \mathbf{R}_+$, we have

$$w(t) - p(t) = w(\theta + nT) - p(\theta + nT) = w_n(\theta) - p(\theta) \quad (0 \le \theta < T)$$

and therefore $w(t) - p(t) \longrightarrow 0$, weakly in H, as $t \longrightarrow \infty$. The proof is complete.Q.E.D.

Remark 2 We mention that if $A = (\partial\varphi)_\lambda$, where $\lambda > 0$ and $\varphi : H \longrightarrow] - \infty, +\infty]$ is a proper, convex and lower–semicontinuous function, then all the assumptions of Theorem 2 are satisfied.

2. Comparison criteria

Theorem 3 *Assume that $A : \underline{D(A)} \subset H \longrightarrow H$ is a maximal monotone operator and $f \in L^1(\mathbf{R}_+; H)$. Let $\{S(t) : \overline{D(A)} \longrightarrow \overline{D(A)}; t \ge 0\}$ be the semigroup generated by $-A$. If*

$$\forall x \in \overline{D(A)}, \; \sigma(t)x := \frac{1}{t} \int_0^t S(s)x \; ds \; converges \tag{14}$$

$$strongly \; (weakly), \; as \; t \longrightarrow \infty,$$

then for every weak solution $z(t)$, $t \geq 0$ of

$$\frac{dz}{dt} + A(z) \ni f(t), \ t > 0, \tag{$\widetilde{E}$}$$

the average $\sigma_z(t) := \frac{1}{t}\int_0^t z(s)ds$ converges strongly (respectively, weakly) as $t \longrightarrow \infty$.

Proof Firstly, we mention that the assumption (14) implies that $F := A^{-1}(0) \neq \emptyset$ and for any $x \in \overline{D(A)}$, $S(t)x$ and $z(t)$, the solution of $(\widetilde{E})$ with $z(0) = x$, are bounded on $\mathbf{R}_+$ (see Moroşanu 1988, p.70). Let $z(t)$ be a weak solution of $(\widetilde{E})$. Define $f_n : \mathbf{R}_+ \longrightarrow H$ as follows:

$$f_n(t) = \begin{cases} f(t), & t < n \\ 0, & t \geq n, \end{cases}$$

where $n \in \mathbf{N}^*$. For each $n \in \mathbf{N}^*$, denote by $z_n(t)$, $t \geq 0$, the weak solution of the Cauchy problem :

$$\begin{cases} z_n'(t) + A(z_n(t)) \ni f_n(t), t > 0 \\ z_n(0) = z(0). \end{cases} \tag{$E)_n$}$$

We have

$$z_n(t) = \begin{cases} z(t), & t < n \\ S(t-n)z(n), & t \geq n. \end{cases}$$

Now, using the assumption (14), we can see that for each $n \in \mathbf{N}^*$, there exists a $p_n \in H$ such that

$$\lim_{t \to \infty} \frac{1}{t} \int_0^t S(s)z(n)ds = p_n, \ strongly \ (respectively \ weakly).$$

After some computations, we obtain

$$\lim_{t \to \infty} \frac{1}{t} \int_0^t z_n(\tau)d\tau = p_n, \ strongly \ (respectively \ weakly). \tag{15}$$

Taking into account the estimate

$$\|z_n(t) - z_m(t)\| \leq \int_m^n \|f(s)\|ds, \ m < n, \ t \geq 0$$

we have

$$\left\| \frac{1}{t}\int_0^t z_n(s)ds - \frac{1}{t}\int_0^t z_m(s)ds \right\| \leq \frac{1}{t}\int_0^t \|z_n(s) - z_m(s)\|ds \leq$$

$$\leq \int_m^n \|f(s)\|ds, \ m < n, \ t > 0$$

Therefore

$$\|p_n - p_m\| \le \int_m^n \|f(s)\| ds.$$

Hence $\{p_n\}$ is a Cauchy sequence in H, hence there exists $p \in H$ such that

$$p_n \longrightarrow p, \, as \, n \longrightarrow \infty, \, strongly \, in \, H. \tag{16}$$

On the other hand, since $z(t)$ and $z_n(t)$ are weak solutions of $(\widetilde{E})$ and $(E)_n$, respectively, we have

$$\|z(t) - z_n(t)\| \le \int_n^t \|f(s)\| ds, \quad t \ge n,$$

and

$$\left\| \frac{1}{t} \int_0^t [z(s) - z_n(s)] ds \right\| = \left\| \frac{1}{t} \int_n^t [z(s) - z_n(s)] ds \right\| \le$$
$$\le \frac{1}{t} \int_n^t \|z(s) - z_n(s)\| ds \le \int_n^t \|f(s)\| ds, \quad t \ge n. \tag{17}$$

Now, using the decomposition

$$\frac{1}{t} \int_0^t z(s) ds - p = \left[\frac{1}{t} \int_0^t z(s) ds - \frac{1}{t} \int_0^t z_n(s) ds \right] + \left[\frac{1}{t} \int_0^t z_n(s) ds - p_n \right] + [p_n - p]$$

and $(15), (16), (17)$ we deduce that

$$\lim_{n \to \infty} \frac{1}{t} \int_0^t z(s) ds = p \, in \, the \, strong \, (respectively \, weak) \, topology.$$

Q.E.D.

In what follows we shall be concerned with the discrete version of Theorem 3. We consider the following first order finite difference equations:

$$\begin{cases} y_n & = (I + c_n A)^{-1}(y_{n-1}), \quad (n = 1, 2, ...) \\ y_0 & = x \in H, \end{cases} \tag{18}$$

where $c_n > 0$ $(n = 1, 2, ...)$ and

$$\begin{cases} x_n & = (I + c_n A)^{-1}(x_{n-1}) + f_n, \quad (n = 1, 2, ...) \\ x_0 & = x, \end{cases} \tag{19}$$

with $f_n \in H$ $(n = 1, 2, ...)$.

If A is a maximal monotone operator, then $\forall x \in H$, $\forall \{c_n\} \subset \,]0, +\infty[$ and $\forall \{f_n\} \subset H$, the problems (18) and (19) have unique solutions $\{y_n\}$, respectively $\{x_n\}$. We also consider the following sequences:

$$\sigma_n = \Big(\sum_{i=1}^{n} c_i \Big)^{-1} \Big(\sum_{i=1}^{n} c_i y_i \Big), \quad n = 1, 2, \dots \tag{20}$$

and

$$w_n = \Big(\sum_{i=1}^{n} c_i \Big)^{-1} \Big(\sum_{i=1}^{n} c_i x_i \Big), \quad n = 1, 2, \dots \tag{21}$$

where $\{y_n\}$ and $\{x_n\}$ are the solutions of (18), respectively (19).

Theorem 4 *Let $A : D(A) \subset H \longrightarrow H$ be a maximal monotone operator. Assume that $\forall x \in H, \forall \{c_n\} \notin l^1$, the sequence $\{\sigma_n\}$ defined by (20) converges strongly (weakly). Then, $\forall x \in H$, $\forall \{c_n\} \notin l^1$, $\forall \{f_n\}$ satisfying $\{\|f_n\|\} \in l^1$, the sequence $\{w_n\}$ defined by (21) converges strongly (respectively weakly).*

The proof of this result is similar to that of Theorem 3. So, for the sake of brevity, it will be omitted.

Remark 3 A comparison criterion for $S(t)x$ and $z(t)$ can be found in Moroşanu (1988, p.77). See also Brezis and Lions (1978) for the discrete case.

3. A convergence result

In this section we give the discrete version of a result due to Pazy (1979) (see also Moroşanu 1988, p.115).

Theorem 5 *Let $A : D(A) \subset H \longrightarrow H$ be an odd, maximal monotone operator. Assume that*

$$\forall x \in H, \ \|y_n - y_{n-1}\| \longrightarrow 0, \tag{22}$$

where $\{y_n\}$ is the solution of (18). Then, $\forall x \in H$, $\{y_n\}$ converges strongly. If, in addition, $\{c_n\} \notin l^1$, then the limit of $\{y_n\}$ belongs to $A^{-1}(0)$.

Proof Let $x \in H$ be fixed. Since A is odd (i.e., $D(A) = -D(A)$ and $A(-x) = -A(x)$, $\forall x \in D(A)$) we have that $0 \in A^{-1}(0)$ and, therefore, $\{\|y_n\|\}$ is nonincreasing. So, there is $a \geq 0$ such that

$$\|y_n\| \longrightarrow a \tag{23}$$

Note also that the sequence $\{\|y_{n+m} + y_n\|\}_n$ is nonincreasing for every m. Hence, taking also into account (22) and (23), we get

$$2a = 2\lim_{k\to\infty}\|y_k\| \leq \lim_{k\to\infty}\|y_k + y_{k+m}\| + \lim_{k\to\infty}\|y_k - y_{k+m}\| \leq$$
$$\leq \|y_n + y_{n+m}\|, \quad \forall n, m \in \mathbf{N}.$$

Therefore, using the equality

$$\|y_{n+m} - y_n\|^2 = 2\|y_n\|^2 + 2\|y_{n+m}\|^2 - \|y_{n+m} + y_n\|^2$$

we deduce that

$$\|y_{n+m} - y_n\|^2 \leq 4(\|y_n\|^2 - a^2)$$

wich implies that $\{y_n\}$ is strongly convergent to some p. If, in addition, $\{c_n\} \notin l^1$ then clearly $\{\sigma_n\}$ defined by (20) converges strongly to p and $p \in A^{-1}(0)$ (see, e.g., Moroşanu 1988, p. 139).

Remark 4 Since

$$\|y_n - y_{n-1}\| = c_n\|A_{c_n}(y_{n-1})\| \leq c_n\|A^0(y_{n-1})\| \leq \|A^0(y_1)\|c_n,$$

we can see that $c_n \longrightarrow 0$ implies (22). For example, if $c_n = n^{-r}, 0 < r \leq 1$, and A is maximal monotone and odd then y_n converges strongly to some $p \in A^{-1}(0)$. Also, (22) holds if $\{c_n\}$ is bounded and $\|A^0(y_n)\| \longrightarrow 0$.

References

Baillon J. B. and Haraux A. (1977), Comportement à l'infini pour les équations d'évolution avec forcing périodique, Archive Rat. Mech. Anal. 67, 101-109.

Barbu V. (1979), Necessary conditions for boundary control problems governed by parabolic variational inequalities, INCREST Technical Report No. 53.

Brézis H. (1973), Opérateurs Maximaux Monotones et Semigroupes de Contractions dans les Espaces de Hilbert, North Holland, Amsterdam.

Brézis H. and Lions P. L. (1978), Produits infinis de resolvents, Israel J. Math. 29, 329-345.

Browder F. and Petryshyn W. V. (1966), The solution by iteration of nonlinear functional equations in Banach spaces, Bull. Amer. Math. Soc. 72, 571-575.

Moroşanu G. (1988), Nonlinear Evolution Equations and Applications, D.Reidel, Dordrecht.

Moroşanu G. (1982), Asymptotic dosing problem for evolution equations in Hilbert spaces, Anal. St. Univ. "Al. I. Cuza" Iasi, Sect. I-a Mat. (N.S.), 28, 127-137.

Pazy A. (1979), Semigroups of nonlinear contractions and their asymptotic behaviour, "Nonlinear Analysis and Mechanics : Heriot–Wat Symposium", vol.III, Research Notes in Math.,30, Pitman, London.

Addresses of the authors:

Rodica Luca   Gheorghe Moroşanu
Department of Mathematics   Faculty of Mathematics
Polytechnical Institute   University of Iaşi
6600 Iaşi, Romania   6600 Iaşi, Romania

International Series of Numerical Mathematics, Vol. 107, © 1992 Birkhäuser Verlag Basel   63

ON SOME. NON LINEAR ELASTIC WAVES BIPERIODICAL, OR ALMOST PERIODICAL IN MECHANICS AND EXTENSIONS HYPERBOLIC NON LINEAR PARTIAL DIFFERENTIAL EQUATIONS

By ROBERT FAURE [*]

(*) Laboratoire de Mecanique de l'Université de Lille I

Laboratoire MEDIMAT - Université Pierre et Marie Curie.

RESUME: In this paper we study the existence of different forms of biperiodical or almost periodical waves. The excitation can be continuous or sequence of impulses; in the peculiar case of almost periodical excitation we examine also the cases of "almost periodical distribution excitation". For the precedent cases we study the solution in the neighbourhood of equilibrium position. We study also the case of the "resonance" and also the case of "bicycles": periodical solution of autonomous non linear equation, and, a biological problem (Fitz Hugh Naguno equation).

Here we study the propagation of an nonlinear elastic wave in homogeneous medium along the x'x axis.

The coordenates of the phenomena are x and t: the deplacement is $u(x,t)$.

The equation of the vibration is here:

$$\frac{\partial^2 u}{\partial x^2} - \frac{1}{c^2}\frac{\partial^2 u}{\partial t^2} + a\frac{\partial u}{\partial x} + b\frac{\partial u}{\partial t} + cu = \eta\, F(x,t,u,\frac{\partial u}{\partial x},\frac{\partial u}{\partial t})$$

where a, b, c are constant, η a little parameter; $\eta > 0$. After a change of variables and notations, we have the partial differential equation (E):

$$\frac{\partial^2 u}{\partial x^2} - \frac{\partial^2 u}{\partial t^2} + a\frac{\partial u}{\partial x} + b\frac{\partial u}{\partial t} + cu = \eta\, F(x,t,u,\frac{\partial u}{\partial x}, \frac{\partial u}{\partial t})$$

$$F(x,t,u,\frac{\partial u}{\partial x}, \frac{\partial u}{\partial t}) = G(x,t) + K(x,t,u,\frac{\partial u}{\partial x}, \frac{\partial u}{\partial t})$$

with $K(x,t,u,\frac{\partial u}{\partial x}, \frac{\partial u}{\partial t}) \equiv 0$, for $u(x,t) \equiv 0$

We use in this paper two real Banach's spaces:

a) the space $B(\omega_1,\omega_2)$; the space of the biperiodical series absolutily convergent:

If $f \in B(\omega_1,\omega_2)$: $f = \sum f_{m,n}\, e^{\,mi\omega_1 x + ni\omega_2 t}$

where $\|f\| = \sum_{m,n} |f_{m,n}|$ for all m,n ; $m,n \in \mathbb{Z}$; $\|f\| < \infty$.

ω_1 and ω_2 are constant with $\omega_1 T_1 = \omega_2 T_2 = 2\pi$.

b) We use also one other Banach's space, generalization of the last $B(\omega_1,\omega_2)$

We let if $f \in B(\Omega_1,\Omega_2)$: $f = \sum f_{m,n}\, e^{\,i\omega_{1,m} x + i\omega_{2,n} t}$

with $\|f\| = \sum_{m,n} |t_{m,n}|$, for all m,n; $\|f\| < \infty$.

$\omega_{1,m} \in \Omega_1$, $\omega_{2,n} \in \Omega_2$, Ω_1 , Ω_2 are the sets of the exponents of f and they are infinite denombrable sets.

I - Solution in the neighbourhood of the equilibrium:

We make here the Hypothesis H_1

$$K = \sum_{\substack{i+j+k=n \\ i+j+k=1}} a_{i,j,k} \, u^i \left(\frac{\partial u}{\partial x}\right)^j \left(\frac{\partial u}{\partial t}\right)^k ; \quad a_{i,j,k} \in B(\omega_1,\omega_2); \quad \text{for all } i,j,k.$$

$G(x,t)$ no constant , $G(x,t) \in B(\omega_1,\omega_2)$; $n \geq 2$.

and $\dfrac{\partial G(x,t)}{\partial x}$, $\dfrac{\partial G(x,t)}{\partial t} \in B(\omega_1,\omega_2)$.

For demonstrade the existence of the solution we write the equation (E) in the form

$$(E_1) \qquad\qquad u = \eta \, T(K+G)$$

$$(E_2) \qquad\qquad \frac{\partial u}{\partial x} = \eta \, T_1(K+G)$$

$$(E_3) \qquad\qquad \frac{\partial u}{\partial t} = \eta \, T_2(K+G)$$

where T, T_1, T_2 are the tranformations in the Banach's space $B(\omega_1,\omega_2)$ which are defined in the following lines:

If u, K, G are defined by their Fourier's developments

$$u = \sum u_{m,n} \, e^{mi\omega_1 x + ni\omega_2 t} ;$$

$$K = \sum K_{m,n} \, e^{mi\omega_1 x + ni\omega_2 t},$$

$$G = \sum G_{m,n} \, e^{mi\omega_1 x + ni\omega_2 t}$$

The equations (E_i) give the three formulas

$$u = \sum t_{m,n} (K_{m,n} + G_{m,n}) \, e^{mi\omega_1 x + ni\omega_2 t}$$

$$\frac{\partial u}{\partial x} = \sum t_{1_{m,n}} (K_{m,n} + G_{m,n}) \, e^{\, mi\omega_1 x \, + \, ni\omega_2 t}$$

$$\frac{\partial u}{\partial t} = \sum t_{2_{m,n}} (K_{m,n} + G_{m,n}) \, e^{\, mi\omega_1 x \, + \, ni\omega_2 t}$$

with
$$t_{m,n} = \left[-m^2\omega_1^2 + n^2\omega_2^2 + c + \left(am\omega_1 + bn\omega_2 \right) i \right]^{-1} *$$

$$t_{1_{m,n}} = t_{m,n} \, m\omega_1 i \, ,$$

$$t_{2_{m,n}} = t_{m,n} \, n\omega_2 i$$

We make then the hypothesis H_2: all $t_{m,n}$, $t_{1_{m,n}}$ and $t_{2_{m,n}}$ have a finite upper bound ρ for T, ρ_1 for T_1, ρ_2 for T_2, we let $d \geq (\rho, \rho_1, \rho_2)$.

For resolve the problem (I), we have the three convergent recurrences defined by

$$u_{n+1} = \eta \, T(K+G) = \eta T \left[K(x, t, u_n, \frac{\partial u_n}{\partial x}, \frac{\partial u_n}{\partial t}) + G(x,t) \right]$$

$$\frac{\partial u_{n+1}}{\partial x} = \eta T_1 (K+G)$$

$$\frac{\partial u_{n+1}}{\partial t} = \eta T_2 (K+G)$$

with
$$u_0 \equiv \frac{\partial u_0}{\partial x} \equiv \frac{\partial u_0}{\partial t} \equiv 0.$$

If $A_{i,j,k} = \| a_{i,j,k} \|$ and

$$\bar{K}(u, v, w) = \sum_{i+j+k=1}^{i+j+k=n} A_{j,k} \, u^i v^j w^k$$

and if l is a commom bound for $\| u_n \|$, $\left\| \frac{\partial u_n}{\partial x} \right\|$, $\left\| \frac{\partial u_n}{\partial t} \right\|$. We have $\| u_{n+1} \|$, $\left\| \frac{\partial u_{n+1}}{\partial x} \right\|$, $\left\| \frac{\partial u_{n+1}}{\partial t} \right\|$ bounded by $d\eta \left[\bar{K}(l, l, l) + \| G(x,t) \| \right]$ and if we have

$$d\eta \left[\overline{K}(1,1,1) + \|G(x,t)\| \right] \leq 1$$

all the three terms of recurrence are bounded by 1.

We know that the three recurances converge if we have

$$d\eta \left[\frac{\partial K(1,1,1)}{\partial u} + \frac{\partial K(1,1,1)}{\partial v} + \frac{\partial K(1,1,1)}{\partial w} \right] < 1.$$

For η sufficiently small the two inequatities are verified, we know also that 1, can be taken arbitrary small provided that η is also sufficiently small.

All the three recurrences u_n, $\dfrac{\partial u_n}{\partial x}$, $\dfrac{\partial u_n}{\partial t}$ which belong to $B(\omega_1, \omega_2)$ have uniform limits: L , $\dfrac{\partial L}{\partial x}$, $\dfrac{\partial L}{\partial t}$.

We can also see that the three others recurrences linked with the last three

$$\frac{\partial^2 u_{n+1}}{\partial x^2} = \eta T_1 \left[\frac{\partial K}{\partial x} + \frac{\partial G}{\partial x} \right](x,t,u_n)$$

$$\frac{\partial^2 u_{n+1}}{\partial x \partial t} = \eta T_2 \left[\frac{\partial K}{\partial x} + \frac{\partial G}{\partial x} \right](x,t,u_n)$$

$$\frac{\partial^2 u_{n+1}}{\partial t^2} = \eta \overline{T}_2 \left[\frac{\partial K}{\partial t} + \frac{\partial G}{\partial t} \right](x,t,u_n)$$

with $u_o(x,t) \equiv 0$, belong for η sufficiently small, to $B(\omega_1, \omega_2)$ and converge uniformily to $\dfrac{\partial^2 L}{\partial x^2}$, $\dfrac{\partial^2 L}{\partial x \partial t}$ and $\dfrac{\partial^2 L}{\partial t^2}$

We have the theorem.

The equation (E) for η sufficiently small have a biperiodical solution. This solution tends to zero wich η.

We can make the remark: the same method can be used when $K(u,k,l)$ is a serie in u.

II- Biperiodical solutions excited by biperiodical impulses.

We study here the equation (E); here:

$$K = \sum_{i=1}^{i=n} a_i(x,t)\, u^i \qquad ; \; a_i(x,t) \in B(\omega_1,\omega_2)$$

But $G(x,t)$ is not continuous: $G(x,t) = s_1 e_1(x) + s_2 e_2(t)$

$$e_1(x) = \sum_m \delta(x-mT_1) \; ; \; e_2(x) = \sum_n \delta(x-mT_2)$$

for all values of m and n $m,n \in \mathbb{Z}$.

The equation here also is written so a transformation in Banach's space. We have the equation:

(E_1) $\qquad\qquad\qquad\qquad\qquad\qquad u = \eta\, T(K+G)$

Here $T(G) = s_1 \varphi_1(x) + s_2 \varphi_2(t)$; were

$$\varphi_1(x) = \sum (-m^2\omega_1^2 + am\omega_1 i + c)^{-1}\, e^{mi\omega_1 t}$$

$$\varphi_2(t) = \sum (-n^2\omega_2^2 + bn\omega_2 i + c)^{-1}\, e^{ni\omega_2 t}$$

were a, b and c are differents on zero; φ_1 and φ_2 belong to $B(\omega_1,\omega_2)$. The recurence is possible. The problem (E) has not a solution, but (E_1) has a solution u with $u \in B(\omega_1,\omega_2)$ for η so small.

III - The case of an excitation by bialmost periodical distribution.

We study the problem analogous that in the previus case. K is the same that in (I), but we take

$$G(x,t) = \sum a_m\, e^{i\omega_{1,m} t} + \sum b_n\, e^{i\omega_{2,m} t}$$

m,n take all values; $m,n \in \mathbb{Z}$, and $\omega_{1,m} \in \Omega_1$, $\omega_{2,n} \in \Omega_2$. Ω_1 and Ω_2 are two denanbrables set.

When we write the transformation

(E_1) $\qquad\qquad\qquad\qquad u = \eta\; T(K+G(x,t))$

we suppose that the series

$$\sum |a_m|\; (-\omega_{1,m}^2 + a\omega_{1,m}i + c)^{-1}$$

and $\qquad\qquad\qquad \sum |b_n|\; (-\omega_{2,n}^2 + b\omega_{2,n}i + c)^{-1}$

are absolutily convergent; Then $T(G(x,t)) \in B(\Omega_1,\Omega_2)$ is bialmost-periodical.

We have seen that the problem: for η sufficiently small (E) have not solution, but (E_1) has a solution bialmost periodical which belong to $B(\Omega_1,\Omega_2)$.

We must say also that here $G(x,t)$ is a distribution but we cannot, in the state of the theory of almost periodical distributions, know if a Fourier's development is in connection with a sequence of impulses or not.

IV - Resonance case.
____

Here we study the PPD (E_1),

$$(E_1): \frac{\partial^2 u}{\partial x^2} - \frac{\partial^2 u}{\partial t^2} + \eta(\; a\frac{\partial u}{\partial x} + b\frac{\partial u}{\partial t} + cu\;) =$$

$$= \eta\left[K(x,t,u,\; \frac{\partial u}{\partial x},\; \frac{\partial u}{\partial t}) + G(x,t)\right]$$

where K have the same form that in the first part (I)

$$K = \sum_{i,j,k} a_{i,j,k}\; u^i\; (\frac{\partial u}{\partial x})^j\; (\frac{\partial u}{\partial t})^k \quad \text{with} \quad 1 \le i+j+k \le n,\; n > 2$$

with the hypothesis H;

1°) $\omega_1 = \omega_2 = \omega$; $T_1 = T_2 = T$

2°) It exists an u_0 no constant $u_0 \in B(\omega,\omega)$; with $\dfrac{\partial^2 u_0}{\partial x^2} - \dfrac{\partial^2 u_0}{\partial t^2} = 0.$

3°) u_0 verifies the conditions

$$\mathcal{M}\left\{\left[K(u_0,x,t,\frac{\partial u_0}{\partial x},\frac{\partial u_0}{\partial t}) - a\frac{\partial u_0}{\partial x} - b\frac{\partial u_0}{\partial t} - cu_0 + G(x,t)\right]e^{mi\omega(x+\varepsilon t)}\right\}$$

$$= 0$$

for all vales of $m \in \mathbb{Z}$ and $\varepsilon = \pm 1$ where

$$\mathcal{M}(f) = T^{-2}\int_0^T\int_0^T f(x,t)dxdt$$

We let $u = u_0 + v$ and we have the new equation (E_2).

$$(E_2): \quad \frac{\partial^2 v}{\partial x^2} - \frac{\partial^2 v}{\partial t^2} + \eta\left(a\frac{\partial v}{\partial x} + b\frac{\partial v}{\partial t} + cv\right) = \eta R$$

where

$$R = K(u_0+v) - a\frac{\partial u_0}{\partial x} - b\frac{\partial u_0}{\partial t} - cu_0 + G(x,t).$$

We can write also R

$$R = K(u_0, x, t) - a\frac{\partial u_0}{\partial x} - b\frac{\partial u_0}{\partial t} - cu_0 + K'_{u_0}v + K'_{\frac{\partial u_0}{\partial x}}\frac{\partial v}{\partial x} +$$

$$+ K'_{\frac{\partial u}{\partial t}}\frac{\partial v}{\partial t} + Q(v,\frac{\partial v}{\partial x},\frac{\partial v}{\partial t}).$$

Q is a polynomial in $v, \frac{\partial v}{\partial x}, \frac{\partial v}{\partial t}$ of degree higher or equal at two.

For resolve (E_2) we define three subspaces of $B(\omega,\omega)$:

with $f = \sum f_{m,n}\, e^{mi\omega x + ni\omega t}$

a) $B_1(\omega,\omega)$ is the space of the Fouries's series where $m = n$.

b) $B_2(\omega,\omega)$ then where $m = -n$.

c) $B_3(\omega,\omega)$ where $m^2 \neq n^2$.

We write $v = v_1 + v_2 + v_3$; $v_i \in B_i(\omega,\omega)$

In the space B_1; v_1 is a function of $X = x + t$ only

B_2; v_2 is a function of $Y = x - t$ only.

If the projections of $\dfrac{\partial K}{\partial u}, \dfrac{\partial K}{\partial p}, \dfrac{\partial K}{\partial q}$ with $p = \dfrac{\partial u}{\partial x}$, $q = \dfrac{\partial u}{\partial t}$ for $v = 0$ on B_1, B_2, B_3 are respectively K'_{11}, K'_{21}, K'_{31}, K'_{12}, K'_{22}, K'_{32}, K'_{13}, K'_{23}, K'_{33}:

We have them for the two first projections of E_2 on B_1, B_2.

(E_3):
$$(a + b - K'_{22} - K'_{23})\frac{dv_1}{dX} + (c - K'_{12})v = \varphi_1$$

φ_1 projection on B_1 of R, and with the hypothesis H 3°).

There is no terms constant in φ_1; polynomial of degree higher two.

For v_2 , we have

(E_4):
$$(a - b - K_{22} - K'_{23})\frac{dv_2}{dY} + (c - K'_{13})v_2 = \varphi_2$$

The third equation is of the form of the first problem

(E_5):
$$\frac{\partial^2 v_3}{\partial x^2} - \frac{\partial^2 v_3}{\partial t^2} + \eta\left(a\frac{\partial v_3}{\partial x} + b\frac{\partial v_3}{\partial t} + cv_3 \right) = \eta\varphi_3$$

Here the method of the first part (I) can be applied: because the divisor is $\left[-m^2\omega^2 + n^2\omega^2 + c + \eta(am + bn)\omega i\right]$, and then $t_{m,n}$ are generally bounded by number ρ, $\eta\rho$ tends to zero with η. Hypothesys H'

We can say that if the four next conditions H" are verified. The equations (E_2) have a solution for η small with H'

1°) $a + b - K'_{22} - K'_{23}$ and $a - b - K'_{22} - K'_{23}$ are ever different of zero.

2°) the numbers of Floquet of the periodical equation (E_3) and (E_4) are the two different of the number one .

Then, we have this theorem with the hypothesis (H),(H') and (H"): The partial differential equation (E_1) has a solution $u(x,t)$ belonging to $B(\omega,\omega)$.

u tends uniformily to u_0 when η tends to zero. u defined by the same recurrence that in (IV)

V - The case of bicycles:

Biperiodical solutions of autonomous partial differential equations.

Here we study the same problem that the precedent case but, we have $G(x,t) \equiv 0$.

We study the equation (E_1)

$$\frac{\partial^2 u}{\partial x^2} - \frac{\partial^2 u}{\partial t^2} + \eta(\, a\frac{\partial u}{\partial x} + b\frac{\partial u}{\partial t} + cu \,) = \eta K(u, \frac{\partial u}{\partial x}, \frac{\partial u}{\partial t})$$

K is independent of x, t and function only of $u,, \frac{\partial u}{\partial x}, \frac{\partial u}{\partial t}$.

We make the hypothesis (H)

1°) There are a function $u_o(x,t)$ and a positive constant ω. $u_o(x,t)$ with its derivatives of first and second order belong to $\mathbf{B}(\omega,\omega)$ which satisfies the equation.

$$2°) \quad \frac{\partial^2 u_o}{\partial x^2} - \frac{\partial^2 u_o}{\partial t^2} = 0 \quad \text{and} \quad u_0 \text{ here must be of the form}$$

$u = u_1 + u_2$, u_1 regressive wave $u(x+t)$

u_2 progressive wave $u(x-t)$

all m,n constants.

$$\text{If } \psi = K(u_o, \frac{\partial u_o}{\partial x}, \frac{\partial u_o}{\partial t}) - a \frac{\partial u_o}{\partial x} - b \frac{\partial u_o}{\partial t} - cu_o.$$

3°) $\psi(x,t) \in \mathbf{B}_3(\omega,\omega)$ defined in the previous part

$\psi(x,t)$ non constant

In this part V the function ψ plays the roll of the function R in part IV.

We suppose then that ψ satisfies all the hypothesis corresponding with these of R in IV.

We can then say the Theorem:

If the Hypothesis H and H' are verified the equation E have a biperiodical autonomous solution for η sufficiently small.

VI - On a case of vibration doubly periodical in Biophysics.

Rinzel indicates a simplified form of the equation for the propagation of the nervous spike (AXONA IMPULSE PROPAGATION).

$$\text{i)} \qquad \frac{\partial v}{\partial t} = \frac{\partial^2 v}{\partial x^2} - f(v) - w + I(x,t)$$

$$\text{ii)} \qquad \frac{\partial w}{\partial t} = \varepsilon\,(v - \gamma w).$$

We can research if the excitation $I(x,t)$ is doubly periodical; the system has a solution also doubly periodical.

We make the hypothesis $I(x,t) \in B(\omega_1,\omega_2)$

$$I(x,t) = \sum I_{mn}\, e^{im\omega_1 x + in\omega_2 t} \qquad \text{with } \| I(x,t) \| = \sum |I_{m,n}| < \infty$$

$$\text{We write } v = \sum v_{mn}\, e^{im\omega_1 x + in\omega_2 t}$$

$$w = \sum w_{mn}\, e^{im\omega_1 x + in\omega_2 t}$$

Equation (ii) gives $w_{mn}(\gamma\varepsilon + n\omega_2 i) = \varepsilon\, v_{mn}$

(i) gives:
$$(n\omega_2 i + m^2\omega_1^2 + (n\omega_2 i + \varepsilon\gamma)^{-1})\, v_{mn} = I_{mn} + f_{mn}$$

where
$$f(v) = \sum f_{mn}\, e^{im\omega_1 x + in\omega_2 t}$$

Thes system (i), (ii) gives the transformation in $B(\omega_1,\omega_2)$

$$\text{(iii)} \qquad v_n = T\left[f(v) + I\right]$$

The T is given by $t_{mn} = (m\omega_2 i + m^2\omega_r^2 + (n\omega_2 i + \varepsilon\gamma)^{-1})^{-1}$

We can use a parameter η .

$$\text{(iv)} \qquad v = \eta\, T\left[f(v) + I\right]$$

In Biophysics $f(v)$ is generaly given by $av(v_T - v)(1 \qquad v)$

where a is a positive constant.

In the theory of Rinzel a $=$ 1, v_T $=$ 0.25, γ $=$ 1 and ε $=$ 0.002.

Here if we take ω_1 and ω_2 $>$ 1.

Then the number ρ maximum of $|t_{mn}|$ is inferior that one.

If we take $\|1\|$ $=$ 1; for exemple; we see that the system have a solution biperiodical for η sufficiently small, we resolve the problem (iv) by recurrence

We can write the theorem.

For η sufficiently small the equation (iv) have a doubly periodical solution and belong to $B(\omega_1,\omega_2)$.

REFERENCES

FAURE, Robert; sur certaines solutions périodiques d'équations dif-férentielles non linéaires. Annali di mathematica pura ed applicata tomo XLII, 1956 - 1957.

FAURE, Robert; Etude de certains systèmes d'équations différen-tielles à coefficients périodiques. Solutio; Mémoire et Publications de la Société des Sciences, des Arts et des Lettres du Hainaut T76 1962.

FAURE, Robert; Percussions en Mecanique non linéaire. Annali di Mathematica pura et applicata (IV) vol. CXL 1985.

RINZEL, John; Integration and propagation of Neuroelectric Signals. U.S. Department ot Health, Education and Welfare. 1979.

The controllability of infinite dimensional and distributed parameter systems

SINGULARITIES IN BOUNDARY VALUE PROBLEMS AND EXACT CONTROLLABILITY OF HYPERBOLIC SYSTEMS

PIERRE GRISVARD

Professor, UNSA

1. The model problem of boundary control

For simplicity we consider control by the Dirichlet data of the solutions of the wave equation. One considers a bounded open subset Ω of $\mathbb{R}^n$ with boundary $\Gamma = \partial\Omega$ (assumed to be at least Lipschitz). We denote by Q the cylinder $\Omega \times]0,T[$ and by $\Sigma = \Gamma \times]0,T[$ its lateral boundary. We denote by a mere ' differentiation in $t \in]0,T[$ and $u(0)$ denotes the function u at time $t=0$.

Our problem is the following.

Given an initial position u_0 and an initial speed u_1 on Ω, we look for a time T and a "control" v on Σ such the solution u of

$$u''=\Delta u \text{ in } Q,$$
$$(1) \quad u(0)=u_0,\, u'(0)=u_1,$$
$$u|_\Sigma=v,$$

fulfils

$$u(T)=u'(T)=0.$$

In other words we want to bring the corresponding system to rest at time T under the action of v. We look for a time T as small as possible. We also endeavor to minimize the support Σ_0 of v.

2. Review of the Hilbert Uniqueness Method HUM

J.L.Lions (1988) has put forward a systemetic approach called HUM, whose justification comes from the fact that one can see this problem as an optimization one in which one tries to minimize

$$\int_\Sigma v^2 \, dsdt$$

under some constraints.

Skipping the details we know that exact boundary controllability is secured for every pair $(u_0, u_1) \in L_2(\Omega) \times H^{-1}(\Omega)$, as soon as the following property holds. There exists a subset Σ_0 of Σ such the inequality

$$(2) \quad \|\varphi_0\|^2_{H^1_0(\Omega)} + \|\varphi_1\|^2_{L^2(\Omega)} \leqslant C \int_{\Sigma_0} |\frac{\partial \varphi}{\partial \nu}|^2 \, dsdt$$

holds for every pair $(\varphi_0, \varphi_1) \in H^1_0(\Omega) \times L^2(\Omega)$, where φ is the (finite energy) solution of

$$\varphi'' = \Delta \varphi \text{ in } Q,$$
$$\varphi(0) = \varphi_0, \varphi'(0) = \varphi_1,$$
$$\varphi|_\Sigma = 0.$$

and where ν denotes the outward normal vector field on Γ. Then the control v is supported by Σ_0.

The solution of (1) is understood in a weak sense which is useless to recall here.

3. Minimal smoothness assumptions

J.L.Lions (1988) applies a multiplier method with integrations by parts to prove (2). This technique was suggested by old works by Rellich. This method applies when Γ is only C^2. C.Bardos-G.Lebeau-J.Rauch (1988) apply results on singularity propagation and this works only when Γ is C^∞. However this allows them to find optimal (i.e. minimal) T_0 such that exact controllability works for every $T > T_0$ and a minimal support Σ_0 too.

In all cases it is obviously necessary that

$$\frac{\partial \varphi}{\partial \nu} \in L^2(\Sigma_0)$$

to make (2) meaningful. More precisely the calculations by J.L.Lions make use, in a basic way, of the integral

$$(3) \quad \int_\Sigma m.\nu |\frac{\partial \varphi}{\partial \nu}|^2 \, dsdt \, .$$

which comes from the integration by parts of

$$\int_Q (\varphi'' - \Delta\varphi)\, m.\nabla\varphi\, dt dx$$

where $m = x - x_0$ with given x_0. The subset Σ_0 is then defined by $m.\nu > 0$.

One can obtain convergence of the integral (3) in two differents ways:

(i) Either one has

$$\frac{\partial\varphi}{\partial\nu} \in L^2(\Sigma).$$

(ii) Or one assumes $m.\nu = 0$ at the points in the neighborhood of which $\dfrac{\partial\varphi}{\partial\nu}$ is not square integrable.

Actually such a question arises even when one considers the corresponding stationary problem

$$(4)\quad \Delta w = f$$

with $w \in H_0^1(\Omega)$ and $f \in L^2(\Omega)$, when one calculates the following integral:

$$\int_\Omega \Delta w\, m.\nabla w\, dx =$$

$$-\int_\Omega \nabla w.\nabla(m.\nabla w)\, dx + \int_\Gamma m.\nu \left|\frac{\partial w}{\partial\nu}\right|^2 ds =$$

$$-\sum_{k,l}\int_\Omega D_k m_l\, D_k w\, D_l \varphi w\, dx - \frac{1}{2}\int_\Omega m.\nabla|\nabla w|^2 dx + \int_\Gamma m.\nu\left|\frac{\partial w}{\partial\nu}\right|^2 ds =$$

$$-\sum_{k,l}\int_\Omega D_k m_l\, D_k w\, D_l w\, dx + \frac{1}{2}\int_\Omega \operatorname{div} m\, |\nabla w|^2 ds + \frac{1}{2}\int_\Gamma m.\nu\left|\frac{\partial w}{\partial\nu}\right|^2 ds.$$

4. Regularity of the solution of the stationary problem

The first way, mentioned above, in (i), to make convergent the integral consists in checking whether $w \in H_0^1(\Omega)$ solution of (4) is such that

$$\frac{\partial w}{\partial\nu} \in L^2(\Gamma).$$

This leads to the natural question whether there exists $s > \dfrac{3}{2}$ such that $w \in H^s(\Omega)$?

The answer is clearly yes when Ω is convex or when its boundary Γ has a bounded curvature. Indeed in these cases one has $w \in H^2(\Omega)$.

Otherwise we recall that according to <u>Grisvard</u> (1985) the solution w may have a singular behavior. If Ω is a plane domain that coincides, near the origin, with the sector $0<\theta<\omega$ (with $\omega>\pi$ to avoid the convex case) there exists a constant c such that

$$(5) \quad w - c r^{\pi/\omega} \sin(\frac{\pi\theta}{\omega}) \in H^2(V)$$

in a suitable neighborhood V of the origin. Here r and θ denote the usual polar coordinates. It follows that $w \in H^s(V)$ for $s < 1 + \frac{\pi}{\omega}$. This upper bound for s is strictly larger than $\frac{3}{2}$ if and only if $\omega < 2\pi$, and this excludes domains with cracks.

Globally this implies that $w \in H^s(\Omega)$ for some $s > \frac{3}{2}$ if Ω is a plane polygon in the strict sense i.e. without crack.

It is easy to extend this last result to polyhedra without cracks by the classical methods of Fourier analysis. One knows nowadays that if Ω is an arbitrary polyhedron without cracks there exists $s(\Omega) > \frac{3}{2}$ such that $w \in H^s(\Omega)$ for every $s < s(\Omega)$. This property is also true in the case of a Neumann problem for the Laplace equation. This implies exact controllability of the solutions of the wave equation either by Dirichlet action or by Neumann action on any polyhedron without crack c.f. <u>Grisvard</u> (1989). In addition one can restrict the Dirichlet action to the subset Σ_0 of Σ defined by $m.\nu > 0$.

If we remplace the usual wave equation by the elastic wave system, i.e. the Laplace operator by the elasticity Lamé system, we obtain similar results. Let us consider for instance $w \in H^1_0(\Omega)^n$ with $n=2$ or 3, solution of

$$(6) \quad \mathrm{div}(\sigma(w)) = f \text{ dans } \Omega.$$

If Ω is a plane domain which coincides with the plane sector $0<\theta<\omega$, with $\omega>\pi$, in a neighborhood of the origin, there exist two constants c_1 and c_2 such that

$$w - c_1 r^{z_1} S_{z_1}(\theta) - c_2 r^{z_2} S_{z_2}(\theta) \in H^2(V)^2$$

in a suitable neighborhood V of 0. Here z_1 and z_2 denote the two roots (which are real) of the equation

$$(7) \quad \sin^2(z\omega) = z^2 \sin^2(\omega) \frac{(\lambda+\mu)^2}{(\lambda+3\mu)^2}$$

in the strip $0 < \mathrm{Re}(z) < 1$ where λ and μ are the two Lamé coefficients and where the vector fields are defined by

$$S_z(\theta) =$$
$$\{z(\lambda+\mu)\sin([z-2]\omega)+[(2-z)(\lambda+\mu)+4\mu]\sin(z\omega)\}\Phi_z(\theta)+$$
$$z(\lambda+\mu)\{\cos(z\omega)-\cos([z-2]\omega)\}\Psi_z(\theta)$$

with

$$\Phi_z(\theta)=\left|\begin{array}{c} [(z+2)(\lambda+\mu)+4\mu]\sin(z\theta)- \\ z(\lambda+\mu)\sin([z-2]\theta) \\ \\ \\ z(\lambda+\mu)\{\cos(z\theta)-\cos([z-2]\theta)\} \end{array}\right|$$

and

$$\Psi_z(\theta)=\left|\begin{array}{c} -z(\lambda+\mu)\{\cos(z\theta)-\cos([z-2]\theta)\} \\ \\ \\ -[4\mu+(2-z)(\lambda+\mu)]\sin(z\theta)- \\ z(\lambda+\mu)\sin([z-2]\theta) \end{array}\right|$$

The analysis of the solutions of the transcendental equation (7) shows that the roots in the strip $0<Re(z)<1$ are actually such that $Re(z)>\frac{1}{2}$ when $\omega<2\pi$ and again this rules out domains with cracks. Consequently we have $w\in H^s(V)^2$ for some $s>\frac{3}{2}$.

Globally this implies $w\in H^s(\Omega)^2$ for some $s>\frac{3}{2}$ if Ω is a plane polygon in the strict sense i.e. excluding cracks.

Exactly as in the case of the Laplace equation, this last result is easily generalized to polyhedra by the methods of Fourier analysis. We know nowadays that if Ω is an arbitrary polyhedron without crack there exists $s(\Omega)>\frac{3}{2}$ such that every solution w of (6) is such that $w\in H^s(\Omega)^3$ for every $s<s(\Omega)$. This property hols also in the case of a Neumann boundary value problem for the Lamé system (stress-free boundary). This implies exact controllability of the solutions of the elastic waves equation in any polyhedron without crack by either a Dirichlet action or a Neumann action on the boundary c.f. <u>Grisvard</u> (1992). This action may be restricted to the subset defined by $m.\nu>0$ in the case of a Dirichlet control.

4. The limit case with cracks

So far we have excluded domains with crack. When Ω is a plane domain with a crack that coincides with the half axis Ox, one has (5) with $\omega = 2\pi$ and therefore

$$(8) \quad w - c\sqrt{r}\sin(\tfrac{\theta}{2}) \in H^2(V).$$

It is clear that $\dfrac{\partial w}{\partial \nu} \notin L^2(\Gamma)$ in the neigborhood of the crack tip. Following the above suggested second way (ii) the integral (3) will make sense if and only if $m.\nu$ vanishes at the crack tip. We recall that $m(x) = x - x_0$ and consequently, in the case of a domain with several cracks, we shall need the existence of a particular point x_0 such that all the cracks are supported by straight lines passing by that point. This a drastic geometric assumption.

Under such restriction the above-mentioned integration by parts has to be performed on the truncated domain Ω_ε obtained from Ω by cutting away a disc of radius ε with center at every crack tip. If we denote by γ_ε the "new frontier" i.e. $\Gamma_\varepsilon \backslash \Gamma$, and if we calculate

$$\int_{\Omega_\varepsilon} \Delta w \, m.\nabla w \, dx ,$$

we obtain the following additional boundary integral

$$\frac{1}{2} \int_{\gamma_\varepsilon} \{ m.\nu [(\frac{\partial w}{\partial \tau})^2 - (\frac{\partial w}{\partial \nu})^2] - m.\tau \frac{\partial w}{\partial \tau} \frac{\partial w}{\partial \nu} \} \, ds,$$

where τ denotes the unitary tangent vector field on the boundary with the direct orientation (note that cracks have two sides with opposite orientation). By (8) the most singular part of the integrand behaves as $\dfrac{1}{\varepsilon}$ and since the length of γ_ε is proportional to ε this integral remains obviously bounded as $\varepsilon \to 0$. It has actually a limit whose sign depends on the sign of $m.\tau$ at the crack tip where the convention is that τ is oriented in the direction toward which the crack points.

Consequently the inequality (2) holds if we assume simultaneously

$$m.\nu = 0 \text{ and}$$

$$m.\tau \geqslant 0$$

at each crack tip. Altogether we need the existence of a point x_0 such that all the cracks are supported by straight lines passing by this point and such that each crack points in the opposite direction to x_0. These very severe geometric restrictions are not quite weakened by using the more general multipliers of <u>Triggiani</u> (1988).

In the case of a Neumann control the situation is reversed since one must assume that

$$m.\nu = 0 \text{ et}$$

$$m.\tau \leqslant 0$$

This means that there exists a point x_0 toward which all cracks point as shown in the following figure.

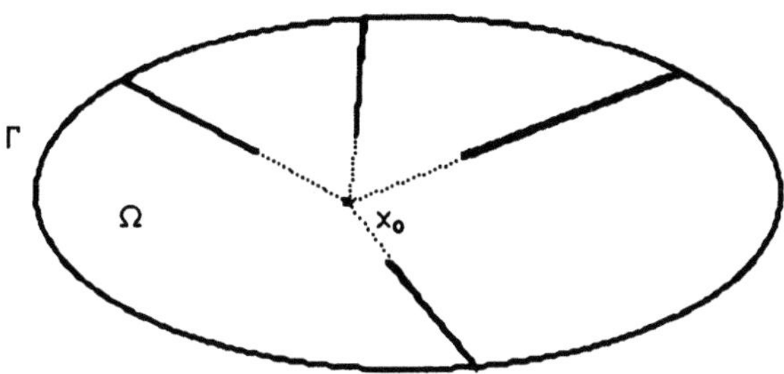

Solutions of the plane elasticity system also have a singular behavior in $\sqrt{r}$ regardless of the boundary condition (Dirichlet or Neumann). The same kind of exact controllability results is therefore plausible although not yet proven. One finds very similar situations in considering boundary controllability with mixed Dirichlet-Neumann action c.f. <u>Grisvard</u> (1989).

5. Remarks

Similar results for the plate equation in a polygon have been derived by <u>M.T.Niane</u> (1988).

Also A.Chaira (not yet published) shows that every (finite energy) solution φ of the wave equation in an arbitrary bounded Lipschitz open subset Ω of $\mathbb{R}^n$, fulfils

$$\frac{\partial \varphi}{\partial \nu} \in L^2(\Sigma).$$

6. References

C.Bardos–G.Lebeau–J.Rauch (1988) Appendix 2 in Lions (1988) below.

P.Grisvard (1985) Elliptic problems in non smooth domains, Monographs and studies in Mathematics, 24, Pitman, London.

P.Grisvard (1989) Contrôlabilité exacte des solutions de l'équation des ondes en présence de singularités, J. de Mathématiques Pures et Appliquées, 68, p.215/259.

P.Grisvard (1992) Singularities in Boundary Value Problems, RMA, Masson, Paris (on print).

J.L.Lions (1988) Contrôlabilité exacte, perturbations et stabilisation de systèmes distribués, tome 1, RMA n°8, Masson, Paris.

M.T.Niane (1988) Contrôlabilité exacte de l'équation des plaques vibrantes dans un polygone, CRAS, Paris, t.307, p.517–521.

R.Triggiani (1988) Exact boundary controllability on $L_2(\Omega) \times H^{-1}(\Omega)$ of the wave equation with Dirichlet boundary control acting on a portion of the boundary $\partial\Omega$ and related problems, Applied Mathematics and Optimisation, 18, p.241/277.

Author's address:

    Laboratoire de Mathématiques, Faculté des Sciences,

    Université de Nice–Sophia–Antipolis, Parc Valrose,

    Nice 06–108 Cedex 2,

    France.

Exact controllability of a shallow shell model

G. GEYMONAT[**], P. LORETI[*], V. VALENTE[*]

Abstract. A coupled reversible system for the vibrations of a shallow spherical shell is given. The exact controllability is studied using HUM method of J.L.Lions and uniqueness results are obtained for small values of the coupling parameter. The efficiency of a spectral numerical approximation method is illustred by the numerical experiments.

1.1 The model

In a previous paper [9] the authors introduce the system (1.1) representing the vibrations of a thin elastic spherical cap [4], [7]

$$(1.1) \quad \begin{cases} du_{tt} - \hat{L}(u) + (1+\nu)w' - \hat{e}\hat{L}(u+w') = 0 \\ dw_{tt} + \frac{\hat{e}}{\sin(\theta)}[\hat{L}(u+w')\sin(\theta)]' - \frac{(1+\nu)}{\sin(\theta)}(u\sin(\theta))' + 2(1+\nu)w = 0 \\ t \in (0,T) \quad \text{and} \quad \theta \in (0,\theta_0) \end{cases}$$

where $\hat{e} = \frac{\varepsilon^2}{3R^2}$, 2ε is the thickness and R the radious of the middle surface; $\nu \in \left(-1, \frac{1}{2}\right)$, $d = d_0 \frac{R^2}{(1-\nu^2)}\hat{E}$ with d_0 the density, $\hat{E}$ Young modulus and

$$\hat{L}(u) = u'' + u'\cot\theta - (\nu + \cot^2\theta)u .$$

Since we are interested to shallow shell approximation we introduce the coordinate $\rho = R\theta$, and we assume that θ_0 is small enough so that $\nu + \frac{1}{\theta^2} \cong \frac{1}{\theta^2}$; $\cot\theta \cong \frac{1}{\theta}$. We have

$$\hat{L}(\varphi) = R^2 \left(\frac{\partial^2\varphi}{\partial\rho^2} + \frac{1}{\rho}\frac{\partial\varphi}{\partial\rho} - \frac{\varphi}{\rho^2}\right) = R^2 L(\varphi)$$

and the system (1.1) becomes

$$(1.2) \quad \begin{cases} \tilde{d}u_{tt} - L(u) + \frac{(1+\nu)}{R}w' - \frac{e}{R}L\left(\frac{u}{R} + w'\right) = 0 \\ \tilde{d}w_{tt} + \frac{e}{\rho}\left[L\left(\frac{u}{R} + w'\right)\rho\right]' - \frac{(1+\nu)}{\rho R}(u\rho)' + \frac{2(1+\nu)}{R^2}w = 0 \end{cases}$$

in $Q = (0,T) \times (0,\rho_0)$; where $\tilde{d} = \frac{d_0}{(1-\nu^2)}\hat{E}$, $e = \frac{1}{3}\varepsilon^2$ and $'$ denotes the first derivative with respect to ρ.

[**]L.M.T., ENS de Cachan/C.N.R.S./Université Paris-VI, 61 avenue du President. Wilson 94230 Cachan – France.
[*]IAC-CNR V.le del Policlinico n. 137 00161 Roma – Italy.

Without loss of generality in the sequel we assume $d = 1$.

We consider the initial conditions

$$(1.3) \qquad u(\rho,0) = u_0 \ , \quad w(\rho,0) = w_0 \ , \quad u_t(\rho,0) = u_1 \ , \quad w_t(\rho,0) = w_1$$

and the simmetry conditions

$$(1.4) \qquad u(0,t) = 0 \ ; \quad \left(\frac{u}{R} + w'\right)(0,t) = 0 \ ; \quad L\left(\frac{u}{R} + w'\right)(0,t) = 0 \ .$$

We are interested to study the exact controllability of the shallow shell (1.2), (1.3), (1.4) assuming

$$(1.5) \qquad u(\rho_0,t) = 0 \ ; \quad \left(\frac{u}{R} + w'\right)(\rho_0,t) = 0 \ ; \quad L\left(\frac{u}{R} + w'\right)(\rho_0,t) = 0$$

respectively

$$(1.5)_0 \qquad u(\rho_0,t) = 0 \ ; \quad \left(\frac{u}{R} + w'\right)(\rho_0,t) = 0 \ ; \quad w(\rho_0,t) = 0 \ .$$

We observe that (1.2) is a coupled reversible system [6, vol. II] with $\frac{1}{R}$ as the coupling parameter. Formally, as R diverges to infinity (and $\theta_0 = \frac{\rho_0}{R} \to 0$) (1.2), (1.3), (1.4), (1.5) (resp. $(1.5)_0$) tends to the uncoupled limit problem

$$(1.1)'_a \qquad\qquad\qquad w_{tt} + \frac{e}{\rho}(L(w')\rho)' = 0$$

$$(1.1)'_b \qquad\qquad\qquad u_{tt} - L(u) = 0$$

with the boundary conditions:

$$(1.4)' \qquad\qquad u(0,t) = 0 \ , \quad w'(0,t) = 0 \ , \quad L(w')(0,t) = 0$$

$$(1.5)' \qquad\qquad u(\rho_0,t) = 0 \ , \quad w'(\rho_0,t) = 0 \ , \quad L(w')(\rho_0,t) = 0$$

respectively

$$(1.5)'_0 \qquad\qquad u(\rho_0,t) = 0 \ , \quad w'(\rho_0,t) = 0 \ , \quad w(\rho_0,t) = 0 \ .$$

1.2 Variational formulation

We set $L^2 = L^2(0,\rho_0;\rho\,d\rho) = \{\varphi : \int_0^{\rho_0} \varphi^2 \rho\,d\rho < +\infty\}$ provided with scalar product $(\varphi,\psi)_0 = \int_0^{\rho_0} \varphi\psi\rho\,d\rho$ and $\mathbf{H} = [L^2(0,\rho_0); \rho\,d\rho)]^2$. For any vector $\mathbf{v} = \{u,w\}$ and $\hat{\mathbf{v}} = \{\hat{u},\hat{w}\}$ in $\mathbf{H}$ we put

$$(\mathbf{v}, \hat{\mathbf{v}}) = (u,\hat{u})_0 + (w,\hat{w})_0 \ .$$

We introduce the bilinear symmetric forms

$$(1.6) \qquad a_{0,R}(\mathbf{v}, \hat{\mathbf{v}}) = \left(u' - \frac{w}{R}, \hat{u}' - \frac{\hat{w}}{R} \right)_0 + \left(\frac{u}{\rho} - \frac{w}{R}, \frac{\hat{u}}{\rho} - \frac{\hat{w}}{R} \right)_0$$
$$+ \nu \left(\hat{u}' - \frac{\hat{w}}{R}, \frac{u}{\rho} - \frac{w}{R} \right)_0 + \nu \left(u' - \frac{w}{R}, \frac{\hat{u}}{\rho} - \frac{\hat{w}}{R} \right)_0$$

$$(1.7) \; a_{1,R}(\mathbf{v}, \hat{\mathbf{v}}) = \left(\frac{u'}{R} + w'', \frac{\hat{u}'}{R} + \hat{w}'' \right)_0 + \left(\left(\frac{u}{R} + w' \right) \frac{1}{\rho}, \left(\frac{\hat{u}}{R} + \hat{w}' \right) \frac{1}{\rho} \right)_0 +$$
$$+ \nu \left(\left(\frac{u'}{R} + w'' \right), \left(\frac{\hat{u}}{R} + \hat{w}' \right) \frac{1}{\rho} \right)_0$$
$$+ \nu \left(\left(\frac{\hat{u}'}{R} + \hat{w}'' \right), \left(\frac{u}{R} + w' \right) \frac{1}{\rho} \right)_0$$

and consequently we define the following functional spaces:

$$(1.8) \quad U = \left\{ \varphi : \int_0^{\rho_0} \left(\varphi'^2 \rho + \frac{\varphi^2}{\rho} \right) d\rho < +\infty \; ; \quad \varphi(0) = \varphi(\rho_0) = 0 \right\}$$

$$(1.9) \quad W = \left\{ \psi : \int_0^{\rho_0} \psi^2 \rho \, d\rho < +\infty \; , \quad \psi' \in U \right\}$$

$$(1.10) \; W_0 = \{ \psi : \psi \in W \, , \quad \psi(\rho_0) = 0 \} \qquad \text{and}$$

$$(1.11) \;\; \mathbf{V} = U \times W \; ; \; \mathbf{V}_0 = U \times W_0 \; .$$

The variational formulation of the problem (1.2), (1.3), (1.4), (1.5) $((1.5)_0)$ is now: find $\mathbf{v} = \{u, w\}$ such that

$$(1.14) \qquad\qquad \mathbf{v} \in C([0,T]; \mathbf{V}) \cap C^1([0,T]; \mathbf{H})$$

resp.

$$\mathbf{v} \in C([0,T]; \mathbf{V}_0) \cap C^1([0,T]; \mathbf{H})$$

$$(1.15) \qquad (\mathbf{v}_{tt}, \hat{\mathbf{v}}) + a_{0,R}(\mathbf{v}, \hat{\mathbf{v}}) + e a_{1,R}(\mathbf{v}, \hat{\mathbf{v}}) = 0 \qquad \forall \hat{\mathbf{v}} \in \mathbf{V} \text{ (resp. } \mathbf{V}_0)$$

and

$$\mathbf{v}_0 = \{u_0, w_0\} \in \mathbf{V} \text{ (resp. } \mathbf{V}_0) \, , \quad \mathbf{v}_1 = \{u_1, w_1\} \in \mathbf{H} \, .$$

To prove the well-posedness of the model we need the following results

LEMMA 1.1. *If $u \in U$ then $\|u\|_u^2 \geq \ \|u\|_0^2$. If $w \in W_0$ then $\|w'\|_0^2 \geq \frac{4}{\rho_0^2}\|w\|_0^2$ and so $\|w\|_0^2 \leq \frac{\rho_0^4}{4\pi^2}\|w'\|_u^2$.*

We observe that if $w \in W$ we have

$$(1.16) \qquad \|w\|_0^2 \leq \rho_0^2 u'(\rho_\ldots)^2 + \frac{\rho_0^4}{2\pi^2}\|w'\|_u^2$$

Let $\mathcal{R} \subset V$ be the space of displacement ($u = 0, w = \text{cost}$) then for $R \to +\infty$ we have

$$a_{0,R}(\mathbf{v},\mathbf{v}) + e a_{1,R}(\mathbf{v}.\mathbf{v}) \to 0 \quad \text{if} \quad \mathbf{v} \in \mathcal{R} .$$

PROPOSITION 1.1. *The form $a_{0,R} + e a_{1,R}$ is V-elliptic with coerciveness constant $\alpha = \alpha(R)$ and*

$$\lim_{R \to \infty} \alpha(R) = 0 .$$

Proof. From (1.6) we have

$$(1.17) \quad a_{0,R}(\mathbf{v},\mathbf{v}) = \|u\|_u^2 - \frac{2(1+\nu)}{R}\int_0^{\rho_0} (u\rho)'w\,d\rho + \frac{2(1+\nu)}{R^2}\int_0^{\rho_0} w^2\rho\,d\rho$$

$$\geq \|u\|_u^2 - \frac{2(1+\nu)}{R}\|u\|_u\|w\|_0 + \frac{2(1+\nu)}{R^2}\|w\|_0^2 .$$

We have also

$$(1.18) \qquad a_{1,R}(\mathbf{v},\mathbf{v}) = \left\|\frac{u}{R} + w'\right\|_u^2 \geq \frac{1}{2}\|w'\|_u^2 - \frac{1}{R^2}\|u\|_u^2$$

and the proposition follows immediately.

PROPOSITION 1.2. *The form $a_{0,R} + e a_{1,R}$ is V_0-elliptic with coerciveness constant $\alpha_0 = \alpha_0(R)$ with*

$$\lim_{R \to \infty} \alpha_0(R) = \overline{\alpha} > 0$$

Proof. The proof follows from inequalities (1.17), (1.18) and from lemma 1.1. We introduce

$$(1.18) \qquad E(\mathbf{v},t) = \frac{1}{2}\left\{\|\mathbf{v}_t\|^2 + a_{0,R}(\mathbf{v},\mathbf{v}) + e a_{1,R}(\mathbf{v},\mathbf{v})\right\}$$

so the conservation of energy is expressed by $E(\mathbf{v},t) = E(\mathbf{v},0) = \frac{1}{2}\left\{\|\mathbf{v}^1\|^2 + a_{0,R}(\mathbf{v}^0,\mathbf{v}^0) + e a_{1,R}(\mathbf{v}^0,\mathbf{v}^0)\right\}$.

Remark. We can study the behaviour of $\{u(R), w(R)\}$ when $\frac{1}{R} \to 0$. It follows from (1.18) and Proposition 1.2 that

$$\{u(R), w(R)\} \to \{u, w\} \text{ in } L^\infty(0, T; V_0) \text{ weak star}$$

$$\{\dot{u}(R), \dot{w}(R)\} \to \{\dot{u}, \dot{w}\} \text{ in } L^\infty(0, T; H) \text{ weak star}$$

where $\{u, w\}$ is the weak solution of the uncoupled system $(1.1)'_a, (1.1)'_b, (1.4)', (1.5)'_0$

2. Some a priori estimates

We now derive the a priori estimates for solutions to (1.2) (1.3) (1.4) $(1.5)_0$ which will be the basis for application of H.U.M. [5] [6] to the boundary exact controllability problem. In an analogous way we can obtain a similar result for problem (1.2), (1.3) (1.4),(1.5) by using inequality (1.16)

THEOREM 2.1. *If $\frac{1}{R}$ is small enough there exists two positive constants γ_1, γ_2 such that, for any sufficiently smooth solution $\{u, w\}$ of (1.2) (1.3) (1.4) $(1.5)_0$ one has*

$$(2.1) \quad \frac{1}{2} \int_0^T u'(\rho_0)^2 \rho_0^2 \, dt + \frac{1}{2} e \int_0^T \left[\left(\frac{u}{R} + w' \right)' \right]^2_{\rho=\rho_0} \rho_0^2 \, dt \geq (\gamma_1 T - \gamma_2) E(0) .$$

Proof. We multiply the first equation of (1.2) by $(u\sqrt{\rho})' \rho \sqrt{\rho}$ and the second equation of (1.2) by $\left(w'\rho - \frac{w}{2} \right) \rho$.

We obtain

$$x_R + y_R + z = 0$$

where

$$x_R = \frac{1}{2} \int_0^T \left\{ \|\dot{u}\|_0^2 + 3\|\dot{w}\|_0^2 + \|u\|_u^2 + e \left\| \frac{u}{R} + w' \right\|_u^2 \right\} dt$$

$$+ \frac{1+\nu}{R} \int_0^T \int_0^{\rho_0} w(u\rho)' \, d\rho \, dt - 3\frac{(1+\nu)}{R^2} \int_0^T \int_0^{\rho_0} w^2 \rho \, d\rho \, dt$$

$$z = \int_0^{\rho_0} \dot{u}(u\sqrt{\rho})' \rho\sqrt{\rho} \, d\rho \bigg|_0^T + \int_0^{\rho_0} \dot{w} \left(w'\rho - \frac{w}{2} \right) \rho \, d\rho \bigg|_0^T$$

$$y_R = -\frac{1}{2} \int_0^T u'(\rho_0)^2 \rho_0^2 \, dt - \frac{1}{2} e \int_0^T \left[\left(\frac{u}{R} + w' \right)' \right]^2_{\rho=\rho_0} \rho_0^2 \, dt$$

$$- \frac{1}{2} \int_0^T \dot{w}^2(\rho_0)\rho_0^2 \, dt + \frac{(1+\nu)}{R^2} \int_0^T w^2(\rho_0)\rho_0^2 \, dt .$$

From Lemma 1.1 we have

$$
\begin{aligned}
x_R \geq \int_0^T &\left(\frac{1}{2}\|\dot{u}\|_0^2 + \frac{1}{2}\|\dot{w}\|_0^2 + \frac{1}{2}\|u\|_u^2 - \frac{(1+\nu)}{2R}\|u\|_u^2 \right. \\
&\left. - \frac{(1+\nu)}{2R}\|w\|_0^2 - \frac{3(1+\nu)}{R^2}\|w\|_0^2 + \frac{1}{2}e\|\frac{u}{R} + w'\|_u^2 \right) dt \\
\geq \int_0^T &\left\{ \frac{1}{2}\|\dot{u}\|_0^2 + \frac{1}{2}\|\dot{w}\|_0^2 + \left(\frac{1}{2} - \frac{(1+\nu)}{2R} \right)\|u\|_u^2 + \frac{(1+\nu)}{R}\left(\frac{1}{2} + \frac{3}{R} \right)\|w\|_0^2 \right. \\
&\left. - \frac{2(1+\nu)}{R}\left(\frac{1}{2} + \frac{3}{R} \right)\frac{\rho_0^4}{4\pi^2}\|w'\|_u^2 + \frac{1}{2}e\|\frac{u}{R} + w'\|_u^2 \, dt \right\} \\
\geq \int_0^T &\left\{ \left(\frac{1}{2}\|\dot{u}\|_0^2 + \frac{1}{2}\|\dot{w}\|_0^2 + \left(\frac{1}{2} - \frac{(1+\nu)}{2R} \right)\|u\|_u^2 + \frac{(1+\nu)}{R}\left(\frac{1}{2} + \frac{3}{R} \right) \right)\|w\|_0^2 \right. \\
&- \frac{4(1+\nu)}{R}\left(\frac{1}{2} + \frac{3}{R} \right)\frac{\rho_0^4}{4\pi^2}\|\frac{u}{R} + w'\|_u^2 - \frac{4(1+\nu)}{R}\frac{\rho_0^4}{4\pi^2}\left(\frac{1}{2} + \frac{3}{R} \right)\left\|\frac{u}{R}\right\|_u^2 \\
&\left. + \frac{1}{2}e\left\|\frac{u}{R} + w'\right\|_u^2 \, dt \right\} \\
\geq \int_0^T &\left\{ \frac{1}{2}\|\dot{u}\|_0^2 + \frac{1}{2}\|\dot{w}\|_0^2 + \left(\frac{1}{2} - \frac{(1+\nu)}{2R} - \frac{4(1+\nu)}{R^3}\frac{\rho_0^4}{4\pi^2}\left(\frac{1}{2} + \frac{3}{R} \right) \right)\|u\|_u^2 \right. \\
&+ \frac{(1+\nu)}{R}\left(\frac{1}{2} + \frac{3}{R} \right)\|w\|_0^2 + \left(1 - \frac{1}{e}\left(\frac{8(1+\nu)}{R}\frac{\rho_0^4}{4\pi^2} \right)\left(\frac{1}{2} + \frac{3}{R} \right) \right)\cdot\frac{1}{2}e \\
&\left. \cdot\left\|\frac{u}{R} + w'\right\|_u^2 \right\} dt
\end{aligned}
$$

since

$$
a_{0,R}(\mathbf{v},\mathbf{v}) \leq \left(1 + \frac{2}{R} \right)\left\{ \|u\|_u^2 + \frac{(1+\nu)}{R}\|w\|_0^2 \right\}
$$

we have

$$
x_R \geq \int_0^T \left\{ \frac{1}{2}\|\dot{u}\|_0^2 + \frac{1}{2}\|\dot{w}\|_0^2 + \frac{1}{2}\varepsilon_1 a_{0,R}(\mathbf{v},\mathbf{v}) + \frac{1}{2}e\varepsilon_2 a_{1,R}(\mathbf{v},\mathbf{v}) \right\} dt
$$

with

$$
\varepsilon_1 = 2\left[\frac{1}{2} - \frac{(1+\nu)}{2R} - \frac{4(1+\nu)}{R^2}\frac{\rho_0^4}{4\pi^2}\left(\frac{1}{2} + \frac{3}{R} \right) \right] \Big/ \left(1 + \frac{2}{R} \right)
$$

and

$$
\varepsilon_2 = 1 - \frac{1}{e}\left[\frac{8(1+\nu)}{R}\frac{\rho_0^4}{4\pi^2}\left(\frac{1}{2} + \frac{3}{R} \right) \right]
$$

we take

$$
\gamma_1 = \min(1, \varepsilon_1, \varepsilon_2) \, .
$$

From the following inequalities (2.2) (2.3)

$$(2.2) \qquad \int_0^{\rho_0} \dot{w}\left(w'\rho - \frac{w}{2}\right)\rho\,d\rho \leq \|\dot{w}\|_0 \|w'\rho - \frac{w}{2}\|_0$$

$$\leq \sqrt{\frac{2\rho_0^3}{\pi^2} + \frac{5\rho_0^4}{16\pi^2}} \|\dot{w}\|_0 \|w'\|_u$$

$$\leq \frac{1}{\sqrt{e}}\sqrt{\frac{2\rho_0^3}{\pi^2} + \frac{5\rho_0^4}{16\pi^2}}\left\{\|\dot{w}\|_0^2 + e\|\frac{u}{R} + w'\|_u^2 + e\|\frac{u}{R}\|_u^2\right\}$$

$$(2.3) \qquad \int_0^{\rho_0} (u\sqrt{\rho})'\dot{u}\rho\sqrt{\rho}\,d\rho \leq \rho_0 \|u\|_u \|\dot{u}\|_0 \leq \frac{\rho_0}{2}\left\{\|\dot{u}\|_0^2 + \|u\|_u^2\right\}$$

we have

$$(2.4) \qquad z \geq -2\rho_0\left\{\frac{\|\dot{u}\|_0^2}{2} + \frac{1}{2}\|u\|_u^2 + 2\frac{1}{\sqrt{e}}\sqrt{\frac{2\rho_0}{\pi^2} + \frac{5\rho_0^2}{16\pi^2}}\left(\frac{1}{2}\|\dot{w}\|_0^2\right.\right.$$

$$\left.\left. + \frac{1}{2}e\|\frac{u}{R} + w'\|_n^2 + \frac{1}{2}\frac{e}{R^2}\|u\|_u^2\right)\right\} .$$

Since $a_{0,R}(\mathbf{v},\mathbf{v}) \geq \left(\frac{1-\nu}{2}\right)\|u\|_u^2$ [see (1.16)] the proof of the Theorem 2.1 follows taking

$$\gamma_2 = 2\rho_0\max\left[\frac{2}{1-\nu}\left(1 + \frac{2}{R^2}\sqrt{e}\sqrt{\frac{2\rho_0}{\pi^2} + \frac{5\rho_0^2}{16\pi^2}}\right), \frac{1}{\sqrt{e}}\sqrt{\frac{2\rho_0}{\pi^2} + \frac{5\rho_0^2}{16\pi^2}}\right] .$$

If we take $\frac{1}{R} \to 0$ we have

$$\gamma_1 \to 1 \quad \text{and} \quad \gamma_2 \to 2\rho_0\max\left[\frac{2}{1-\nu}, \frac{2}{\sqrt{e}}\sqrt{\frac{2\rho_0}{\pi^2} + \frac{5\rho_0^2}{16\pi^2}}\right] .$$

THEOREM 2.2. *Let $\{u,w\}$ be the weak solution of (1.2), (1.3),(1.4), (1.5)$_0$, then there exists a positive constant C_T such that the following estimate holds*

$$(2.5) \qquad \frac{1}{2}\int_0^T u'(\rho_0)^2\rho_0^2\,dt + \frac{1}{2}e\int_0^T\left[\left(\frac{u}{R} + w'\right)'\right]_{\rho=\rho_0}^2 \rho_0^2\,dt \leq C_T E(0) .$$

where

$$C_T = (3 + \frac{4(1+\nu)}{R})T + \gamma_2$$

The proof follows from multiplier method as in the proof of the Theorem 2.1

3. Exact controllability

Let $\mathcal{A}_R$ be the Lax-Milgram operator associated to the form $a_{0,R} + e a_{1,R}$. Given $T > 0$ and an initial state $\phi^0 = \{\varphi^0, \psi^0\}$ and $\phi^1 = \{\varphi^1, \psi^1\}$, find the controls $g_i(i = 1, 2(3))$ such that the unique solution of the problem

$$(3.1) \qquad \ddot{\phi} + \mathcal{A}_R \phi = 0 \quad \text{in} \quad (0, \rho_0) \times (0, T)$$

$$(3.2) \qquad \phi(\rho, 0) = \phi^0 \quad \dot{\phi}(\rho, 0) = \phi^1 \quad \rho \in (0, \rho_0)$$

$$(3.3) \qquad \varphi(0, t) = \psi'(0, t) = L(\tfrac{\varphi}{R} + \psi')|_{\rho=0} = 0$$

$$(3.4) \qquad \varphi(\rho_0, t) = g_{1,R}(t) \quad \tfrac{\varphi(\rho_0, t)}{R} + \psi'(\rho_0, t) = g_{2,R}(t)$$

$$(3.5) \qquad \psi(\rho_0, t) = 0$$

respectively

$$(3.5)' \qquad\qquad L\left(\frac{\varphi}{R} + \psi'\right)_{\rho_0} = g_{3,R}(t)$$

satisfies the further condition

$$\phi(\rho, T) = \dot{\phi}(\rho, T) = 0 .$$

The problem is attacked by H.U.M. method [see J.L. Lions] in conjugation with the a priori estimate given by Theorem 2.1, 2.2.

THEOREM 3.1. *If $\frac{1}{R}$ is small enough there exists a $T^0 > 0$ such that we have exact controllability for any $\phi^0, \phi^1 \in (\mathbf{H} \times \mathbf{V}_0)'$.*

The proof follows from the Theorem 2.1 when $T^0 \geq \frac{\gamma_2}{\gamma_1}$ with

$$g_1(t) = u'(\rho_0) - (1 + \nu)\frac{w(\rho_0)}{R} \quad \text{and} \quad g_2(t) = \left(\frac{u(\rho_0)}{R} + w'(\rho_0)\right)$$

We can also give the following uniqueness result.

THEOREM 3.2. *If $\{u, w\}$ is the solution of the problem (1.2) (1.3) (1.4) (1.5) and $\{u, w\}$ satisfies the following conditions: $w(\rho_0) = u'(\rho_0) = \frac{u}{R}(\rho_0) + w'(\rho_0) = 0$ for any $t \in]0, T^0[$, then $\mathbf{v} = \{u, w\} \equiv 0$ in Q.*

Remark. As in [6, Vol II], from the Theorem 2.1 and 2.2 we have

$$g_{i,R} \to g_i \quad \text{when} \quad \frac{1}{R} \to 0 \, in \, L^2(0, T) weakly$$

where g_i are the controls of the uncoupled system, leading respectively φ, ψ to the rest.

4. A spectral numerical approximation method

The controllability problem (3.1) (3.2) (3.3) (3.4) (3.5) (resp. (35)') is equivalent to following problem

$$
P \begin{cases}
find \ \inf J(\mathbf{v}^0, \ \mathbf{v}^1) \\
\mathbf{v}^0, \mathbf{v}' \in \mathbf{V}_0 \times \mathbf{H} \quad (resp. \ \mathbf{V} \times \mathbf{H}) \\
J(\mathbf{v}^0, \mathbf{v}') = \frac{1}{2} \left\{ \int_0^T \left(u'(\rho_0) - (1+\nu)\frac{w}{R} \right)^2 \rho_0 \, dt + \int_0^T w(\theta_0)^2 \rho_0 \, dt + \right. \\
\left. + e \int_0^T \left[\left(\frac{u}{R} + w' \right)' \right]^2_{\rho=\rho_0} \rho_0 \, dt \right\} - \int_0^{\rho_0} (\varphi' u^0 + \psi' w^0)\rho \, d\rho + \\
+ \int_0^{\rho_0} (\varphi^0 u' + \psi^0 w')\rho \, d\rho \ .
\end{cases}
$$

This problem can be solved by conjugate gradient method as in [8]. In order to avoid numerical instabilities, we compute the first N eigenvalues and eigenfunctions of the operator $\mathcal{A}_R$ to derive the approximate problem.

For this let $v_j = \{u_j, w_j\}$ and λ_j the solutions of the eigenvalue problem:

$$
(4.2) \quad \begin{cases}
-L(u_j) + \frac{(1+\nu)}{R}w_j' - \frac{e}{R}L\left(\frac{u_j}{R} - w_j' \right) = \lambda_j u_j \\
\frac{e}{\rho}\left[L\left(\frac{u_j}{R} + w_j \right)\rho \right]' - \frac{(1+\nu)}{\rho R}(u_j\rho)' + 2\frac{(1+\nu)}{R^2}w_j = \lambda_j w_j
\end{cases}
$$

$$
(4.3) \quad \begin{cases}
u_j(0) = 0 & u_j(\rho_0) = 0 \\
u_j(0) + w_j'(0) = 0 & (u_j + w_j')_{\rho_0} = 0 \\
L(u_j + w_j')|_0 = 0 & L(u_j + w_j)(\rho_0) = 0 \ .
\end{cases}
$$

We set $u_j = \sigma_j'$, and we arrive to the following eigenvalue problem

$$(4.4) \quad \begin{cases} -(\nabla \sigma_j)' + \frac{(1+\nu)}{R} w_j' - \frac{\epsilon}{R} \nabla \left(\frac{\sigma_j}{R} + w \right)' = \lambda_j \sigma_j' \\ \frac{\epsilon}{\rho} \left[\nabla \nabla \left(\frac{\sigma_j}{R} + w_j \right) \right] - \frac{(1+\nu)}{R} \nabla \sigma_j + 2 \frac{(1+\nu)}{R^2} w_j = \lambda_j w_j \end{cases}$$

$$(4.5) \quad \begin{cases} \sigma_j'(0) = 0 \qquad\qquad \sigma_j'(\rho_0) = 0 \\ \left(\frac{\sigma_j}{R} + w_j \right)_0' = 0 \qquad \left(\frac{\sigma_j}{R} + w_j \right)_{\rho_0} = 0 \\ \left[\nabla \left(\frac{\sigma_j}{R} + w_j \right) \right]_0' = 0 \quad \left[\nabla \left(\frac{\sigma_j}{R} + w_j \right) \right]'(\rho_0) = 0 \end{cases}$$

where

$$\nabla \psi = \psi'' + \frac{\psi'}{\rho} \ .$$

If we denote with φ_j^h and $\mu_j^h (J = 1, 2 \ldots N)$ the solutions of the discrete eigenvalue problem:

$$(4.6) \quad \begin{cases} \nabla^h \varphi_j^h = -\mu_j^h \varphi_j^h \\ D^h \varphi_j^h |_0 = D^h \varphi_j^h |_{\rho_0} = 0 \qquad \left(D\varphi = \frac{d\varphi}{d\rho} \right) \end{cases}$$

we have:

$$(4.7) \quad \begin{cases} \nabla^h \nabla^h \varphi_j^h = (\mu_j^h)^2 \varphi_j^h \\ D^h \varphi_j^h |_0 = D^h \varphi_j^h |_{\rho_0} = 0 \\ D^h (\nabla^h \varphi_j^h)_0 = D^h (\nabla^h \varphi_j^h)_{\rho_0} = 0 \ . \end{cases}$$

Moreover taking

$$\sigma_N^h = \sum_{j=1}^N A_j \varphi_j^h \ , \quad w_N^h = \sum_{j=1}^N B_j \varphi_j^h$$

we obtain

$$u_j^h = D^h \varphi_j^h \ , \quad (w_j^h)^\pm = \frac{(\lambda_j^h)^\pm - \mu_j^h \left(1 + \frac{1}{3} \frac{\epsilon^2}{R^2} \right)}{\frac{1+\nu}{R} + \frac{1}{3} \frac{\epsilon^2}{R} \mu_j^h} \varphi_j^h$$

where:

$$(\lambda_j^h)^\pm = \frac{1}{2} \Bigg\{ \left[\mu_j^h \left(1 + \frac{1}{3} \frac{\epsilon^2}{R^2} \right) + \frac{2(1+\nu)}{R^2} + \frac{\epsilon^2}{3} (\mu_j^h)^2 \right] \pm$$

$$\pm \sqrt{ \left[\mu_j^h \left(1 + \frac{1}{3} \frac{\epsilon^2}{R^2} \right) + \frac{2(1+\nu)}{R^2} + \frac{\epsilon^2}{3} (\mu_j^h)^2 \right] - 4 \left[\mu_j^h \left(1 + \frac{1}{3} \frac{\epsilon^2}{R^2} \right) \right] \cdot }$$

$$\cdot \left(\frac{2(1+\nu)}{R^2} + \frac{\epsilon^2}{3} \mu_j^h)^2 \right) - \mu_j^h \left(\frac{1+\nu}{R} + \frac{1}{3} \frac{\epsilon^2}{R} \right)^2 \Bigg\} \ .$$

Then the initial state of the homogeneous problem can be represented by

$$(\mathbf{v}^0)_N^h = \sum_{j=1}^{N} S_j^0 \mathbf{v}_j^h \quad (\mathbf{v}^1)_N^h = \sum_{j=1}^{N} S_j^1 \mathbf{v}_j^h$$

and the problem P_h can be formulated in term of Fourier coefficients.

$$P_h \begin{cases} Find\ (s_j^0, s_j^1)_{(j=1,\dots N)} \in \mathbf{R}^2\ such\ that \\ J_h^N(s_j^0, s_j^1,\ J = 0,1,\dots N) \le J_h^N(\tau_j^0, \tau_j^1,\ J = 0,1,\dots N) \\ \forall(\tau_j^0, \tau_j^1) \in \mathbf{R}^2\ j = 0,1,\dots N\ . \end{cases}$$

We observe that the numerical approximation of the problem follows from the solution of the discrete eigenvalue problem (4.6) + conjugate gradient method + Fourier series.

From the uniqueness results of §3 and regularity properties of the solution $\{u, w\}$ as well as an accurate discretization of the eigenvalue problem (4.6), we deduce the convergence of the algorithm (see [8], [11]).

Some preliminary numerical experiments are shown in picture 4.1 (initial configuration of the shell), picture 4.2 (shell deformation at $t = \frac{9}{10}T^0$) and picture 4.3 at $t = T^c$) the parameters are: N (number of eigenvectors)= 9, $h = \frac{1}{300}$ and $R = 100$.

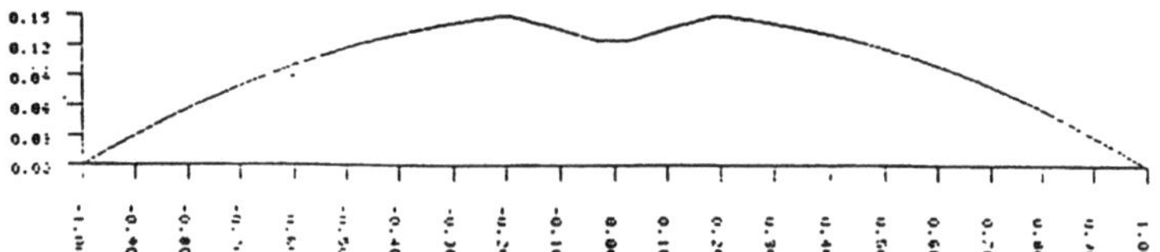

Figure 4.1

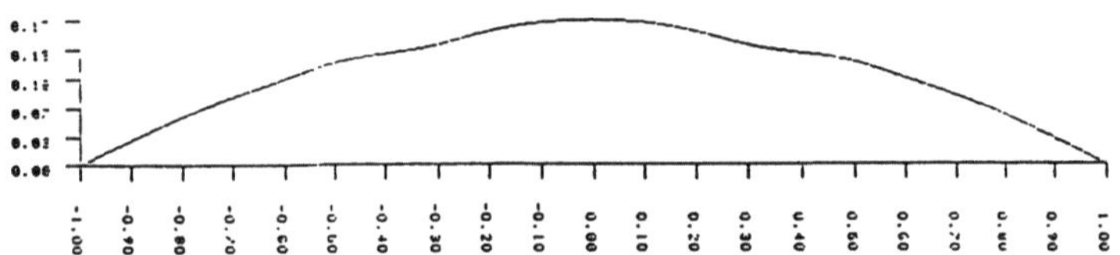

Figure 4.2

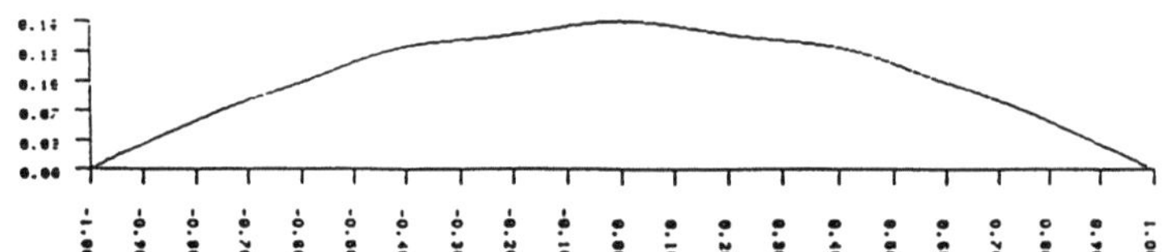

5. Bibliography

[1] L. BAUER, E.L. REISS, H.B. KELLER: *Axisymmetric buckling of hollow spheres and hemisphere*, Comm. Pure and Applied Math., XXIII, (1970), 529–568.

[2] P. PODIO GUIDUGLI, M. ROSATI, A. SCHIAFFINO, V. VALENTE: *Equilibrium of an elastic spherical cap pulled at the rim*, SIAM. J. Math. Anal., vol. 20, n. 3 (1989), 643-663.

[3] G. GEYMONAT, M. ROSATI, V. VALENTE: *Numerical analysis for eversion of elastic spherical caps*, Computer Methods in Appl. Mech. and Eng. 75, (1989) 39-52.

[4] A.E.H. LOVE: *A Treatise on the mathematical theory of elasticity*, 4th editions, Cambridge Univ. Press, (1927).

[5] J.E. LAGNESE, J.L. LIONS: *Modelling analysis and control of thin plates*, Masson (1988).

[6] J.L. LIONS: *Contrôlabilité exacte perturbations et stabilisation de systèmes distribués*,Vol.I et II, Masson, (1988).

[7] S.TIMOSHENKO : *Theory of elastic stability*, Mc Graw–Hill (1963).

[8] C.I. LI, R GLOWINSKY, J.L. LIONS: *A numerical approach to exact boundary controllability of the wave equation*, Japan J. Applied Math., 7, (1990) 1-76.

[9] G. GEYMONAT, P. LORETI, V. VALENTE, *Controlabilité exacte d'un modèle de coque mince*, C.R.A.S. t313 série I, pp. 81-86, 1991.

[10] G. GEYMONAT, P. LORETI, V. VALENTE: *En Preparation*.

[11] P.A. RAVIART, J.M. THOMAS, *Introduction à l'analyse numérique des équations aux dérivées portreller*. Masson (1988).

International Series of Numerical Mathematics, Vol. 107, © 1992 Birkhäuser Verlag Basel

Inverse Problem : Identification of a melting front in the 2D case

Xin-Fang Wang, Marie-Minerve Rosset-Louërat and Christine Bénard

Abstract. The moving solid-liquid interface of a melting solid in a rectangular enclosure is identified from temperature and flux measurements performed on the cold side of the solid only . The lack of measurement in the liquid phase prevents the interface to be recovered by straightforward use of the direct Stefan solution . The identification of this melting front is an inverse problem that requires particular solving methods . An algorithm is used, based on a regularized least square approach, that is extended to this nonlinear case by a predictive-sliding-horizon technique .

1.Introduction

The Stefan Problem is a particular moving boundary problem (Carslow and Jaeger 1959) : the isothermal interface between the solid phase and the liquid phase is driven by the diffusive heat transfer in the two connected phases . Studies on the Stefan problem have started with the direct solution . Then inverse situations have been examined. Among phase change problems, two inverse cases can arise : the *control* problem or the *tracking* one. The control problem consists in computing boundary conditions on the solid boundaries for the solid to melt or solidify at a prescribed velocity . Several studies have been devoted to this problem (Colton 1974, Colton and Reemtsen 1984, Jochum 1980 a and b, Jochum 1982, Knaber 1985, Moreno and Saguez 1978, Reemtsen and Kirsch 1984, Saguez 1976, Saguez 1978) . In the particular 1D space case, Jochum (1980 a and b) has studied existence and unicity of the optimal control solution and has given a numerical solution . He has extended his numerical method to the 2D space case (Jochum 1982) . Colton (1974), Moreno and Saguez (1978) have also studied the 2D space control problem . Fewer results are available for the inverse tracking problem . This inverse problem is considered here . It consists in identifying the time dependent position of the melting front from temperature and flux measurements performed on the solid phase boundary alone (the accessible cold side of the solid) . Indeed, direct computation of the interface is not possible as soon as some of the boundary conditions lack, as it often happens in real situations when no measurement is performed in the liquid phase . Results are given here in the 2D case that extend the results given in the 1D case by Afshari (1990), Afshari et al. (1989,1991), Bénard (1990) and Bénard and Afshari (1991) .

2. Definition of the problem

A rectangular enclosure of aspect ratio A (height H along $0y$, width $L = H/A$ along $0x$) is filled with a material at initial temperature T_i, which is below T_f, the melting temperature of the material . From time t_0 the temperature $T(x = 0, y, t)$ is set to $T_0(y, t)$, higher than T_f; therefore the solid begins to melt (Fig. 1) and the solid-liquid interface $\Sigma(t)$ begins to move .

Using dimensionless variables (κ_{sol} is the diffusivity of the solid) :

$$\frac{y}{H} \longrightarrow y \qquad\qquad \frac{x}{H} \longrightarrow x$$

$$\frac{\kappa_{sol}}{H^2}t \longrightarrow t \qquad\qquad \frac{T - T_f}{T_f - T_i} \longrightarrow T$$

one gets the heat conduction equation in the time dependent solid domain:

$$\frac{\partial^2 T}{\partial x^2} + \frac{\partial^2 T}{\partial y^2} = \frac{\partial T}{\partial t} \tag{1}$$

The boundary conditions are known on three walls of the enclosure : Γ_0, Γ_1 , Γ_2 ; and an extra data is known on Γ_0. They are defined as follows :

- on Γ_1 et Γ_2:

$$\frac{\partial T}{\partial y} = 0 \tag{2}$$

- on Γ_0:

$$T(1/A, y, t) = \theta(y, t) \tag{3}$$

$$-\left.\frac{\partial T(x,y,t)}{\partial x}\right|_{x=1/\Lambda} = \phi(y,t) \qquad (4)$$

- on the moving interface $\Sigma(t)$:

$$T = T_f. \qquad (5)$$

The initial temperature is known :

$$T(x,y,t_0) = T_i \qquad (6)$$

The inverse problem consists in identifying the interface position $s(y,t)$ ($s(y,t)$ is the liquid thickness) from the measurements performed on Γ_0, with the adiabaticity hypothesis (Equation 2).

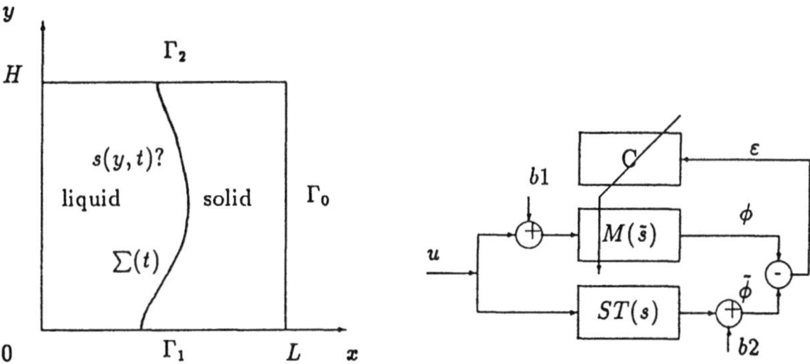

Figure 1: Identification problem   Figure 2: Identification process

3. Computational method

The model approach is used to solve this inverse problem . It is shown in Fig.2 where :

- $ST(s)$ is the real thermal system
- $M(\tilde{s})$ is the direct model that simulates $ST(s)$
- ϕ is the model output

- $\bar{\phi}$ is the output measurement
- u is the model input and the real system input
- the interface position $s(y,t)$ is a parameter of M .

The identification of the parameter s is a nonlinear problem that is solved by the minimization of a criterion (algorithm C of Fig.2) based on the distance between ϕ and $\bar{\phi}$.

Due to the fact that a diffusive system behaves as a low pass filter, very quick variations of $s(y,t)$ will produce vanishingly small output at the solid boundary Γ_0 . As a consequence, this inverse problem is an ill-posed problem (Guerrier 1989, Hadamard 1923) . In order to restore the continuous dependence of the solution on the data, a priori information on the solution is added . Here the regularization method (Tikhonov 1963 a and b, Tikhonov and Arsenin 1977) is used : in the criterion, a quadratic term is added to the output error function that is a weighted stabilizing function of the unknown to be identified .

Another difficulty in our problem is that, even after time and space discretisation, the dimension of the unknown $s(y,t)$ is still large . This is the reason why $s(y,t)$ is estimated by a sequential method : the criterion is defined on a sliding time horizon τ, the length of which is chosen so as to obtain a maximum sensitivity of the optimization of the criterion with respect to the first values of $s(y,t)$ in the considered time horizon . The criterion is the following :

$$J_\alpha = \int_{t_1}^{t_2} \int_0^1 (\phi(y,t) - \bar{\phi}(y,t))^2 dy dt + \Omega(\alpha, s) \qquad (7)$$

where the first term in the right handside of Equation 7 is the distance between the model output and the measurement and the second term is the regularization term depending on a weighting parameter α .

The choice of the sub horizon $\tau = t_2 - t_1$ is based on previous studies by Afshari (1990) and Bénard and Afshari (1989) . They showed that the sensivity function : output $\phi(t + t_i)$ versus parameter $s(t)$ has a maximum for a positive value of t_i and vanishes for large values of t_i after a value of about t_r . Since a good value of τ should be large enough to garantee sensitivity of J_α to s and small enough not to increase the bias resulting of the prediction of $s(y,t)$ on the horizon τ (Cf. Section 3.2, Eq.9), the value t_2 is set to t_r .

3.1 Direct model . ϕ is the model output corresponding to a given $s(y,t)$. It is computed with the help of a 2D direct model that simulates heat diffusion through the solid domain with one moving boundary (liquid-solid interface) . The numerical scheme is the one proposed by Gobin (1984) and Zanoli (1984) . It is based on Patankar (1980) finite volume method . Since the solid domain is nonrectangular, a nonorthogonal coordinate transformation is used to map the irregular physical cavity onto a fixed rectangular computational space . This transformation gives a transformed heat conduction equation with additional non diagonal terms , that can be solved, at each discretization time step δt by a usual ADI method . Either the temperature ($\theta(y,t)$ on $x = 1/A$) or the flux ($\phi(y,t)$ on $x = 1/A$) can be taken as input to the direct model, together with the assumed $s(y,t)$. In the 1D case, without measurement noise, the flux has been shown to be more sensitive to the interface position $s(y,t)$ than the temperature (Afshari 1990), and thus the flux ($\phi(y,t)$ on $x = 1/A$) is chosen as the model output and the temperature ($\theta(y,t)$ on $x = 1/A$) as the input.

3.2 Resolution of the inverse problem . The discretisation parameters are the following :

- measurement sampling time interval Δt

- direct model internal time step δt : we know (Afshari 1990) that $\delta t \leq 0.25\Delta t$ is a good choice

- direct model internal space steps δx and δy : we know (Afshari 1990) that $\delta x^2/\delta t \simeq 1$ is a good choice

- space discretisation of $s(y,t)$ with space step equal to : H/ns and time discretisation of $s(y,t)$ which is taken to be the sampling time Δt . In order to reduce the number of unknowns, the space discretisation of $s(y,t)$ can be different from the space discretisation of the direct model . A spline interpolation is used to develop $s(y,t)$ in space from its values identified on the $ns + 2$ nodes along $0y$ (including 2 fictitious nodes, see Gobin 1984 and Zanoli 1984) .

This allows us to express the unknown $s(y,t)$ as $s_{i,j}$ where : $i = 1,2, \cdots,ns$ and j is the time index : $t = j\Delta t$. The values of $s_{i,j}$ are calculated successively in time at each sampling instant : $s_{i,m+1}$, the value of $s_{i,m+1}$ at time $t_{m+1} = (m + 1)\Delta t$, is identified through minimization of JD_α a discretized version of J_α defined on the interval $[t_{m+1}, t_{m+r}]$:

$$JD_\alpha = \frac{1}{r\,ny}\sum_{i=1}^{ny}\sum_{j=1}^{r}\left(\phi(i,m+j) - \tilde{\phi}(i,m+j)\right)^2 + \alpha\sum_{i=1}^{ns}(s_{i,m+1} - s_{i,m})^2 \qquad (8)$$

where α is the regularization term coefficient .

The length of the observation horizon is $\tau = r.\Delta t$. The identification algorithm is iterative and requires, at each iteration, the resolution of the direct problem (i.e. the calculation of the model output) on the horizon $[t_{m+1}, t_{m+r}]$. When JD_α is optimized for $[t_{m+1}, t_{m+r}]$, $s_{i,m+1}$, the optimized value at t_{m+1}, is the only one kept . The optimization process is resumed on the next time interval $[t_{m+2}, t_{m+r+1}]$.

To reduce the number of unknowns of the minimization problem the following linear assumption is made on $s_{i,j}$:

$$s_{i,m+j} = s_{i,m+1} + \beta(s_{i,m+1} - s_{i,m})(j-1) \qquad (9)$$
$$i = 1, 2, \cdots ns$$
$$j = 1, 2, \cdots r$$

The meaning of Equation (9) is that at a time step t_{m+1}, s is temporally assumed to depend on $s_{i,m+1}$ and β only, on the time horizon $[t_{m+1}, t_{m+r}]$, so that β can be considered as some prediction parameter . This parameter is chosen a priori in $[0, 1]$. The only unknowns left are $s_{i,m+1}$, $(i = 1, \cdots, ns)$, so that the criterion is a function of ns unknowns only . Its minimization is performed by a Gauss Newton algorithm .

Here the criterion to be minimized is JD_α :

$$JD_\alpha = J_0 + J_1 \qquad (10)$$

with :

$$J_0 = \frac{1}{r\,ny}\sum_{i=1}^{ny}\sum_{j=1}^{r}(\phi_{i,m+j} - \tilde{\phi}_{i,m+j})^2 \qquad (11)$$

$$J_1 = \alpha\sum_{i=1}^{ns}(s_{i,m+1} - s_{i,m})^2$$

Let :

$$S_{m+1} = \begin{pmatrix} s_{1,m+1} \\ s_{2,m+1} \\ \vdots \\ s_{ns,m+1} \end{pmatrix} \quad \Phi_i = \begin{pmatrix} \phi_{1,m+i} \\ \phi_{2,m+i} \\ \vdots \\ \phi_{ny,m+i} \end{pmatrix} \quad \tilde{\Phi}_i = \begin{pmatrix} \tilde{\phi}_{1,m+i} \\ \tilde{\phi}_{2,m+i} \\ \vdots \\ \tilde{\phi}_{ny,m+i} \end{pmatrix} \tag{12}$$

$$X_i^T = \begin{pmatrix} \dfrac{\partial \phi_{1,m+i}}{\partial s_{1,m+1}} & \dfrac{\partial \phi_{2,m+i}}{\partial s_{1,m+1}} & \cdots & \dfrac{\partial \phi_{ny,m+i}}{\partial s_{1,m+1}} \\[2mm] \dfrac{\partial \phi_{1,m+i}}{\partial s_{2,m+1}} & \dfrac{\partial \phi_{2,m+i}}{\partial s_{2,m+1}} & \cdots & \dfrac{\partial \phi_{ny,m+i}}{\partial s_{2,m+1}} \\[2mm] \vdots & \vdots & \vdots & \vdots \\[2mm] \dfrac{\partial \phi_{1,m+i}}{\partial s_{ns,m+1}} & \dfrac{\partial \phi_{2,m+i}}{\partial s_{ns,m+1}} & \cdots & \dfrac{\partial \phi_{ny,m+i}}{\partial s_{ns,m+1}} \end{pmatrix}$$

$$A(S_{m+1}^k) = X^T X = \sum_{i=1}^{r} X_i^T X_i + \alpha \cdot r \cdot ny \cdot I$$

$$B(S_{m+1}^k) \;=\; \alpha \cdot r \cdot ny \cdot (S_{m+1}^k - S_m) - \sum_{i=1}^{r} X_i^T (\Phi_i^k - \tilde{\Phi}_i)$$

where the superscript k stands for the iteration number . The resulting equation for the unknown S_{m+1}^{k+1} (S_{m+1} at the iteration number k) that minimizes JD_α :

$$A(S_{m+1}^k) \cdot (S_{m+1}^{k+1} - S_{m+1}^k) = B(S_{m+1}^k) \tag{13}$$

is solved by L.U. decomposition of Matrix A .

4. Example

4.1 Definition of our reference .   The measurements used to solve the inverse problem are obtained by direct simulation : a noiseless output $\tilde{\phi}(y,t)$ is obtained from noiseless input $s(y,t)$ and $\theta(y,t)$. Then Gaussian white noise is added to $\tilde{\phi}$ (standard deviation σ_ϕ) and θ (standard deviation σ_θ) to simulate the noisy measurements that will be used to get $s(y,t)$ back .

Our reference is a smooth symmetrical interface (Fig. 3):

$$s_1(y,t) = l(1 - e^{-t})\frac{1 - \cos 2\pi y}{2}$$

where $l = 1/A = 0.4$, $s(y,t)$ is the liquid thickness and the temperature $\theta(y,t)$ on the cold face of the solid $x = 1/A$ is constant and equal to the initial temperature .

4.2 Parameters of the numerical problem . Previous analyses (Afshari 1990) allow us to choose the following values of the discretisation parameters :

- 12x12 grid for the direct model

- number of discretization points for $s(y,t)$ along $0y$: $ns = 10$

- number of measurement points along $0y$: $ny = 10$

- time step of the direct model $\delta t = 0.0025$

- measurement samplig time $\Delta t = 0.01$

- sub time interval for minimization $r = 10$.

To figure out the quality of the identification, a relative distance is defined :

$$ER = \sqrt{\frac{\sum_{i=1}^{ns} \sum_{j=0}^{nt}(s_{i,j} - \tilde{s}_{i,j})^2}{\sum_{i=1}^{ns} \sum_{j=0}^{nt} s_{i,j}^2}} \tag{14}$$

where $nt.\Delta t$ is the whole time interval where the identification is performed ($nt = 80$) . The behavior of ER, as a function of α , is typical of regularization methods : when α increases starting from a small value, ER decreases thanks to the filtering effect of $\Omega(\alpha, s)$; it increases again when this filtering becomes too important .

4.3 Results . In the case of noiseless measurements, Figures 4 and 6 show the results of identification with $\alpha = 1$ and $\beta = 1$; the relative distance ER is 0.024 . In the case of noisy measurements, a larger value of α has to be used to stabilize the solution . Figures 5 and 7 show the results of the identification with parameter values set to $\alpha = 10$ and $\beta = 1$ ($ER = 0.156$).

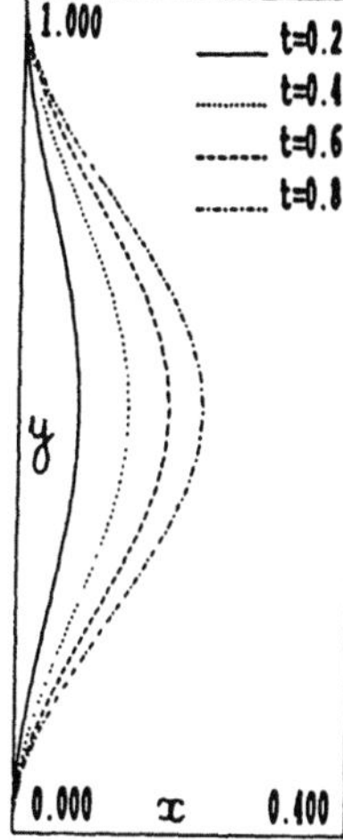

Fig.2
Melting front
Reference.

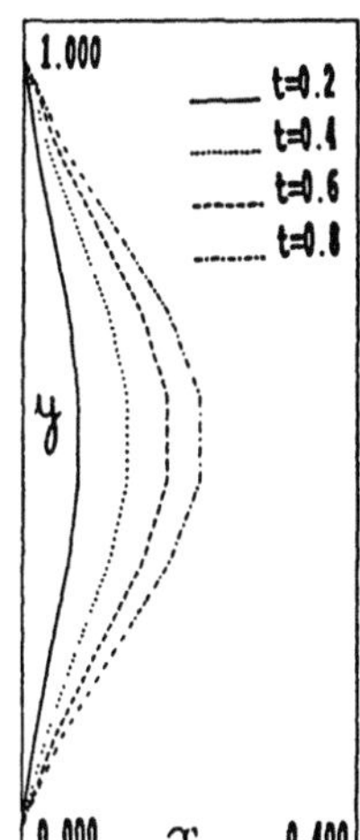

Fig.3
Identified interface
without noise.

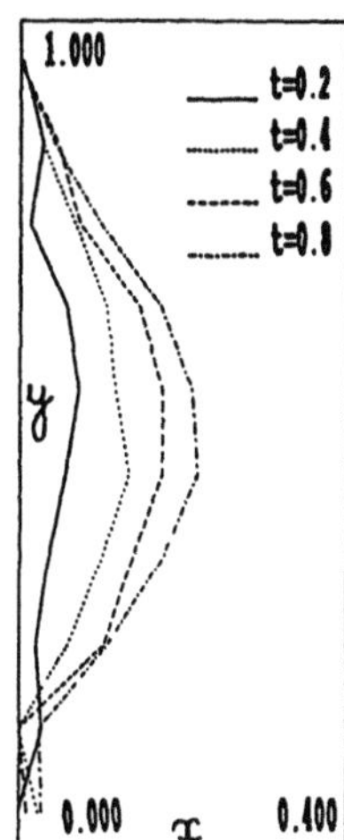

Fig.5
Identified interface
with noise.

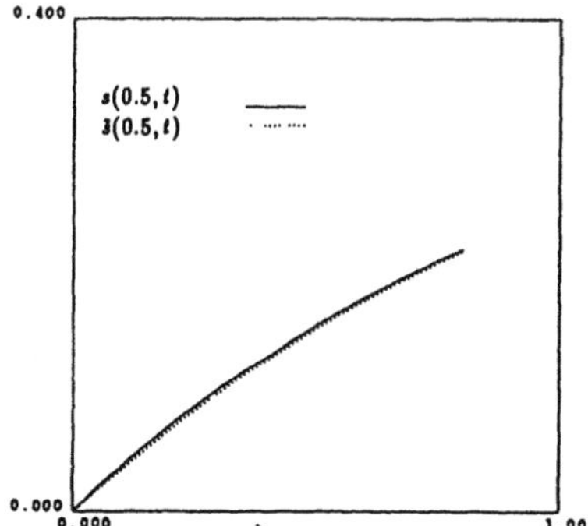

Fig.6
Liquid thickness at $y=0.5$
as a function of time.
Identification without noise .

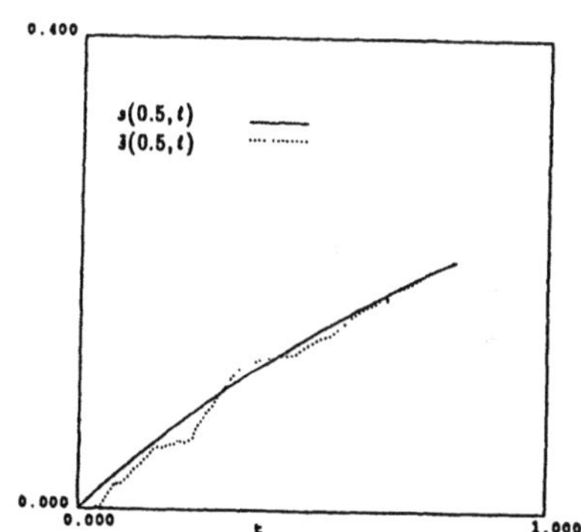

Fig.7
Liquid thickness at $y=0.5$
as a function of time.
Identification with noise .

Here a regularization process together with a predictive-sliding-time technique has allowed to recover a moving interface . Let us note however that the information present in $\bar{\phi}$ on the cold face of the solid contained enough information due to the number of measurements and the small enough width of the solid .

5. Conclusion

A numerical solution of the 2D inverse Stefan problem has been proposed that gives satisfactory results . Measurement noise is handled more easily than by other methods (Colton and Reemtsen 1984, Jochum 1982) . On a Sun 4/110 computer, typical CPU time for Fig. 5 is of the order of 4 hours . Improvement of the optimisation algorithm as well as identification of various shapes of interfaces will be the next steps of this work .

References

Afshari, A. (1990)*Identification de l'évolution d'un front de fusion/solidification par résolution inverse de l'équation de la chaleur dans le domaine solide.* Thèse de Doctorat, Université de Paris-Sud, Orsay, .

Afshari,A. , Bénard,C. ,Duhamel,C. and Guerrier, B. (1989) 'On-line identification of the state of the surface of a material undergoing thermal processing' *Proc. 5th IFAC symposium on control of distributed parameter systems*, A.El Jai Ed., Perpignan, pp. 459-463 .

Afshari,A., Bénard,C. Rosset-Louërat M.-M., Wang X.F. (1991), Problème inverse : suivi de l'évolution de la position d'un front de fusion à l'aide de mesures sur la face solide , *Journée S.F.T. Modélisation en thermique industrielle.*

Bénard, C.,(1990), Méthodes inverses : application au suivi d'un front de changement d'état , *Cours de l'Ecole d'Eté du Groupement Universitaire de Thermique* .

Bénard, C. and Afshari, A. (1991) 'Front tracking for the control of solid-liquid phase change processes' *Proc. 7th Int. Conf. on Numerical Methods in Thermal Problem,* Stanford , Vol. VII. Part 1, pp.186-198 .

Carslaw, H.S. and Jaeger, J.C. (1959)*Conduction of heat in solids*, Oxford University Press.

Colton, D. (1974) 'The Inverse Stefan Problem for the Heat Equation in Two Space Variables' *Mathematika*, 21, pp.282-286.

Colton, D. and , Reemtsen, R. (1984) 'The Numerical Solution of the Inverse Stefan Problem in Two Space Variables' *SIAM. J. Appl. Math.*, Vol.44, no.5, pp.996-1013.

Gobin, D. (1984) *Changement d'état solide-liquide : Evolution temporelle du couplage entre la convection naturelle dans la phase liquide et la conduction dans la phase so-lide . Etude numérique et expérimentale* , Thèse de Doctorat d'Etat, Université de Paris 6 .

Guerrier, B., (1989) *Relation entrées-sorties dans les systèmes diffusifs : Réduction du modèle d'état, Identification de conditions aux limites* , Thèse de Doctorat d'Etat, Université de Paris 6 .

Hadamard, J. (1923) *Lectures on the Cauchy problem in linear partial differential equations* , Yale University Press .

Jochum, P. (1980a) 'The Numerical Solution of the Inverse Stefan Problem' *Numer. Math.* Vol.34, pp.411-42.

Jochum, P. (1980b) 'The Inverse Stefan Problem as a Problem of Nonlinear Approximation Theory' *J. Approximation Theory*, Vol.30, pp.81-98 .

Jochum, P. (1982) 'To The Numerical Solution of an Inverse Stefan Problem in Two Space Variables' *Numerical Treatement of Free Boundary Value Problem*, Ed Albrecht J. et al, ISNM 58, Birkhäuser-Verlag, Basel.

Knaber, P. (1985) 'Control of Stefan Problem by Means of Linear-Quadratic Defect Minimization' *Numer. Math.* 46, pp. 429-442.

Moreno, C. and Saguez, C. (1978) 'Dépendance par rapport aux données de la frontière libre associée à certaines inéquations variationnelles d'évolution' *Rapport de recherche 298, IRIA*.

Patankar, S.V. (1980) *Numerical heat transfer and fluid flow* , Hemisphere.

Reemtsen, R. and Kirsch, A. (1984) 'A Method for the Numerical Solution of the One Dimensional Inverse Stefan Problem' *Numer. Math.* 45, pp.253-273.

Saguez, C. (1976) 'Contrôle optimal de systèmes gouvernés par des inéquations variationelles. Applications à des problèmes de frontière libre' *Rapport de Recherche 191, IRIA.*

Saguez, C. (1978) 'Contrôle optimal d'inéquations variationelles avec observation de domaines' *Rapport de Recherche 286, IRIA.*

Tikhonov, A.N. (1963a) 'Solution of Incorrectly Formulated Problem and The Regularization Method' *Soviet Math. Dokl.*, Vol.4, pp.1035-1038.

Tikhonov, A.N. (1963b) 'Regularization of incorrectly posed problems' *Soviet Math. Dokl.*, Vol.4, pp.1624-1627.

Tikhonov, A.N. and Arsenin, V.Y. (1977) *Solutions of ill-posed problems*, Translation ed. by F. John, Winston/Wiley .

Zanoli, A. (1984) *Stockage thermique par chaleur latente : Couplage de la conduction dans la phase solide avec le déplacement de l'interface et la convection naturelle dans la phase liquide* , Thèse de Doctorat-Ingénieur, Université de Paris 6 .

Authors' address:
Xin-Fang Wang, Marie-Minerve Rosset-Louërat, Christine Bénard
Laboratoire FAST Unité Associée 871, Paris 6-Paris 11-CNRS
Bât. 502
Campus Universitaire
91405 Orsay Cedex FRANCE

MICRO-LOCAL APPROACH TO
THE CONTROL FOR THE PLATES EQUATION

Nicolas Burq and Gilles Lebeau

Abstract: We present recent results in the theory of control for the plates equation in an open subset of $\mathbf{R}^n$. These results were obtained by G. Lebeau using semi-classical micro-local analysis and the results of propagation of singularities of R.B. Melrose and J. Sjöstrand. They give natural sufficient conditions for the exact controllability in an arbitrary small time. We also present further developments obtained by T. Hargé and N. Burq using technics of the same kind.

1. Introduction

The problem of the control for the plates equation has been studied by J.L. Lions (1988). The basic tool for the study of this problem is his H.U.M. method. Using this method and multiplier technics, J.L. Lions obtains results of exact controllability of the following type:

Theorem (Lions 1988). Let Ω be an open bounded subset of $\mathbf{R}^n$ with a C^3 boundary, let $x_0 \in \mathbf{R}^n$, $m(x) = x - x_0$, $R(x) = \max\limits_{x \in \overline{\Omega}} |m(x)|$ and $\Sigma(x_0) = \{x \in \partial\Omega \ ; \ m(x) \cdot n(x) > 0\}$ where $n(x)$ is the unit outer normal of $\partial\Omega$. Let λ_0 be the smallest eigenvalue of the Laplacian with Dirichlet boundary conditions on Ω.

Then for all $T > T_0 = \frac{R(x_0)}{\lambda_0}$ and for all $(u_0, u_1) \in L^2(\Omega) \times H^{-2}(\Omega)$ there exists a control $v \in L^2(\Sigma(x_0) \times [0, T_0])$ such that the solution of the system:

$$(\frac{\partial^2}{\partial t^2} + \Delta^2)u = 0 \ in \ \Omega \times]0, +\infty[,$$

$$u \mid_{\partial \Omega} = 0,$$

$$\frac{\partial u}{\partial n} \mid_{\partial \Omega} = v \times 1_{\Sigma(x_0) \times [0,T]},$$

$$u \mid_{t=0} = u_0,$$

$$\frac{\partial u}{\partial t} \mid_{t=0} = u_0,$$

verifies $u \equiv 0$ for all $t \geq T$.

We can make two remarks about this result. The first one is that since the plates equation is not hyperbolic we might think that in fact the control in an arbitrary small time holds, whereas here, the time is fixed by the geometry. The second remark is that the hypothesis on the geometrical structure of the part of the boundary on which we impose a non-zero control is not quite natural.

An answer to the first remark was given by E. Zuazua (1988) who proved that in certain cases, with the same geometrical hypothesis as Lions', we do have exact controllability in an arbitrary small time.

In the more particular case of a rectangle, A.Haraux (1989) has solved the "semi-internal" control problem in which we allow the control to act in a strip parallel to one side of the rectangle, and afterwards S. Jaffard (1988) proved that the internal control in any subset of a rectangle, with non empty interior, holds. However their proofs, using harmonic analysis do not seem to be generalizable to other geometric situations.

After the results of C. Bardos, G. Lebeau and J. Rauch (1988) on the exact controllability of the wave equation obtained using micro-local analysis technics and propagation results, G. Lebeau has studied the application of such methods to the plates equation. In this text we will show how these ideas can give natural geometric conditions and powerful results.

2 Main theorem

2.1 Definition. Let Ω be an open bounded subset of $\mathbf{R}^n$ with a C^∞ boundary such that $\partial \Omega$ has not contact of infinite order with his tangent lines. Let Γ_0 be an open subset of $\partial\Omega$, we say that Γ_0 controls geometrically Ω in time $T_0 > 0$ if every generalized ray, covered at speed 1, starting from Ω in any direction and reflecting on $\partial\Omega$ according to the optical laws hits Γ_0 in a non diffractive point and in a time smaller than T_0 (we say that a point (x_0, ρ_0) is diffractive if in a neighborhood of x_0 the line $(x_0 + s\rho_0)$ is in $\overline{\Omega}$).

2.2 Theorem (Lebeau 1989). *Let Γ_0 controlling geometrically Ω in finite time, let T_1 be any strictly positive number, then for all $(u_0, u_1) \in H_0^1(\Omega) \times H^{-1}(\Omega)$ there exists a control $v \in L^2(\Gamma_0 \times [0, T_1])$ such that the solution of the system:*

$$(\frac{\partial^2}{\partial t^2} + \Delta^2)u = 0 \ in \ \Omega \times]0, +\infty[,$$

$$u \mid_{\partial\Omega} = 0,$$

$$\Delta u \mid_{\partial\Omega} = v \times 1_{\Gamma_0 \times [0, T_0]},$$

$$u \mid_{t=0} = u_0,$$

$$\frac{\partial u}{\partial t} \mid_{t=0} = u_0,$$

verifies $u \equiv 0$ for all $t \geq T_1$.

The results of A.Haraux and S.Jaffard show that contrarily to the case of the wave equation, the geometric controllability gives here only a sufficient condition. This phenomena will appear in the proof of this result that we are going to outline.

3. Reduction of the problem

Applying the H.U.M. method of J.L. Lions we reduce the problem to the proof of the fact that the solution of the system

$$(\frac{\partial^2}{\partial t^2} + \Delta^2)u = 0 \ in \ \Omega \times]0, +\infty[,$$

$$u \mid_{\partial\Omega} = 0,$$

$$\Delta u \mid_{\partial\Omega} = 0,$$

$$u \mid_{t=0} = u_0,$$

$$\frac{\partial u}{\partial t} \mid_{t=0} = u_0,$$

verifies

$$E_0(u) = \|u_1\|_{H^{-1}}^2 + \|u_0\|_{H_0^1}^2 \le C \int_0^{T_1} \int_{\Gamma_0} |\frac{\partial u}{\partial n}|^2 \, d\sigma(x) dt.$$

But the solution of this system can be written, if e_ν is the orthonormal basis of $L^2(\Omega)$ of eigenvectors for the Laplacian with Dirichlet boundary conditions, associated with the eigenvalues λ_ν, as $u = u_+ + u_-$ with

$$(\pm i \partial_t - \Delta)u_\pm = 0,$$

$$u_\pm \mid_{\partial\Omega} = 0,$$

$$u_\pm = \sum_\nu e^{\pm it\lambda_n u} a_{\pm,\nu} e_\nu(y),$$

$$u_0 = \sum_\nu (a_{+,\nu} + a_{-,\nu}) e_\nu,$$

$$u_1 = \sum_\nu \lambda_n u (a_{+,\nu} - a_{-,\nu}) e_\nu,$$

and

$$E_0(u) = \|u_+\|_{H^1}^2 + \|u_-\|_{H^1}^2 = E(u_+) + E(u_-).$$

Since u_+ and u_- do not interact it is sufficient to prove that

$$E(u_+) \le C \int_O^{T_1} \int_\Gamma |\partial_n u_+|^2,$$

and

$$E(u_-) \le C \int_O^{T_1} \int_\Gamma |\partial_n u_-|^2.$$

We'll show how we can prove for example the first inequality.

4. h-pseudo-differential operators and semiclassical wave fronts.

In this part we present the main tools needed in the following.

4.1 h-pseudo-differential operators. For $m \in \mathbf{R}$ and $d \in \mathbf{N}^*$, let $S^m(\mathbf{R}^d)$ be the vector space of the symbols $q(x, \xi, h)$ of class C^∞ in $(x, \xi) \in \mathbf{R}^d \times \mathbf{R}^d$, defined for $h \in]0, h_0[$ $(h_0 > 0)$ and such that

$$\forall \alpha, \beta \in \mathbf{N} \; \exists C_{\alpha,\beta}; \; |\partial_x^\alpha \partial_\xi^\beta q| \le C_{\alpha,\beta}(1 + |\xi|)^{m-|\beta|}.$$

For $q \in S^m(\mathbf{R}^d)$ let $Op(q) = Q(x, hD_x, h)$ be the operator defined by

$$Op(q)f(x) = \frac{1}{(2\pi h)^d} \int e^{i(x-y)\cdot\xi/h} q(x, \xi, h) f(y) dy d\xi.$$

We recall that if $Q \in Op(S^0)$ then Q is bounded in L^2 uniformly with respect to $h \in]0, h_0[$.

If near a point $x_0 \in \partial\Omega$, we have a coordinate system $x = (x', x_n)$ such that Ω is locally defined by $x_n > 0$ then we can define tangential operators S_t^m of order m by

$$Op(q)f(x) = \frac{1}{(2\pi h)^{n-1}} \int e^{i(x'-y')\cdot\xi'/h} q(x', x_n, \xi', h) f(y', x_n) dy' d\xi',$$

$$|\partial_x^\alpha \partial_{\xi'}^\beta q| \le C_{\alpha,\beta}(1 + |\xi'|)^{m-|\beta|}.$$

4.2 Semi-classical wave fronts. In the following we'll note $h_k = 2^{-k}$. Let $\alpha < 1/2$ and $\beta > 2$ be two reals. For $k \in \mathbf{N}$ we set

$$F_k = \{\nu;\ \alpha 2^K \le \sqrt{\lambda_n u} \le \beta 2^k\},$$

$$E_k = \{u = \sum_{\nu \in F_k} u_\nu e_n u\},$$

and

$$B = \{U = (u_k)_{k \in \mathbf{N}};\ \forall k u_k \in E_k \text{ and } \|u_k\|_{L^2} \le 1\}.$$

We will not distinguish between elements of E_k, $u_{k,0}$ and the associated solutions of the system

$$(ih_k\partial_t - h_k^2\Delta)u_k = 0,$$

$$(A) \qquad u_k \mid_{\partial\Omega} = 0,$$

$$u_k \mid_{t=0} = u_{k,0}.$$

For $U \in B$ and $\rho_0 \in T^*\mathbf{R}_t T^*\Omega \cup T^*\partial\Omega$ we will say that ρ_0 is not in the wave front of U if there exists a symbol $q \in S^0(\mathbf{R}_t \times \mathbf{R}_x^n)$ (tangential if $\rho_0 \in T^*\partial\Omega$) such that $p(\rho_0) \ne 0$ and a function $\phi \in C_0^\infty(\mathbf{R}_t \times \mathbf{R}_x^n)$ such that for all $N \in \mathbf{N}$ there exists a constant C_N such that

$$\|Op(p) \mid_{h=h_k} u_k\|_{L^2}^2 \le C_N h_k^N$$

We will note $\rho_0 \notin WF(U)$.

5 Estimates

Solutions of the system (A) behave like solutions of a wave equation. Indeed, we took initial data with spacial frequencies of order h_k^{-1} and the equation $ih\partial_t - h^2\Delta$ is taken so that we'll have a finite speed for the propagation of singularities. We are going to precise this heuristic idea.

5.1 Elliptic estimates. There exists $C_1 > 0$ such that if $\phi \in C_0^\infty(\Omega)$ and $q \in S^0(\mathbf{R}_x^n)$ with support in $|\xi| \geq C_1$ then for all $\gamma \in \mathbf{N}^n$, $M \in \mathbf{N}$ there exists $C_{\gamma,M}$ such that for all $U \in B$ we have

$$\|\partial_x^\gamma Op(q)\,|_{h=h_k}\,(\phi(x)u(x))\|_{L^2}^2 \leq C_{\gamma,M} h_k^M.$$

We have a similar result near the boundary using tangential operators. This result means that if you take initial data in E_k, the part of its L^2-norm in the zone $|\xi| \geq C$ will be negligible.

Let $\Psi \in C^\infty(\mathbf{R})$, $\Psi = 0$ in a neighborhood of $[\alpha^2, \beta^2]$, $\Psi = 1$ in a neighborhood of the infinite, let $\chi \in C_0^\infty(\mathbf{R}_t)$, let $q \in S^0(\mathbf{R}_x^n)$ (S_t^0 near the boundary) such that $supp(q) \subset \{|\xi| \leq C\}$ ($\{|\xi'| \leq C\}$ near the boundary), and let $\phi \in C_0^\infty(\mathbf{R}_x^n)$ then for all $N \in \mathbf{N}$ there exists C_N such that

$$\|\Psi(h_k D_t)\chi(t)Op(q)\,|_{h=h_k}\,\phi u_k\|_{L^2(\mathbf{R}\times\Omega)}^2 \leq C_N h_k^N.$$

This result means that the part of the energy outside the zone $\tau \in h_k^{-1}[\alpha^2, \beta^2]$ is negligible. It is due to the fact that if $|\xi| \in h_k^{-1}[\alpha, \beta]$ and $\tau \notin h_k^{-1}[\alpha, \beta]$ then the symbol of the operator $ih_k\partial_t - h_k^2\Delta$, $h_k^2|\xi|^2 - h_k\tau$ does not cancel. This is why we call these estimates "elliptic estimates".

5.2 Propagation estimates. To obtain more precise estimates we are going to make a transformation introducing a new variable s, and suppressing the parameter h_k. For $U \in B$ we set

$$\Theta(U)(s, t, x) = \sum_k e^{ish_k^{-1}} u_k(t, x).$$

This new distribution verifies

$$\frac{\partial^2 \Theta(U)}{\partial s \partial t} = \Delta_x \Theta(U),$$
$$\Theta(U)\mid_{\mathbf{R}_s \times \mathbf{R}_t \times \partial\Omega} = 0.$$

It is possible to define the usual wave front of $\Theta(U)$, $WF_b(\Theta(U))$ which is a closed subset of $T^*\mathbf{R}_s \times T^*\mathbf{R}_t \times (T^*\Omega \cup T^*\partial\Omega)$. This wave front is, according to the theorem of Melrose-Sjöstrand (1982), invariant by the generalized bicaracteristic flow of $\sigma\tau - |\xi|^2$ and if we note $\theta(s,\rho) = (s, \sigma = 1, \rho)$ then we have the following result

$$\forall s_0, \ \forall U \in B, \forall \rho_0 \in T^*\Omega \cup T^*\partial\Omega,$$

$$\rho_0 \in WF(U) \iff \theta(s_0,\rho_0) \in WF_b(\Theta(U)).$$

We deduce from this that

i) $WF(U) \subset \{\tau \in [\alpha,\beta], \ \tau = |\xi|^2\} = \Sigma_b$ ($WF(U) \subset \{\tau \in [\alpha,\beta], \ |\xi'|^2\} \le \beta$ near the boundary)

i i) $WF(U)$ in invariant by the generalized flow $g(u,\cdot)$ in $T^*\mathbf{R} \times (T^*\Omega \cup T^*\partial\Omega)$.

From this we deduce the following estimate: let $e \in S^0(\mathbf{R}_t \times \mathbf{R}_x^n)$ (S_t^0 near the boundary of $\mathbf{R}_t \times \Omega$) be elliptic in $\rho_0 \in \Sigma_b$ and ϕ_0 with support near (t_0, x_0), $\phi_0 = 1$ near (t_0, x_0), then for all $u \in \mathbf{R}$ there exists $f \in S^0(\mathbf{R}_t \times \mathbf{R}_x^n)$ (S_t^0 near the boundary) elliptic in $\rho_1 = g(u, \rho_0)$ and ϕ_1 with support near (t_1, x_1) such that for all $U \in B$ and all $N \in \mathbf{N}$ there exists $C_N > 0$ such that for all $k \in \mathbf{N}$

$$\|Op(f)\mid_{h=h_k} \phi_1 u_k\|_{L^2} \le \|Op(e)\mid_{h=h_k} \phi_0 u_k\|_{L^2} + C_N h_k^N.$$

This result means that we can estimate the micro-local L^2-norm near a point ρ_1 by the micro-local L^2-norm near a point ρ_0 if starting from the point x_0 in the direction ξ_0 at the time t_0, travelling at speed $2|\xi|$ and reflecting on $\partial\Omega$ according to the laws of optics, you reach the point x_1 with the direction ξ_1 at the time $t_1 = t_0 + u$ ($\rho_i = (t_i, \tau_i, x_i, \xi_i)$ with $\tau_0 = \tau_1 = |\xi_0|^2 = |\xi_1|^2$).

5.3 Lifting estimate. Let $\rho_0 \in T^*\mathbf{R}_t \times T^*\partial\Omega$, $\rho_0 \in \Sigma_b$, be a non-diffractive point such that $t_0 \in]0, 2\beta T_0[$. Then there exists a symbol $e \in S_t^0(\mathbf{R}_t \times \mathbf{R}_x^n)$, elliptic in ρ_0, and $\phi \in C_0^\infty$, $\phi = 1$ near (t_0, x_0), such that for all $N \in \mathbf{N}$ there exists $C_N > 0$ such that for all $U \in B$ and $k \in \mathbf{N}$

$$\|Op(e)\mid_{h=h_k} \phi_0 u_k\|_{L^2} \le \int_0^{2\beta T_0} \int_{\Gamma_0} |h_k \partial_n u_k|^2.$$

This result means that near a non diffractive point the micro-local L^2-norm is bounded by h_k times the local L^2-norm of the normal derivative.

5.4 Conclusion. Putting together these results we can show that there exists $\phi \in C_0^\infty(\mathbf{R})$; $\phi = 1$ near 0 such that for all $N \in \mathbf{N}$ there exists $C_N > 0$ such that for all $U \in B$

$$\int_{\mathbf{R}} \int_{\Omega} |\phi(t)u_k(t,x)|^2 dx dt \leq C \int_0^{2\beta T_0} \int_{\Gamma_0} |\partial_n h_k u_k(x,t)|^2 d\sigma(x) dt + C_N h_k^N.$$

Indeed, to estimate the left term, we only have to estimate it in a compact subset of $T^*\Omega \cup T^*\partial\Omega$ according to the elliptic estimates, then using the part i) of the propagation estimates, we only have to estimate it near the set $t = 0$, $\tau = |\xi|^2$, which can be done, according to the propagation estimate and the hypothesis of geometrical control, by the micro-local L^2-norm near a point $\rho = (t, \tau, x, \xi) \in T^*\partial\Omega$ with $0 < t < \beta T_0$. Finally this norm can be estimated by the local L^2-norm of the normal derivative, the term $C_N h_k^N$ being used to control the errors made at each step.

We see that since the L^2-norm of $u_k(t, \cdot)$ is independent of the time, this result means that if $\|u_k\|_{L^2(\Omega)} \leq 1$

$$\|u_k\|_{L^2(\Omega)}^2 \leq C \int_0^{\alpha^{-1}T_0} \int_{\Gamma_0} |h_k \partial_n u_k(x,t)|^2 d\sigma(x) dt + C_N h_k^N.$$

If we set $v_k(x,t) = u_k(x, h_k^{-1}t)$ then we have

$$(i\partial_t - h_k^2 \Delta)v_k = 0,$$

$$v_k \mid_{\partial\Omega} = 0,$$

$$v_k \mid_{t=0} = u_{k,0},$$

and

$$\|v_k(\cdot, 0)\|_{L^2(\Omega)}^2 \leq C \int_0^{\alpha^{-1}h_k T_0} h_k^{-1} \int_{\Gamma_0} |h_k \partial_n v_k(x,t)|^2 d\sigma(x) dt + C_N h_k^N.$$

Applying this result at the instants ph_k; $0 \leq p \leq [T_1 h_k^{-1}]$, using the fact that the L^2-norm of $u_k(t, \cdot)$ is independent of t and summing the inequalities obtained, we get

$$h_k^{-2}\|u_k\|_{L^2(\Omega)}^2 \leq C \int_0^{T_1} \int_{\Gamma_0} |\partial_n u_k(x,t)|^2 d\sigma(x) dt + C_N h_k^N,$$

and since on E_k the norms $\| \ \|_{H^1}$ and $h_k^{-1}\| \ \|_{L^2}$ are equivalent, we get for initial data in a subspace E_k and $s < 0$

$$\|u_k\|^2_{H^1(\Omega)} \le C \int_0^{T_1} \int_{\Gamma_0} |\partial_n u_k(x,t)|^2 d\sigma(x)dt + C_s\|u_k\|_{H^s}.$$

Finally, since $u \in H^1(\Omega)$ can be written as $u = \sum_{k \in \mathbb{N}} u_k$ with $u_k \in E_k$ and since each u_k interact with only a finite number of $u_{k'}$, we obtain with $s < 1$

$$\|u\|^2_{H^1(\Omega)} \le C \int_0^{T_1} \int_{\Gamma_0} |\partial_n u(x,t)|^2 d\sigma(x)dt + C_s\|u\|^2_{H^s}.$$

The term $\|u\|_{H^s}$ can be removed by a unicity argument, and this proves the estimation of §3, and hence the theorem of §2.

6. Further developments

6.1 Factorization. While looking at the structure of the proof of the previous result, we might think that it is closely related with the factorization of the solution of this plates equation into two terms u_+ and u_-. In fact it is not at all the case and the method can be modified to apply in other cases. This was done by T.Hargé in his thesis of the université d'Orsay (1991). He proved the following result under the same hypotheses of geometrical controllability and the additional hypothesis that the boundary of Ω is analytic.

Theorem (Hargé). Let $T > 0$ be a real. Then for all $(u_0, u_1) \in L^2(\Omega) \times H^{-2}(\Omega)$ there exists a control $g \in L^2(]0,T[\times\Gamma_0)$ such that the solution of the system

$$(\frac{\partial^2}{\partial t^2} + \Delta^2)u = 0 \ in \ \Omega \times]0, +\infty[,$$

$$u \mid_{\partial\Omega} = 0,$$

$$\frac{\partial u}{\partial n} \mid_{\partial\Omega} = g \times 1_{\Gamma_0 \times]0,T[},$$

$$u \mid_{t=0} = u_0,$$

$$\frac{\partial u}{\partial t} \mid_{t=0} = u_1,$$

verifies $u \equiv 0$ for all $t \ge T$.

The hypothesis of analyticity of the boundary is used here to obtain the unicity argument of the end of §5. Without this hypothesis we only control a subspace of $L^2(\Omega) \times H^{-2}(\Omega)$.

The main idea of the demonstration of this theorem is that even thought you can't factorize globally the solution into two solutions of the Schrödinger equation, this can be done at least micro-locally and hence you can apply the propagation technics.

6.2 Non-geometric controllability. As we saw at the beginning, the hypothesis of geometrical controllability is not a necessary condition for the exact controllability. In fact the results of A.Haraux can be deduced from the results of G.Lebeau in the more general case of $\Omega = \Omega_1 \times \Omega_2$, if you allow the control to act in a strip $\Omega_1 \times \omega_2$ with $\omega_2 \subset \Omega_2$. An other case where the geometrical controllability does not hold has been studied by N.Burq (1992). The geometric situation is the following:

Let $\widehat{\Omega}$ be a bounded connected open set of $\mathbf{R}^3$ with C^∞ boundary and such that $\partial\Omega$ has not contact of infinite order with his tangent lines. Let $\theta_1, \cdots, \theta_N$ be N strictly convex closed subset of $\widehat{\Omega}$ with C^∞ boundary and non-empty interior such that the convex hull $\mathrm{conv}(\underset{i}{\cup} \theta_i)$, of the union of the obstacles θ_i is included in $\widehat{\Omega}$. Let $\Omega = \widehat{\Omega} \setminus (\underset{i}{\cup} \theta_i)$, and let $\Gamma_0 = \partial\widehat{\Omega}$.

We suppose that

$$H_1) \forall 1 \leq j_1, j_2, j_3 \leq N, \ j_i \neq j_k \Rightarrow \mathrm{conv}(\theta_{j_1} \cup \theta_{j_2}) \cap \theta_{j_3} = \emptyset,$$

and we also make another hypothesis on the strongly hyperbolic structure of the billiard which is for example always verified if we have only two obstacles or if the obstacles are N balls of diameter smaller than $r > 0$ separated each one from the others by a distance greater than $N \times r$.

We see now that Γ_0 does not control geometrically Ω since you have captive trajectories (one if $N = 2$ and an infinite number if $N > 2$) which do not hit Γ_0. However we have the following result:

Theorem (Burq 1992). Let T_1 be any strictly positive number, then for all $\epsilon > 0$ and all $(u_0, u_1) \in H_0^{1+\epsilon}(\Omega) \times H^{-1+\epsilon}(\Omega)$ there exists a control $v \in L^2(\Gamma_0 \times [0, T_1])$ such *that the solution of the system:*

$$(\frac{\partial^2}{\partial t^2} + \Delta^2)u = 0 \; in \; \Omega \times]0, +\infty[,$$

$$u\,|_{\partial\Omega} = 0,$$

$$\Delta u\,|_{\partial\Omega} = v \times 1_{\Gamma_0 \times [0, T_0]},$$

$$u\,|_{t=0} = u_0,$$

$$\frac{\partial u}{\partial t}\,|_{t=0} = u_0,$$

verifies $u \equiv 0$ *for all* $t \geq T_1$.

The main idea of the proof of this result is that in §5 we saw that we did not need to control the L^2-norm of u_k in time T_0 but only to control h_k^{-1} times the L^2-norm in time $T_0 \times h_k^{-1}$. Using a parametrix construction inspired from the works of M.Ikawa (1988), It is possible to show that you control the L^2-norm of u_k in time $T_0 \times -\log(h_k)$ and hence you control $\frac{h_k^{-1}}{-\log(h_k)}$ times the L^2-norm in time $h_k^{-1}T_0$. This means that you control more than the $H^{1-\epsilon}$-norm of u_k which, by the duality of the H.U.M method will give the announced result.

6.3 Conclusion. We have seen in this part that the micro-local approach does not only give convenient geometric conditions for the exact controllability, but that it also seems to be applicable to various geometric data where the usual methods failed.

References

Bardos C., Lebeau G., Rauch J. (1988), appendix 2 of the book of J.L. Lions undermentioned.

Burq N. (1992), Contrôle de l'équation des plaques en présence d'obstacles strictement convexes, Prépublications de l'Ecole Polytechnique.

Haraux A. (1989), Series lacunaires et contrôle semi-interne des vibrations d'une plaque rectangulaire. Journal de Mathématiques pures et appliquées 68, 457-465.

Hargé T. (1991), Thèse du l'Université de Paris-Sud (Orsay).

Ikawa M. (1988), Decay of solutions of the wave equation in the exterior of several convex bodies. Annales de l'institut Fourier, Grenoble, 38, 2, 113-146.

Jaffard S. (1988), Note aux Comptes Rendus de l'Academie des Sciences de Paris, Tome 307, serie 1, 759-762.

Lebeau G. (1989), Contrôle de l'équation de Schrödinger, to be published in the Journal de Mathématiques Pures et Appliquées.

Lions J.L. (1988), Contrôlabilité exacte, pertubations et stabilisation de systèmes distribués, tome 1, Collection R.M.A.,Masson.

Melrose R.B., Sjöstrand J. (1982), Singularities of boundary value problems II, Communications on Pure and Applied Mathematics, Vol 35.

Zuazua E. (1988), Contrôlabilité exacte en temps arbitrairement petit de quelques modèles de plaques (1988), appendix 1 of the book of J.L.Lions above-mentioned

Authors' address:

Nicolas Burq
Centre de Mathématiques (CMAT)
Ecole Polytechnique
91128 PALAISEAU CEDEX
FRANCE

Gilles Lebeau
Université de Paris-Sud (Orsay)
Département de Mathématiques
91405 ORSAY CEDEX
FRANCE

International Series of Numerical Mathematics, Vol. 107, © 1992 Birkhäuser Verlag Basel

Bounded solutions for controlled hyperbolic systems

C.Vårsan

Abstract. *The existence of bounded or periodic solution is proved for hyperbolic systems when the right hand side is perturbed by a control fulfiling a weak controllability property.*

1. Introduction.

It is well known that a control $g(t,x)=A(t)x+b(t)$, $t>0$, added to the right hand side $f(t,x)$ of a differential equation insures the existence of a bounded or periodic solution provided $A(t)$ fulfills a stability property $\leq A(t)x,x>\leq-L|x|^2$, $t\geq0$, $x\in E$, for some $L>0$; the strong stability condition can be replaced by a weaker controllability property and the result is still true.

In this paper we consider hyperbolic systems

$$(*) \quad \frac{\partial x_1}{\partial t_1}=f_1(t,x), \quad \frac{\partial x_2}{\partial t_2}=f_2(t,x), \quad x=(x_1,x_2), \quad t=(t_1,t_2)\in R_t \times [0,a],$$
$$x_1(0,t_2)=\rho_1(t_2), \quad x_2(t_1,0)=\rho_2(t_1)$$

and the existence of bounded or periodic solutions for $(*)$ is proved if f_1 is perturbed by a control $\sum_1^m u_i(t)g_i(x_1)$, where the vector fields g_i fulfil a controllability property assumed for the Lie algebra generated by them.

Second order hyperbolic system could be derived from $(*)$ and, adding a control variable, the existence of periodic solutions is considered in [1] but from a different point of view.

The main result is stated in Theorem 2 and it can be extended easily to the case of unbounded $t_2>0$ if along with f_1 one perturbs f_2 by a control $\sum_j v_j(t)h_j(x_2)$ with h_j fulfilling a similar controllability property.

The case of Ito's stochastic differential equations replacing (*) bring no major difficulties in proving that similar perturbations of the drifts generate the corresponding bounded or periodic solutions. Related properties for diffusion and hyperbolic differential equations one may find in [2] and [3].

2. Formulation of the problem and main results. Let
$f_i:R_+\times J\longrightarrow R^{n_i}$, $i=1,2$, $n_1+n_2=n$, $J=\left[0,T_2\right]\subset R_+$, be given and consider the system

1) $\partial x_1/\partial t_1=f_1(t,x)$, $\partial x_2/\partial t_2=f_2(t,x)$, $x_1(0,t_2)=\varphi_1(t_2)$, $x_2(t_1,0)=\varphi_2(t_1)$ for $t=(t_1,t_2)\in R_+\times J$, $x=(x_1,x_2)\in R^n$,   where $\varphi_1,\varphi_2,f_1,f_2$ are supposed to be continuous functions. A standard procedure allow one to obtain a unique global solution in (1) if f_1 and f_2 fulfil the following property: for any compact set $K\subset R^n$ and $T>0$ there is $L(T,K)>0$ such that

a) $\|f_i(t,x'')-f_i(t,x')\|<L\|x''-x'\|$, $t\in[0,T]\times J$, $x',x''\in K$.

b) $\|f_i(t,x)\|<C(1+\|x\|)$, $t\in R_+\times J$, $x\in R^n$, for some $c>0$.

A solution $x(t)$ in (1) is called bounded if

α) $\max_{t_2\in J}\|x(t_1,t_2)\|\leq M<\infty$, $t_1\geq 0$, for some $M>0$;

$x(t)$ is periodic if there exists $T_1>0$ such that

β) $x(t_1+kT_1,t_2)=x(t_1,t_2)$, $k\geq 1$, k integer, $t\in R_+\times J$.

In particular (α) and (β) are suitable for second order systems
$\partial y/\partial t_1\partial t_2=f(t,y,y_1,y_2)$, $y_1=\partial y/\partial t_1$, $y_2=\partial y/\partial t_2$, $t=(t_1,t_2)$
defining $x_1=y_2$ and $x_2=(y,y_1)$.

More complicated system as
$\partial^3 y/\partial t_1\partial t_2\partial t_3=f(t,y,y_1,y_2,y_3,y_4,y_5,y_6),t=(t_1,t_2,t_3)$,

$y_i=\partial y/\partial t_i$, $i=1,2,3$, $y_4=\partial^2 y/\partial t_1\partial t_2$, $y_5=\partial^2 y/\partial t_1\partial t_3$,

$y_6=\partial^2 y/\partial t_2\partial t_3$, can be written in the form

1') $\partial x_1/\partial t_1=f_1(t,x)$, $\partial x_2/\partial t_2=f_2(t,x)$, $\partial x_3/\partial t_3=f_3(t,x)$,
$x=(x_1,x_2,x_3)$ where $x_1=(y_6)$, $x_2=(y_3,y_5)$, $x_3=(y,y_1,y_2,y_4)$.

Write $J=[0,T_2]\times[0,T_3]\subseteq R_+^2$ and (α), (β) can be rewritten for a solution in (1'); the results for (1) are easily translated for (1') or more general.

The problem we consider for (1) is the following. We are given $g_1,...,g_m : R^{n_1} \longrightarrow R^{n_1}$ of class C^∞ and we are looking for bounded $u_i : R_t \times J \longrightarrow R$, $i=1,...,m$, such that the solution of the controlled system

$$2) \quad \frac{\partial x_1}{\partial t_1} = f_1(t,x) + \sum_1^m u_i(t) g_i(x_1), \quad x_1(0,t_2)=\varphi_1(t_2), \quad \|\varphi_1(t_2)\| \le \gamma$$

$$\frac{\partial x_2}{\partial t_2} = f_2(t,x) \qquad\qquad x_2(t_1,0)=\varphi_2(t_1)$$

fulfills either (α) or (β).

Write $\mathcal{L}(g_1,...,g_m)$ for the Lie algebra over R generated by $g_1,...,g_m$ using the usual bracket

$$[g_i,g_j](y) = \left(\frac{\partial g_i}{\partial z} g_i - \frac{\partial g_i}{\partial z} g_i \right)(y),$$

Let $T_1>0$, $b_1,...,b_{m_1} \in \mathcal{L}(g_1,...,g_m)$ and continuous functions $v_j : [0,T_1] \times J \longrightarrow R$, be fixed.

It is useful to consider the enlarged system

$$3) \quad \frac{\partial z_1}{\partial t_1} = f_1(t,z) + \sum_{j=1}^{m_1} v_j(t) b_j(z_1), \quad z_1(0,t_2)=\varphi_1(t_2)$$

$$\frac{\partial z_2}{\partial t_2} = f_2(t,z) \qquad\qquad z_2(t_1,0)=\varphi_2(t_1)$$

where $z=(z_1,z_2)$ and φ_1,φ_2 are those in (2).

Write $u(t)=(u_1(t),...,u_m(t))$ and $v(t)=(v_1(t),...,v_{m_1}(t))$

It holds the following

Theorem 1.

Assume that f_1,f_2 fulfil (a) and (b), and $b_1,...,b_{m_1} \in \mathcal{L}(g_1,...,g_m)$ fulfil (b).

Assume that $v(t)$, $\frac{\partial v}{\partial t_1}(t)$, $\frac{d\varphi_2}{dt_1}(t_1)$ are continuous and let $z(t)$, $t \in [0,T_1] \times J$, be the corresponding solution in (3). Then, for each $h \in (0,1)$, there exist a continuous control $u_h(t)=\varphi_h(t,v(t))$ such that the corresponding solution $x_h(t)$ in

(2) fulfils $\max\limits_{t\in[0,T_1]\times J} \|x_h(t)-z(t)\| \leq M\sqrt{h}$, for some $M>0$.

In addition, $\varphi_h(t,v)$ is continuous (even linear in v) and doesn't depend on f_i, ξ_j, φ_i; the constant M depends only on r for any φ_1, φ_2, v fulfilling

$$\|\varphi_1(t_2)\| + \|\varphi_2(t_1)\| + \left\|\frac{d\varphi_2}{dt_1}(t_1)\right\| + \|v(t)\| + \left\|\frac{\partial v}{\partial t_1}(t)\right\| \leq r, \quad t\in[0,T_1]\times J.$$

The above theorem is the main tool for geting bounded or periodic solutions in (2) provided the solution in (3) is bounded and the estimate in Theorem 1 is uniformly with respect to compact sets $[t_1, t_1+T_1], t_1>0$. This goal is accomplished assuming

c) there exist $b_1,...,b_n \in \mathcal{L}(\xi_1,...,\xi_m)$ such that

$$\det(b_1,...,b_n)(x_1) \neq 0 \quad (V) \quad x_1 \in S(0,\gamma) \triangleq \{y\in R^{n_1} : \|y\|\leq\gamma\} \quad \text{for some}$$

$\gamma>0$;

d) f_1, f_2 and φ_2 are C^1 and for any compact $K\subset R^n$ there exists $M_K>0$ such that

$$\left\|\frac{\partial f_1}{\partial t_1}(t,x)\right\| + \left\|\frac{\partial f_2}{\partial x}(t,x)\right\| + \left\|\frac{\partial f_2}{\partial t_1}(t,x)\right\| + \|\varphi_2(t_1)\| + \left\|\frac{d\varphi_2}{dt_1}(t_1)\right\| \leq M_K$$

$(\forall)\ t\in R_t\times J,\ x\in K$.

Theorem 2.

Assume that f_1, f_2 fulfil (b) and $\xi_1,...,\xi_m$ fulfil (c). Let f_1, f_2, φ_2 verify (d). Then there exists a bounded control $\tilde{u} : R_t \times J \longrightarrow R^m$ such that the corresponding solution $\tilde{x}$ in (2) is bounded. In addition, assume that f_1, f_2 and φ_2 are periodic functions with respect to t_1 and let T_1 be the period. Then

there exists a continuous function $\tilde{\varphi}_1(t_2)$, $t_2\in J$, and a periodic control $\tilde{u} : R_t \times J \longrightarrow R^m$ such that the corresponding solution $\tilde{x}$ in (2) is periodic with the same period T_1.

3. Proofs of Theorem 1 and 2.

Proof of Theorem 1.

The statement is similar to that of Theorem 1 in [2] and the proof repeats the same scheme as in the case of one dimensional parameter t as it is given in the above cited work provided $g_1, \ldots, g_m$ are bounded.

Using a standard argument one replaces $g_1, \ldots, g_m$ by bounded as follows. The solution $z(t)$, in (3), exists and is bounded, $\|x(t)\| \le \varrho/2$ $\forall$ $t \in [0, T_1] \times J$. Define C^∞, bounded functions $\tilde{g}_j(x_1) : R^{n_1} \longrightarrow R^{n_1}$ such that $\tilde{g}_j(x_1) = g_j(x_1)$ for $\|x_1\| \le \varrho$ and $\tilde{g}_j(x_1) = 0$, for $\|x\| \ge 2\varrho$. The bounded functions $\tilde{g}_j, j=1, \ldots, m$, fulfil the hypothesis in Theorem 1 and the conclusion shows that the corresponding solution $\tilde{x}_h(t)$ in (2) satisfies $\|\tilde{x}_h(t)\| \le \varrho$ $(\forall)$ $t \in [0.T_1] \times J$, for h sufficiently small, which allow one to replace $\tilde{g}_1, \ldots, \tilde{g}_m$ in (2) and $\tilde{b}_1, \ldots, \tilde{b}_m \in \mathcal{L}(\tilde{g}_1, \ldots, \tilde{g}_m)$ in

(3) by the original ones.

Proof of Theorem 2.

The hypothesis (c) for $b_1, \ldots, b_{n_1}$ allow one to define $\bar{v}_j(t,z)$, $j=1, \ldots, n_1$, of class C^1, such that

$$4) \quad f_1(t_1,z) + \sum_{j=1}^{n_1} \bar{v}_j(t,z) b_j(z_1) = -z_1$$

Write $\bar{z}_1(t) = (\exp{-t_1}) \varphi_1(t_2)$ and define $\bar{z}_2(t)$ as the solution of the following system

$$5) \quad \frac{\partial z_2}{\partial t_2} = f_2(t, \bar{z}_1(t), z_2) \qquad z_2(t_1, 0) = \varphi_2(t_1)$$

Using the hypothesis (b) for f_2 and the boundedness of $\bar{z}_1(t)$ and $\varphi_2(t_1)$ for $t_1 \ge 0$, from (5) we get the boundedness of $\bar{z}_2(t)$.

On the other hand, $y(t) = \partial \bar{z}_2 / \partial t_1$ is the solution of the following system

$$y(t) = \frac{d\varphi_2}{dt_1}(t_1) + \int_0^{t_2} \frac{\partial f_2}{\partial x_2}(t_1, s, \bar{z}(t_1, s)) y(t_1, s) + \int_0^{t_2} F(t_1, s) ds$$

where $\bar{z} = (\bar{z}_1, \bar{z}_2)$, $F(t) = \dfrac{\partial f_2}{\partial t_1}(t, \bar{z}(t)) + \dfrac{\partial f_2}{\partial x_1}(t, \bar{z}(t)) \dfrac{\partial \bar{z}_1}{\partial t_1}(t)$, and

using (d) and the boundedness of $\bar{z}(t)$ we get that $\dfrac{\partial \bar{z}_2}{\partial t_1}(t)$, $t_1 \ge 0$, is bounded too.

Write $\bar{v}_j(t)=\bar{v}_j(t,\bar{z}(t))$, $j=1,...,n_1$, and using (b) and (d) for

f_1 in (4) we get that $\bar{v}_j(t)$ and $\dfrac{\partial \bar{v}_j}{\partial t_1}(t)$, $t_1 \geq 0$, are bounded. It

is easy to see that $\bar{z}(t)$ is the unique solution of the
following system

$$6)\quad \frac{\partial z_1}{\partial t_1} = f_1(t,z) + \sum_{j=1}^{n_1} \bar{v}_j(t)b_j(z_1), \qquad z_1(0,t_2)=\varphi_1(t_2)$$

$$\frac{\partial z_2}{\partial t_2} = f_2(t,z) \qquad\qquad\qquad z_2(t_1,0)=\varphi_2(t_1)$$

and (2) and (6) fulfil the hypothesis in Theorem 1. Therefore,
a continuous control $u^0(t)$, $t \in [0,T_1] \times J$, will exists such that
the corresponding solution $x^0(t)$ in (2) fulfils

$$7)\quad \|x^0(t)-\bar{z}(t)\| \leq \delta, \ (\forall)\ t \in [0,T_1]\times J, \ \text{for some}\ \delta \in (0,1)$$
$$\|x^0(T_1,t_2)\| \leq \|\varphi_1(t_2)\| \quad (\forall)\ t_2 \in J$$

On the next interval $t_1 \in [T_1,2T_1|$, $t_2 \in J$, one uses $x_1^0(T_1,t_2)$,
$t_2 \in J$, and $\varphi_2(t_1)$, $t_1 \in [T_1,2T_1]$, as the new Cauchy conditions for
(2) and (6). Repeating the above algorithm we define
$\bar{z}_1^1(t)=(\exp(T_1-t_1))x_1^0(T_1,t_2)$ and let $\bar{z}_2^1(t)$ be the solution of
the system

$$\frac{\partial z_2}{\partial t_2} = f_2(t,z_1^1(t),z_2), \quad z_2(t_1,0)=\varphi_2(t_1), \quad t_1 \geq T_1$$

Write $\bar{v}_j^1(t)=\bar{v}_j(t,\bar{z}^1(t))$, where $\bar{v}_j(t,z)$, $j=1,...,n_1$, are defined
in (4) and $\bar{z}(t)=(\bar{z}_1^1(t),\ \bar{z}_2^1(t))$.

One notices that $\bar{z}^1(t)$ is the unique solution of the following
system

$$8)\quad \frac{\partial z_1}{\partial t_1} = f_1(t,z) + \sum_{j=1}^{n_1} \bar{v}_j^1(t)b_j(z_1), \qquad z_1(T_1,t_2)=x_1^0(T_1,t_2)$$

$$\frac{\partial z_2}{\partial t_2} = f_2(t,z) \qquad\qquad\qquad z_2(t_1,0)=\varphi_2(t_1)$$

Using Theorem 1 for (2) and (8) we get a bounded and continuous $u^1:[T,2T]\times J\longrightarrow R^m$ such that the corresponding solution $x^1(t)$ in

$$9)\quad \begin{cases} \dfrac{\partial x_1}{\partial t_1}=f_1(t,x)+\sum_1^m u_i^1(t)g_i(x_1) & x_1(T_1,t_2)=x_1^0(T_1,t_2),\, t_2\in J\\[4mm] \dfrac{\partial x_2}{\partial t_2}=f_2(t,x) & x_2(t_1,0)=\varphi_2(t_1),\, t_1\geq T_1 \end{cases}$$

fulfils

$$10)\quad \begin{cases} \|x^1(t)-\bar{z}^1(t)\|\leq\varrho\ ,\ t\in[T_1,2T_1]\times J,\ \Big(\delta\in(0,1)\text{ as in }(7)\Big)\\[4mm] \|x_1^1(2T_1,t_2)\|\leq\|x_1^0(T_1,t_2)\|\leq\|\varphi_1(t_2)\|,\ t_2\in J. \end{cases}$$

On each interval $[kT_1,\ (k+1)T_1]\times J$ we get $(\bar{v}^k(t),\bar{z}^k(t))$ and $(u^k(t),x^k(t))$ such that $u^k(t)=\psi(\tilde{h},\ \bar{v}^k(t))$, $(\bar{v}^k(t),\bar{z}^k(t)),k\geq1,$ are uniformly bounded and

11) $\|x^k(t)-\bar{z}^k(t)\|\leq\delta,\ t\in[kT_1,(k+1)T_1]\times J,\ k\geq1,$ for some $\delta\in(0,1)$.

Define $\tilde{u}:R_+\times J\longrightarrow R^m$ by $\tilde{u}(t)=u^k(t)$, $t\in[kT_1,(k+1)T_1]\times J$, and let $\tilde{x}(t)$ be the corresponding continuous solution in (2).
It holds $\tilde{x}(t)=x^k(t)$ for $t\in[kT_1,(k+1)T_1]\times J$, $k\geq0,$ and using (11) one concludes that $\tilde{x}(t)$ and $\tilde{u}(t)$ are bounded and

12) $\|\tilde{x}(kT_1,t_2)\|=\|\tilde{x}^{k-1}(kT_1,t_2)\|\leq\|\varphi_1(t_2)\|$ $(\forall)\ t_2\in J,\ k\geq1.$
To prove the second part of the theorem using the periodicity property we may consider that the sequence $(u^k(t),x^k(t))$, $t\in[kT_1,(k+1)T_1]\times J$, $k\geq1,$ satisfy the following equations

$$13)\quad \tilde{x}_1^{k+1}(t)=\tilde{x}_1^k(T_1,t_2)+\int_0^{t_1}\tilde{f}_1^{k+1}(\tau,t_2,\tilde{x}^{k+1}(\tau,t_2))d\tau,\ t_1\in[0,T_1]$$

$$\tilde{x}_2^{k+1}(t)=\varphi_2(t_1)+\int_0^{t_2}f_2(t_1,\tau,x^{k+1}(t_1,\tau))d\tau$$

where $\tilde{f}_1^{k+1}(t,x)=f_1(t,x)+\sum_1^m\tilde{u}_i^{k+1}(t)g_i(x_1)$, and

$(\tilde{u}^k(t),\tilde{x}^k(t))$, $t\in[0,T_1]\times J$, is defined by

$(\tilde{u}^k(t),\tilde{x}^k(t))=(u^k(t_1+kT_1,t_2),\ x^k(t_1+kT_1,t_2))$, for any $k\geq1$

By definition $\tilde{u}^k(t)=\psi(\tilde{h},\tilde{v}^k(t))$, where

$\tilde{v}^k(t)=\tilde{v}(t,\tilde{z}^k(t))$, $\tilde{z}^k(t)=(\tilde{z}_1^k(t), \tilde{z}_1^k(t))$ and $\tilde{z}_1^k$, $\tilde{z}_2^k$ fulfill

$$14)\quad \begin{cases} \tilde{z}_1^k(t)=(\exp-t_1)\tilde{x}_1^{k-1}(T_1,t_2) \\[2em] \tilde{z}_2^k(t)=\varphi_2(t_1)+\int_0^{t_2} f_2(t_1,\tau,\tilde{x}^k(t_1,\tau))d\tau \qquad t\in[0,T_1]\times J. \end{cases}$$

Using (12) in (13) we get that $\|\tilde{x}_1^k(T_1,t_2)\|\leq\|\varphi_1(t_2)\|$ (V) $k\geq 1$,
and a subsequence $\{\tilde{x}_1^k(T_1,.)\}$ exists such that

$$15)\quad \lim_{k\to\infty} x_1^k(T_1,t_2)=\tilde{\varphi}_1(t_2) \text{ uniformly in } t_2\in J.$$

Using (15) in (14) one concludes the uniform convergence of $\{\tilde{z}^k(t)\}_{k\geq 1}$ and as a result $\{\tilde{v}^k(t)\}_{k\geq 1}$ and $\{\tilde{u}^k(t)\}_{k\geq 1}$

converge uniformly to the corresponding continuous $\tilde{v}(t)$, $\tilde{u}(t)$, $t\in[0,T_1]\times J$, and

$$16)\quad \lim_{k\to\infty} \tilde{f}_1^k(t,x)=f_1(t,x)+\sum_1^m \tilde{u}_i(t)g_i(x_1)\overset{\Delta}{=} \tilde{f}_1(t,x) \qquad \text{uniformly in}$$

$t\in[0,T_1]\times J$ and for x in compact sets in R^n.
Using (15) and (16) one obtains the uniform convergence of $\{\tilde{x}^k(t)\}_{k\geq 1}$ in (13) to a solution $\tilde{x}(t)$ such that

$$17)\quad \tilde{x}_1(t)=\tilde{\varphi}_1(t_2)+\int_0^{t_1} \tilde{f}_1(\tau,t_2,\tilde{x}(\tau,t_2)) \, d\tau$$

$$\tilde{x}_2(t)=\varphi_2(t_1)+\int_0^{t_2} f_2(t_1,\tau,\tilde{x}(t_1,\tau))d\tau$$

and $\tilde{x}_1(T_1,t_2)=\tilde{\varphi}_1(t_2)=\tilde{x}_1(0,t_2)$, $t_2\in J$.

Noticed that $\tilde{x}_2(T_1,t_2)$ and $\tilde{x}_2(0,t_2)$, $t_2\in J$ are the solution for

the equation $y(t_2)=y_0+\int_0^{t_2} f_2(0,\tau,\tilde{x}_1(0,\tau), \quad y(\tau))d\tau$, the

uniqueness property implies $\tilde{x}_2(0,t_2)=\tilde{x}_2(t_1,T_2)$, $T_2\in J$. The proof
is complete.

References.

[1] Lamberto Cesari, Periodic Solutions of Hyperbolic Partial Diff.Eq., pp.33-57, Int.Symp. on Nonlinear Diff. Eq. and Nonlinear Mechanics, Eds.J.P. La Salle, S.Lefschetz, Academic Press 1963.

[2] C.Vârsan, Approximation and Existence of Periodic Solutions for Controlled Diffusion Equations, Stochastics and Stochastics Reports, vol.30, pp.130-150, 1990.

[3] C.Vârsan, Approximation of Goursat problem for hyperbolic controlled stochastic differential equations, to appear in Revue Roum.Math.Pures Appl.

Author's address:

C.Vârsan
Institute of Mathematics
Romanian Academy
P.O.Box 1-764
RO-70700 Bucharest
Romania

International Series of Numerical Mathematics, Vol. 107, © 1992 Birkhäuser Verlag Basel

CONTROLLABILITY AND TURBULENCE

J.F. POMMARET

ABSTRACT: The purpose of this communication is to make an elementary study of various nonlinear phenomena by using a new formal approach to controllability instead of the usual functional one, in the control theory of partial differential equations.

We recall briefly how the concept of controllability in the *ordinary differential* ("*classical*") *control theory* has been extended to the case of distributed systems. We follow closely the exposition of Lions (1990).

If u and y respectively denote the (possibly multi dimensional) input and output (or state) of a control system:

$$\dot{y} = f(y; u)$$

we shall say that this control system is *controllable* if for any couple of initial output y_0 and final output y_T , we may find a convenient input $u(t)$ such that the corresponding output $y(t)$ with $y(0) = y_0$ also satisfies $y(T) = y_T$ for a certain given $T < \infty$.

Having thus treated the case of ordinary differential equations (ODE) it seems therefore natural to treat the case of partial differential equations (PDE) in the following manner.

Let now u and y be functions of time t and space x (we use 1-dimensional space for simplicity) satisfying the control system:

$$\frac{\partial y}{\partial t} = \mathcal{F}(y; u)$$

where $\mathcal{F}$ is a nonlinear operator involving only space derivatives. The problem of *exact controllability* amounts to find a convenient input $u(x, t)$ such that, for any couple of initial output $y_0(x)$ and final output $y_T(x)$, the corresponding output $y(x, t)$ solution of the preceding evolution equation with initial condition $y(x, 0) = y_0(x)$ also satifies $y(x, T) = y_T(x)$ for a certain given $T < \infty$. One can for instance consider the problem of how to cut off the vibration of a drum by an external mechanical action. Of course, one can consider the set $y_T(x)$ when $y_0(x)$ and T are given but when u varies, in order

to define *approximate controllability* whenever the "reachable set" is dense.
For example, let ω be an open subset of space and χ_ω be its characteristic ("bump")
function. It can be proved that the control system:

$$\frac{\partial y}{\partial t} = \frac{\partial^2 y}{\partial x^2} + \chi_\omega u$$

is approximately controllable, while the control system:

$$\frac{\partial y}{\partial t} = \frac{\partial^2 y}{\partial x^2} - y^3 + \chi_\omega u$$

is not approximately controllable (Henry 1977 and Lions 1990).

Though the preceding definitions of controllablity seems pertinent and useful for
applications, we shall prove that functional analysis is not the only good framework
for understanding controllability.

For this, let us consider a linear control system of the form:

$$\dot{y} = Ay + Bu$$

Of course, in this simple case, people must answer by "yes" or "no" to the question:"Is
such a control system controllable ?". However, it is known from any textbook that
such a control system with n outputs is controllable if and only if :

$$rank(B, AB, ..., A^{n-1}B) = n$$

Therefore it becomes clear that CONTROLLABILITY IS A FORMAL PROPERTY
OF THE CONTROL SYSTEM. Indeed, it is a property that only depends on the
equations but not on their solutions as it is independent of the initial data and not
related to functional analysis through the study of trajectories. As a byproduct, CON-
TROLLABILITY CAN BE TESTED BY MEANS OF A SYMBOLIC COMPUTER.

It remains to answer to the 3 following questions on controllability:
1) Does it admit any *formal definition* ?
2) Is it depending on the state representation ?
3) Is it possible to extend it from ODE to PDE ?
Of course, if such a formal definition may exist, it must be tested. Hence, a first idea
is to look, through the literature, at the so-called nonlinear control systems of the form:

$$\dot{y} = a(y) + b(y)u$$

It is known (Isidori 1985) that one can introduce the *weak control distribution* Δ_W
spanned by the vector fields a, b and their brackets on one side and the *strong control
distribution* Δ_S such that $\Delta_W = a + \Delta_S$. However, any definition of controllability
based on the rank of Δ_W is meaningless as one just needs to consider the control systems $\dot{y} = 0$ and $\dot{y} = 1$ which are certainly not controllable. The exclusive importance

of Δ_S has been stressed out definitively in Haddak (1989) as it is the only way to generalize the linear situation.

REMARK: The two simple examples above also prove that it is wrong to understand controllability as a way to fill in the whole tangent space at a point of the output space just by changing the input in the dynamical system determined by the control system.

Nevertheless, not a word can be said for PDE through this approach and the possibility to answer to the three last questions at the same time will provide a new unexpected link between differential algebra, computer algebra, control theory and turbulence !.

Indeed, *partial differential control theory* will study input/output relations defined by linear or even nonlinear systems of PDE. Examples of such systems can be found in all branches of engineering sciences (mechanics, hydrodynamics,...). In that case, an *observable* is any physical scalar quantity that can be measured by means of a local apparatus, that is, always a (most of the time rational) function of the input,output and their derivatives (pressure, temperature, vertical component of the vortex vector,...). It follows that turbulence or instability can be seen on the measurement through a fluctuation or an explosion of the numerical value. Hence, this quantity cannot be fixed arbitrarily when the boundary values are regular because we could choose it to be regular. As a byproduct, turbulence or instability can only happen for an observable which is not *free*, that is satisfies at least one PDE (eventually highly nonlinear and of very high order) for itself.

The following formal definition, though valid for ODE or PDE and independent of the state representation, seems, surprisingly, completely disconnected from the previous one:

DEFINITION: A control system is said to be (totally) controllable if any observable is free. It will be said to be (totally) uncontrollable if any observable must satisfy at least one PDE for itself. Any intermediate situation may exist.

Of course, if an observable does satisfy a PDE for itself, the control system cannot be controlled because there is no way to control this PDE ... which does not depend on any input !. The fact that the converse is also true is not evident at first sight and has been proved in Pommaret (1988). (Also see Fliess 1991).

REMARK: Testing the preceding definition involves delicate methods and concepts from the formal theory of systems of PDE (diagram chasing, Spencer cohomology, formal integrability). A detailed example of the new type of computations involved can be found in Pommaret (1990).

In actual practice of engineering sciences, the control systems are usually made up by algebraic PDE (polynomials in the derivatives) and an observable is, most of the time, a rational function of input, output and their derivatives: this is the basic reason for using differential algebra.

Let us therefore suppose that the coefficients of the PDE defining the control system belong to a certain *ground differential field K* (numerical constants, rational funstions of space and time,...). The reader can refer to Kolchin (1973) for a classical exposition and Pommaret (1988) for a modern intrinsic formulation. The rational functions of input, output and their derivatives satisfying the control system determine the *input/output differential field N* that contains K as a subdifferential field. Equivalently, one can say that the control system determines a differential extension N/K. At that time, controllability has just to do with the *differential algebraic closure K'* of K in N. In particular, the control system will be said to be totally differentially algebraically controllable if $K' = K$ or uncontrollable if $K' = N$.

After a few remarks, we review, in the light of this new approach, a certain number of examples in engineering sciences.

REMARK: The set of C^∞ differential functions on a manifold is a differential ring which is not an integral domain because the product of two bump functions with different supports is zero though each bump function is not zero. Such an argument is crucial in the proof of the counterexample already produced where no ground field can be defined (Henry 1977).

REMARK: Differential algebra can only be a guide towards new concepts because certain non-algebraic observables can be constrained by algebraic PDE, even in the case of differentially algebraic control systems. For example, $u\dot{y} - \dot{u} = 0$ is an algebraic ODE and $z = y - log\, u$ is constrained by $\dot{z} = 0$ though K is differentially algebraically closed in N. However, we notice that the linearized system $u\dot{Y} + \dot{y}U - \dot{U} = 0$ is not controllable because, setting $Z = Y - (1/u)U$, we get $\dot{Z} = 0$ (Compare with Fliess 1991, appendix).

REMARK: As the differential algebraic closure of K' in N is again K', the definition is *optimum* in some sense, leading automatically to the concept of "*minimum representation*" in classical control theory.

Let us start with a very simple academic example proving that controllability can be understood in a new intrinsic fashion, even in the classical setting (Pommaret 1990).

EXAMPLE: With $n = 2$ and $K = Q$, let us consider:

$$\dot{y}^1 = y^2 + u, \dot{y}^2 = \alpha y^1 + u \qquad with \qquad \alpha = cst$$

The classical test shows that the system is controllable if and only if $\alpha \neq 1$. Indeed, if $\alpha = 1$, then $z = y^2 - y^1$ does satisfy $\dot{z} + z = 0$ and one cannot escape from the diagonal in the (y^1, y^2) space. Setting $y = y^1$, we obtain $\ddot{y} - \alpha y - u - \dot{u} = 0$ and $y^2 = \dot{y} - u$, but no test of controllability can be given on the latter ODE alone. Let us choose an observable $z = h(y, \dot{y}; u, \dot{u}, \ddot{u}, ...)$ as $\ddot{y}, y^{(3)}, ...$ can be known by using the control second order ODE and its various derivatives. The condition $\dot{z} + \Phi(z) = 0$ implies that $z = h(y, \dot{y}; u)$ because, if $u^{(r)}$ is the highest derivative of u appearing in in h, then $u^{(r+1)}$ must appear in $\dot{z}$ and cannot be eliminated, unless $r = 0$. Let us set $z = \dot{y} + \lambda y + \mu u$ with $\dot{z} = \ddot{y} + \lambda \dot{y} + \mu \dot{u} = \lambda \dot{y} + \alpha y + u + (\mu + 1)\dot{u} \equiv \nu z$. The only possibility is $\lambda = \mu, \lambda \nu = \alpha, \mu \nu = 1, \mu = -1$, that is $\lambda = -1, \mu = -1, \nu = -1 \Rightarrow \alpha = 1$, a fact proving that controllability does not depend on the (state) representation of the second order ODE as a system of two first order ODE.

The following example establishes a striking link with another approach to turbulence (Lions 1990) and future research must explain the relations existing between *formal analysis* and *functional analysis*.

EXAMPLE: Bénard problem (Landau et al. 1971, Pommaret 1988,1990).
After convenient dimensional simplifications, the Bénard control system in the Boussinesq linearized approximation can be written:

$$\vec{\nabla}.\vec{v} = 0, \Delta \vec{v} - \theta \vec{g} - \vec{\nabla}\pi = 0, \Delta \theta - \vec{v}.\vec{g} = 0$$

where $\vec{v}$ is the speed, θ and π the respective perturbations of temperature and pressure, $\vec{g} = (0, 0, -g)$ is gravity. The two only known approaches to the stability problem involve isolating (decoupling !) θ or the vertical speed which, surprisingly, satisfy the same 6th order PDE though having completely different physical meaning:

$$\Delta \Delta \Delta \theta - g^2(\partial_{11} + \partial_{22})\theta = 0$$

One can prove by using computer algebra that any observable is constrained by at least one PDE. For example, $\zeta = \partial_1 v^2 - \partial_2 v^1$ satisfies $\Delta \zeta = 0$. More generally, it seems important for future research to notice that all the examples of the excellent reference (Chandrasekhar 1970) dealing with hydrodynamic or hydromagnetic instabilities enter the same scheme, *even in the nonlinear time depending case*, and are totally uncontrollable (Pommaret 1988,1990).

The following example, on the contrary, is totally controllable.

EXAMPLE: Maxwell equations with sources (Pommaret 1990).

$$\vec{\nabla} \wedge \vec{H} - \frac{\partial \vec{D}}{\partial t} = \vec{j} \qquad , \qquad \vec{\nabla}.\vec{D} = \rho$$

The input $(\rho, \vec{j})$ is not free as one must have $\frac{\partial \rho}{\partial t} + \vec{\nabla}.\vec{j} = 0$. However, any observable can be expressed as a function of $(\vec{H}, \vec{D})$ and their derivatives because $(\rho, \vec{j})$ are too. But

$(\vec{H}, \vec{D})$ can be given arbitrarily in order to provide $(\rho, \vec{j})$. Hence, any observable is free.

Though this exposition is very elementary, we would like to emphasize that the corresponding mathematical framework is based on the formal theory of systems of PDE created by D.C. Spencer during the years 1960-1975 in the USA. However, such a theory is not so well known and its application to differential algebra is recent (Pommaret 1988,1989,1990). Therefore, we hope that this lecture will prove that there is an alternative to functional or numerical analysis for studying physical phenomena. As the formal theory of PDE furnishes the algorithms needed for using computer algebra in the study of these formal problems, we also hope that the brief sketch that we presented will provide a new domain for applied mathematics in a near future.

BIBLIOGRAPHY

CHANDRASEKHAR S.(1961,1970), Hydrodynamic and hydromagnetic stability, Dover, 650p.

FLIESS M.(1991), The Influence of R.E. Kalman, Edited by A.C. Antoulas, Springer-Verlag, Berlin, 463-474.

HADDAK A.(1989), Differential algebra and controllability, Proceedings IFAC Symposium, Nonlinear Control and Systems Design, 14-16 june, Capri, Italy, 434-437.

HENRY J.(1977), Etude de la contrôlabilité de certaines équations paraboliques non-linéaires, Thèse, Paris.

ISIDORI A.(1985), Nonlinear control systems, an introduction, Lecture Notes in Control and Information Science 92, Springer Verlag, Berlin, 297p.

KOLCHIN E.R.(1973), Differential algebra and algebraic groups, Academic Press, New York, 446p.

LANDAU L., LIFSCHITZ E.(1971), Hydrodynamics, MIR, Moscow, 670p.

LIONS J.L.(1990), Are there connections between turbulence and controllability ?, IX Int. Conf. Analysis and Optimization of Systems, INRIA, 12-15 june, Antibes, France.

POMMARET J.F.(1988), Lie pseudogroups and mechanics, Gordon and Breach, New York, 560p.

POMMARET J.F.(1989), Differential algebra and partial differential control theory, Proceedings IFAC Symposium, Nonlinear Control and Systems Design, 14-16 june, Capri, Italy, 430-433.

POMMARET J.F.(1990), Equations aux dérivées partielles et théorie des groupes, perspectives nouvelles en informatique et en automatique, cours, 19-23 novembre, INRIA, Rocquencourt,France.

Author's address

J.F. Pommaret
Centre de Mathématiques Appliquées (CERMA)
Ecole Nationale des Ponts et Chaussées (ENPC)
La Courtine, 93167 Noisy-le-Grand Cedex, France

The H$_\infty$ control problem

International Series of Numerical Mathematics, Vol. 107, © 1992 Birkhäuser Verlag Basel

H_∞ BOUNDARY CONTROL WITH STATE FEEDBACK; THE HYPERBOLIC CASE

Viorel BARBU

Abstract

We solve the H_∞-control problem with state feedback for infinite dimensional boundary control systems of hyperbolic type with distributed disturbances.

1   Introduction and problem formulation

Let X, U, W be real Hilbert spaces and let A be the infinitesimal generator of C_0 semigroup e^{At} on H, $B_1 \in L(W, X)$, $B_2 \in L(U, (D(A^*))')$, $C_1 \in L(X, Z)$, $D_{12} \in L(U, Z)$. Consider the input-output system defined by

$$x'(t) = Ax(t) + B_1 w(t) + B_2 u(t), \quad t \in R^+ = [0, \infty] \tag{1.1}$$

$$z(t) = C_1 x(t) + D_{12} u(t), \quad t \in R^+. \tag{1.2}$$

Here $x(t) \in X$ is the state of the system, $u(t) \in U$ is the control input, $w(t) \in W$ is an exogeneous input and $z(t) \in Z$ is the controlled output.

We shall assume

i) $A^{-1} B_2 \in L(U, X)$ and for every $T > 0$ the operator $B_2^* e^{A^* t}$ admits a continuous extension from X to $L^2(0, T; U)$ i.e.

$$\int_0^T |B_2^* e^{A^* t} x|_U^2 dt \le C_T |x|^2 \quad \forall x \in X. \tag{1.3}$$

We have denoted by A^* the adjoint of A and by $B_2^* \in L(D(A^*), U)$ the adjoint of $B_2 \in L(U, (D(A^*))')$, $((D(A^*))'$ is the dual of $D(A^*)$). We shall denote by $|\cdot|_X$, $|\cdot|_Z$, $|\cdot|_U$, $|\cdot|_W$ the norms of X, Z, U, W and by $(\cdot, \cdot)$, $(\cdot, \cdot)_Z$, $(\cdot, \cdot)_U$, $(\cdot, \cdot)_W$ the corresponding scalar products. By assumption (i) it follows (see e.g. [9]) that for every $T > 0$, system (1.1) with initial condition $x(0) = x_0 \in X$ and inputs $u \in L^2(0, T; U)$, $w \in L^2(0, T; W)$ has a mild solution $x \in C([0, T]; X)$ given by

$$x(t) = e^{At} x_0 + \int_0^t e^{A(t-s)}(B_1 w(s) + B_2 u(s)) ds, \quad t \in [0, T]. \tag{1.4}$$

This is the sense in which we shall view the system (1.1) in this paper.

This is the abstract formulation of a large class of boundary control systems with distributed disturbance input including that governed by the wave equation with Dirichlet and Neumann boundary control, first order hyperbolic systems, Euller-Bernoulli equations (see [8], [9]). For instance the input-output system

$$\begin{aligned} y_{tt} - \Delta y &= w \quad in \quad \Omega \times R^+ \\ y &= u \quad in \quad \partial\Omega \times R^+ \end{aligned} \tag{1.5}$$

where $u \in L^2(R^+; L^2(\partial\Omega))$, $w \in L^2(R^+; L^2(\Omega))$ can be written in the form (1.1) when $X = L^2(\Omega) \times H^{-1}(\Omega)$, $U = L^2(\partial\Omega)$, $W = L^2(\Omega)$ and

$$A = \left\|\begin{matrix} 0 & I \\ -\Delta & 0 \end{matrix}\right\|, \quad B_1 w = \left\|\begin{matrix} 0 \\ w \end{matrix}\right\|, \quad B_2 u = \left\|\begin{matrix} 0 \\ -\Delta Du \end{matrix}\right\|.$$

Here Δ is the Laplace operator with the domain $H_0^1(\Omega) \cap H^2(\Omega)$ (Ω is an open bounded subset of R^n with a smooth boundary $\partial\Omega$) and $D \in L^2(L^2(\partial\Omega), L^2(\Omega))$ is the Dirichlet map, i.e., $\Delta Du = 0$ in Ω, $Du = u$ in $\partial\Omega$.

For a given feedback controller $F \in L(D(A), U)$ denote by $S_F : L^2(R^+; W) \to L^2(R^+; Z)$ the closed loop operator

$$z = S_F w = (C_1 + D_{12}F) \int_0^t e^{(A+B_2F)(t-s)} B_1 w(s) ds. \tag{1.6}$$

The standard problem of H_∞-control theory is the following (see e.g. [2], [3]) : given $\gamma > 0$ find $F \in L(D(A), H)$ which internally stabilizes the system (1.1) and makes $S_F \in L(L^2(R^+; W), L^2(R^+; Z))$ with $\|S_F\| < \gamma$ ($\|S_F\|$ is the operational norm of S_F. The main result of this paper, Theorem 1 below, gives a necessary and sufficient condition for existence of a suboptimal solution F to this H_∞-control problem. This criterium is expressed in terms of an algebraic Riccati equation associated with system (1.1), (1.2) and resembles the standard finite dimensional results. In the case of systems with distributed input controller u i.e., $B_2 \in L(U, X)$, it was established in [6], [7], [8], (for other related results in infinite dimensional spaces see also [5], [11]).

A standard approach to the suboptimal H_∞-controllers is to associate with system (1.1), (1.2) the differential game

$$\sup_{w \in L^2(R^+; W)} \inf_{u \in L^2(R^+; U)} \int_0^\infty (|z(t)|_Z - \gamma^2 |w(t)|_W^2) dt \tag{1.7}$$

and to write the closed loop strategies in terms of an algebraic Riccati equation. However, the extension of [7] to the present situation is not trivial and it relies on same recent results of Flandoli et al. (1988) on synthesis of quadratic optimal control problems with infinite horizon governed by boundary control systems of hyperbolic type. The H_∞-control problem for parabolic system of the form (1.1), (1.2) (in the sense of [4], [9]) can be treated analogously.

2   The main result
Beside assumption (i) we shall assume here that

 ii) The pair (C_1, A) is exponentially detectable i.e., there exists $K \in L(Z, X)$ such that $A + KC_1$ generates an exponentially stable semigroup.

 iii) $D_{12}^*[C_1, D_{12}] = [0, I]$.

Theorem 1 *Suppose that $\gamma > 0$ and hypotheses (i), (ii), (iii), hold. Then there exists an $F \in L(D(A), X)$ such that $A + B_2F$ generates an exponentially stable semigroup and $\|S_F\| < \gamma$ if and only if there exists $P \in L(X, X)$ with $B^*P \in L(D(A), U)$, $P = P^* \geq 0$ satisfying*

$$(Ax, Py) + (Px, Ay) - (P(B_2 B_2^* - \gamma^{-2} B_1 B_1^*) Px, y) + (C_1^* C_1 x, y) = 0 \quad \forall x, y \in D(A) \tag{2.1}$$

and such that $A - (B_2 B_2^* - \gamma^{-2} B_1 B_1^*)P$ generates an exponentially stable semigroup. Moreover, in this case the state feedback $F = -B_2^* P$ is exponentially stabilizing and $\|S_F\| < \gamma$.

Remark Assumptions (ii), (iii) are made in order to simplify the statements and the proofs but they can be relaxed as in [9] to

(ii') there exists $\varepsilon > 0$ such that for all $(\omega, x, u) \in R \times D(A) \times D(B_2)$ satisfying $i\omega x = Ax + B_2 u$ there holds

$$|\omega x + D_{12}u|_Z^2 \geq \varepsilon |x|^2$$

(iii') $D_{12}^* D_{12}$ is coercive.

Now coming back to system (1.5) we note that assumption (ii) is satisfied and so Theorem 1 is in particular applicable with the controlled output z given by

$$z(t) = (y(t), my_t, u(t)) \in Z = L^2(\Omega) \times H^{-1}(\Omega) \times L^2(\partial\Omega)$$

where $m = m(x)$ is a smooth non-negative function with compact support in Ω (see [1]).

3 Proof of Theorem 1 *Only if* part. Let $K : L^2(R^+; U) \times L^2(R^+; W) \to [-\infty, +\infty]$ be defined by

$$K(u, w) = \int_0^\infty (|C_1 x + D_{12}u|_Z^2 - \gamma^2 |w|_W^2) dt \tag{3.1}$$

where x is the solution to system (1.1) with $x(0) = x_0$.

Consider the minimization problem

$$\inf\{K(u, w) \; ; \; u \in L^2(R^+; U)\} \tag{3.2}$$

where $w \in L^2(R^+; W)$ is arbitrary but fixed. Since by assumption (iii),

$$|C_1 x + D_{12}u|_Z^2 = |C_1 x|_Z^2 + |u|_U^2 \quad \forall(x, u) \in X \times U$$

it is readily seen that problem (3.2) has a unique solution $\bar{u} = \Gamma w$. Moreover, we may represent $\Gamma : L^2(R^+; w) \to L^2(R^+; U)$ as

$$\Gamma w = \Gamma_0 w + f_0 \quad \forall w \in L^2(R^+; W) \tag{3.3}$$

where $\Gamma_0 \in L(L^2(R^+; W), L^2(R^+; U))$ and $f_0 \in L^2(R^+; U)$ ($f_0 = \arg\inf K(u, 0)$).

Lemma 1 *There is $p \in C(R^+; X) \cap L^2(R^+; X)$ such that*

$$p' = -A^* p + C_1^* C_1 \bar{x} \text{ in } R^+ \tag{3.4}$$

$$B_2^* p(t) = \bar{u}(t) \text{ a.e. } t > 0. \tag{3.5}$$

Here $\bar{x}$ is the solution to (1.1) with $u = \bar{u} = \Gamma w$. The equation (3.4) should be understood of course in the following mild sense

$$p(t) = e^{A^*(T-t)} p(T) + \int_t^T e^{A^*(s-t)} C_1^* C_1 \bar{x}(s) \tag{3.4}'$$

for all $0 \leq t \leq T < \infty$.

Proof Under our assumptions system (1.1) is internally exponentially stabilizable by a state feedback $u = -B_2^* P_0 x$ where $P_0 \in L(X, X)$, $P_0 = P_0^* \geq 0$ and (see [4])

$$B_2^* P_0 e^{(A - B_2 B_2^* P_0)t} x_0 \in L^2(R^+; U) \quad \forall x_0 \in X_0. \tag{3.6}$$

Let $p \in C(R^+; X) \cap L^2(R^+; X)$ be the solution to

$$p' = (A - B_2 B_2^* P_0)^* p - P_0 B_2 \bar{u} + C_1^* C_1 \bar{x}, \quad t \geq 0 \tag{3.7}$$

i.e.,

$$p(t) = \int_t^\infty e^{(A - B_2 B_2^* P_0)^*(s-t)} (C_1^* C_1 \bar{x}(s) - P_0 B_2 \bar{u}(s)) ds, \tag{3.8}$$

In virtue of (3.6) it is clear that (3.8) is well defined and $p \in L^2(R^+; X)$ (p can be equivalently defined by (3.4)'). Now by an easy calculation involving (3.1) and (3.7) we get

$$\int_0^\infty (\bar{u}(t) - B_2^* p(t), v(t) + B_2^* P_0 x(t))_U dt = 0 \tag{3.9}$$

for all $(x, v) \in C(R^+; X) \cap L^2(R^+; U)$ satisfying $x' = Ax + B_2 v$ in R^+ ; $x(0) = 0$.

We note that $B_2^* P_0 x \in L^2(R^+; U)$. Indeed $y = B_2^* P_0 x$ is the solution to integral equation

$$y(t) = \int_0^t L(t - s) y(s) ds + \int_0^t L(y - s) v(s) ds \tag{3.10}$$

where $L(t) = B_2^* P_0 e^{A_{P_0} t} B_2$, $A_{P_0} = A - B_2 B_2^* P_0$. Noticing that

$$(L(t) u_0, v_0)_U = (e^{\frac{1}{2} A_{P_0} t} B_2 u_0, (B_2^* P_0 e^{\frac{1}{2} P_0 t})^* v_0)_U,$$

for all $u_0, v_0 \in U$ and $t \geq 0$ it follows by assumption (i) and (3.6) that $L(t) \in L(U, U)$ a.e. $t > 0$ and

$$\int_0^\infty \|L(u)\|_{L(U,U)} dt \leq \infty$$

This implies by a standard device that for every $v \in L^2(R^+; U)$, eq. (3.11) has a unique solution $y \in L^2(R^+; U)$ as claimed. Now if in (3.9) we take $v = u - B_2^* P_0 x$ where u is arbitrary in $L^2(R^+; U)$ and x is the solution to equation

$$x' = A_{P_0} x + B_2 u, \quad x(0) = 0.$$

we get (3.5) as claimed.

Now consider the function $\varphi : L^2(R^+; W) \to R$ defined by

$$\varphi(w) = -K(\Gamma w, w). \tag{3.11}$$

By (3.3) we may represent φ as

$$\varphi(w) = \|Dw\|_{L^2(R^+;W)}^2 + ((Dw, f)) + \delta \tag{3.12}$$

where $D \in L(L^2(R^+; W), L^2(R^+; W))$, $((.,.))$ is the scalar product of $L^2(R^+; W)$ and $\delta \in R$. On the other hand, by assumption there exists $F \in L(D(A), U)$ such that $\|S_F\| < \gamma$. This clearly implies that

$$\varphi(w) \geq \alpha \|w\|_{L^2(R^+;W)}^2 + \beta \quad \forall w \in L^2(R^+; W)$$

where $\alpha > 0$ and $\beta \in R$. Along with (3.12) the latter implies that $D^* D$ is coercive and so φ attains its infimum on $L^2(R^+; W)$ in a unique point w^*, i.e.,

$$w^* = \arg\inf\{\varphi(w); w \in L^2(R^+; W)\}.$$

Lemma 2 *We have*

$$w^*(t) = -\gamma^{-2} B_1^* p(t) \text{ a.e. } t > 0 \tag{3.13}$$

Proof We have

$$(w^*, w)_W - (\Gamma_0^*(\Gamma_0 w^* + f_0), w)_W - (C_1^* C_1 x^*, \tilde{x}))dt = 0$$

for all $w \in L^2(R^+; W)$ where

$$\tilde{x}' = A\tilde{x} + B_2 \Gamma_0 w + B_1 w, \quad \tilde{x}(0) = 0.$$

Then using systems (1.1), (3.4), (3.5), we get

$$\int_0^\infty (\gamma^2 w^*(t) + B_1^* p(t), w(t))_W dt = 0 \quad \forall w \in L^2(R^+; W)$$

which implies (3.13).

To summarise, we have proved so far that the problem

$$\sup_{w \in L^2(R^+;W)} \inf_{u \in L^2(R^+;W)} K(u, w) \tag{3.14}$$

has a unique solution (u^*, w^*) characterised by the Euler-Lagrange system

$$(x^*)' = -Ax^* + B_1 w^* + B_2 u^*, \quad t \in R^+; \quad x^*(0) = x_0 \tag{3.15}$$

$$p' = -A^* p + C_1^* C_1 x^*, \quad p \in L^2(R^+; X) \tag{3.16}$$

$$u^*(t) = B_2^* p(t), \quad w^*(t) = -\gamma^{-2} B_1^* p(t), \text{ a.e. } t > 0 \tag{3.17}$$

We set

$$Px_= - p(0). \tag{3.18}$$

It is readily seen that $P \in L(X, X)$, $P = P^*$ whilst by $(3.15) \sim (3.17)$ we have

$$(Px_0, x_0) = -(p(0), x_0) = \int_0^\infty (|C_1 x^*(t) + D_{12} u^*(t)|_Z^2 - \gamma^2 |w^*(t)|_W^2 dt$$

$$\geq \inf \left\{ \int_0^\infty |C_1 x(t) + D_{12} u(t)|_Z^2 dt; u \in L^2(R^+, U) \right\}.$$

Moreover, since for every $t \geq 0$, (u^*, w^*) is the solution to problem

$$\sup_{w \in L^2(R_t^+;W)} \inf_{u \in L^2(R_t^+;U)} \left\{ \int_t^\infty |C_1 x + D_{12} u|_Z^2 ds \; ; \; x' = Ax + B_2 u + B_1 w \right\}$$

in $R_t^+ = (t, +\infty); x(t) = x^*(t)$, we infer that

$$p(t) = -Px^*(t) \quad \forall t \geq 0. \tag{3.19}$$

Let us denote by $S_P(t) : X \to X$ the family of operators

$$S_P(t)x_0 = x^*(t) \quad \forall t \geq 0 \tag{3.20}$$

where x^* is defined by (3.15) (i.e., x^* is the optimal state in problem (3.14)). It is readily seen that S_P is a C_0-semigroup in X. Denote by A_P the infinitesimal generator of S_P. We collect in Lemma 3 some properties of A_P and $B_1^* P$ (related result have been previously established in [4]).

Lemma 3 *We have*

$$D(A_P) \subset D(B_1^* P) \tag{3.21}$$

$$B_1^* P \in L(D(A_P), U) \cap L(D(A), U) \tag{3.22}$$

$$A_P x = Ax - B_2 B_2^* Px + \gamma^{-2} B_1 B_1^* x \quad \forall x \in D(A_P). \tag{3.23}$$

Proof We shall systematically use assumption (i). By (3.16, (3.18), (3.19) we have

$$Px_0 = -e^{A^*T}Pe^{A_PT}x_0 - \int_0^T e^{A^*t}C_1^*C_1e^{A_Pt}xdt \quad \forall T > 0$$

If $x_0 \in D(A_P)$ this yields

$$\begin{aligned}
B_2^*Px_= &B_2^*e^{A^*T}Pe^{A_PT}x_0 \\
&-B_1^*(A^*)^{-1}(e^{A^*T}C_1^*C_1e^{A_PT}x_0 - C_1^*C_1x_0) \\
&-\int_0^T e^{A^*t}C_1^*C_1e^{A_Pt}A_Px_0dt).
\end{aligned}$$

Since $B_2^*e^{A^*t}Pe^{A_Pt}x \in L^2_{loc}(R^+;U)$ the latter makes sense for almost all $T > 0$ and so $B_2^*Px_0 \in X$. Hence $D(A_P) \subset D(B_2^*P)$. In particular, this implies that B_2^*D is densely defined and is continuous from $D(A_P)$ to U.

On the other hand, for all $x \in H$ and $z \in D(A^*)$ we have

$$\begin{aligned}
\frac{d}{dt}(S_P(t)x_0, z) &= (S_P(t)x_0, A^*z) + (B_2u^*(t) + B_1w^*(t), z) \\
&= (S_P(t)x_0, A^*z) - (B_2B_2^*PS_P(t)x_0 - \gamma^{-2}B_1B_1^*PS_P(t)x_0, z)
\end{aligned}$$

Inasmuch as $B_2^*P \in L(D(A_P), U)$ and B_2 is closed in the latter yields

$$(A_Px_0, z) = (x_0, A^*z) - (B_2B_2^*Px_0 - \gamma^{-2}B_1B_1^*Px_0, z) \quad \forall x_0 \in D(A_P),\ z \in D(A^*)$$

Hence

$$A_Px_0 = Ax_0 - B_2B_2^*Px_0 + \gamma^{-2}B_1B_1^*Px_0 \quad \forall x_0 \in D(A_P).$$

On the other hand, an easy calculation involving (3.16) and (3.18) reveals as above that

$$A^*P \in L(D(A_P), H), \quad A_P^*P \in L(D(A), X)$$

Then by (3.23) we see that

$$\begin{aligned}
|(B_1^*Px, B_1^*Pz)| &\leq |(Ax, Pz)| + |(A_Px, Pz)| \\
&\leq C\|x\|_{D(A)}\|z\|_{D(A)} \quad \forall x \in D(A_P), z \in D(A).
\end{aligned}$$

Hence $B_1^*P \in L(D(A), U)$ thereby completing the proof of Lemma 3.

Proof of Theorem 1 (continued). To prove that P is a solution to the Riccati equation (2.1) we notice first that in virtue of (3.16), (3.17) and (3.18) we have

$$(Px^*(t), x^*(t)) = \int_t^\infty (|C_1x^*(s)|_Z^2 + |u^*(s)|_U^2 - \gamma^2|w^*(s)|_W^2)ds \quad \forall t \geq 0$$

For $x_0 \in D(A_P)$ this yields

$$2(Px^*(t), A_Px^*(t)) + |C_1x^*(t)|_Z^2 + |B_2^*Px^*(s)|_U^2 = -\gamma^{-2}|B_1^*Px^*(t)|_W^2 = 0 \quad \forall t \geq 0$$

and since $B_2^*P \in L(D(A_P), X)$ we get

$$2(Px_0, A_Px_0) + |C_1x_0|_Z^2 + |B_2^*Px_0|_U^2 - \gamma^{-2}|B_1^*Px_0|_W^2 = 0$$

Finally, since $B_2^* P \in L(D(A), X)$ the latter extends to all of $D(A)$,

$$2(Px_0, A_P x_0) + |C_1 x_0|_Z^2 - |B_2^* P x_0|_U^2 + \gamma^{-2}|B_1^* P x_0|_W^2 = 0 \qquad (3.24)$$

for all $x_0 \in D(A)$. Differentiating (3.24) in the space $D(A)$ we get eq. (2.3).

To prove that the semigroup $e^{A_P t}$ is exponentially stable we use detectability assumption (ii). Let $K \in L(Z, X)$ as in this assumption. Then we have

$$x^*(t) = e^{(A+KC_1)^*} x_0 + \int_0^t e^{(A+KC_1)(t-s)}(B_2 u^*(s) + B_1 w^*(s))ds$$

$$- \int_0^t e^{(A+KC_1)(t-s)} KC_1 x^*(s)ds \quad \forall t \geq 0.$$

Since $B_1 w^*, KC_1 x^* \in L^2(R^+; X)$ it remains to be shown that

$$\int_0^t e^{(A+KC_1)(t-s)} B_2 u^*(s)ds \in L^2(R^+; X)$$

The latter follows by assumption (i) arguing as in lemma 5.2 in [4].

It is also easy to see that the operator $A - B_2 B_2^* P$ generates an exponentially stable semigroup. Moreover, by an easy calculation involving eq. (2.1) we see that (see [8])

$$\frac{d}{dt}(Px(t), x(t)) = |Gx(t)|_Z^2 + |B_2^* Px(t)|_U^2 + \gamma^2|\tilde{w}(t)|_W^2 - \gamma^2|w(t)|_W^2 \text{ a.e. } t > 0$$

where

$$w \in L^2(R^+; W) \cap C^1(R^+; W), \quad x' = (A - B_2 B_2^* P)x + B_1 w, \quad x(0) = 0$$

and $\tilde{w} = w - \gamma^{-2} B_1^* Px$. This yields

$$\|Dx\|_{L^2(R^+;Z)}^2 - \gamma^2\|w\|_{L^2(R^+;W)}^2 = \gamma^2\|\tilde{w}\|_{L^2(R^+;W)}^2$$
$$\geq \delta\|w\|_{L^2(R^+;W)}^2$$

where δ is independent of w. Hence $\|S_{B_2 P}\| < \gamma$ which completes the proof of *only if part*.

Suppose now that $P = P^* \geq 0$ is a solution to eq (2.1) such that $A - (B_2 B_2^* - \gamma^{-2} B_1 B_1^*)P$ generates an exponentially stable semigroup. Then clearly $A - B_2 B_2^* P$ generates an exponentially stable semigroup and arguing as above it follows that $\|S_{B_2^* P}\| < \gamma$.

Acknowledgements   This work was done during the stay of the author at INRIA-Rocquencourt.

References

[1] C. Bardos, G. Lebeau, J. Rauch (1990) : Contrôle et stabilisation de l'équation des ondes. (Appendix J.L. Lions, Controllabilité exacte, perturbations et stabilisation de systèmes distribués, Masson, Paris).

[2] R. Curtain (1992) : A synthesis time and frequency domain methods for the control of infinite-dimensional systems ; a system theoretic approach. (To appear in SIAM Frontières in Applied Mathematics).

[3] J. Doyle, K. Glover, P. Khargonekar, B. Francis (1989) : State-space solutions to standard H_2 and H_∞-control problems. IEEE Trans. Aut. Control, Vol AC 34, pp. 831-847.

[4] F. Flandoli, I. Lasiecka, R. Triggiani (1988) : Algebraic Riccati equations with nonsmoothing observation arising in Euler-Bernoulli boundary control problems. Annali Mat. Pura Applicata TCLIII, pp. 307-382.

[5] A. Halanay (1992) : Stabilising compensators with disturbance attenuation (to appear).

[6] A. Ichikawa (1992) : H_∞-control and min-max problems in Hilbert space (to appear).

[7] B. Van Keulen (1991) : H_∞-control with measurement feedback for linear infinite-dimensional systems (to appear).

[8] B. Van Keulen, M. Peters, R. Curtain (1991) : H_∞-control with state feedback : the infinite dimensional case (to appear).

[9] I. Lasiecka, R. Triggiani (1991) : DIfferential and algebraic Riccati equations with applications to boundary/point control problems : continuous theory and approximation theory. Lectures Notes in Control and Information Sciences Vol. 164, Springer Verlag.

[10] A.J. Pritchard, D. Salmon (1987) : A semi-group theoretic approach for systems with unbounded input and output operators. SIAM J. Control and Optimization 25, pp. 121-144.

[11] A.J. Pritchard, S. Townley (1989) : Robustness of linear systems. J. Diff. Eqs. 17, pp. 254-286.

Author's address   : Viorel BARBU - University of Iasi - 6600 Iasi - Romania.

International Series of Numerical Mathematics, Vol. 107, © 1992 Birkhäuser Verlag Basel

Remarks on the theory of robust control

A. Bensoussan
University Paris Dauphine and INRIA

P. Bernhard
INRIA Sophia antipolis

June 9, 1992

1 Introduction

We deal in this article with the theory of robust control for infinite dimensional systems in the case of full and partial observation. The first case corresponds to a theorem of [1]. Here the contribution consists in a simplification of the proof showing that all steps reduce to standard LQ problems for which the theory is well known and can be heavily applied. As a side effect an interesting LQ problem is identified, with a nice interpretation. The problem is equivalent to solving a differential game with full information for both players. The second case corresponds to a theorem of [2]. Here again we provide a simplified proof. In this case the problem is not equivalent to a differential game, but includes the solution of three related differential games with full information. We show the role of the "observer" concept in this context.

2 General Presentation of the Problem

2.1 Notation and Assumptions

Let X be a real separable Hilbert space, and A be the infinitesimal generator of a C_0-semigroup e^{At} in X. We denote by $D(A)$ the domain of the operator A. We say that the semigroup is exponentially stable if one has

$$\|e^{At}\| \leq M e^{-\alpha t}, \alpha > 0.$$

A linear unbounded operator A in X is said *exponentially stable* if it is the infinitesimal generator of a C_0-semigroup e^{At} in X which is exponentially stable. We recall the important result of DATKO [3],namely that A is exponentially stable iff for any h in X the solution of

$$x' = Ax, x(0) = h$$

belongs to $L^2(0, \infty; X)$. Let also U and W be real separable hilbert spaces, and B and D linear bounded operators from respectively U and W to X. Consider the dynamic system governed by the equation

$$(2.\ 1) \qquad \begin{aligned} x' &= Ax + Bv + Dw \\ x(0) &= 0. \end{aligned}$$

In equation (2. 1) w stands for a disturbance and v stands for a control. We now make precise what is meant by *robust control*. Let Z be an addditional Hilbert space and $H \in \mathcal{L}(X; Z)$. Suppose the controller is interested in the cost function

$$K_0(v, w) = \int_0^\infty (|Hx|^2 + |v|^2)\, dt$$

which may take the value $+\infty$. He is interested in minimizing the value of the cost function. The value of the disturbance is not fixed, but it may be observable or not. Because of the linearity the only value that he can hope to control is the ratio

$$\rho(v,w) = \frac{K_0(v,w)}{\displaystyle\int_0^\infty |w|^2\, dt}$$

provided of course the disturbance is square integrable, which will always be assumed in this paper.

Definition 2.1 *We say that the γ^2 robustness property (with full observation) holds for the equation (2. 1) and the cost function $K_0(v,w)$ if one has*

$$\sup_w \inf_v \rho(v,w) < \gamma^2.$$

Another definition will be given in the case of partial information.

2.2   Review of some results

Consider the pair of operators A, B, H as above.

Definition 2.2 *We say that the pair A, B is H stabilizable if for any $h \in X$ there exists $v \in L^2(0,\infty;U)$ such that the solution x of the equation*

$$(2.\ 2) \qquad\qquad\qquad\qquad \begin{aligned} x' &= Ax + Bv \\ x(0) &= h \end{aligned}$$

satisfies $Hx \in L^2(0,\infty;Z)$. We say that the pair A, B is stabilizable if it is I stabilizable.

It is a classical result that the pair A,B is stabilizable iff there exists an operator $F \in \mathcal{L}(X,U)$ such that $A + BF$ is exponentially stable. We now give another

Definition 2.3 *We say that the pair A, H is detectable if the pair A^*, H^* is stabilizable.*

It follows from the characterization of stabilizability that the pair A,H is detectable iff there exists an operator $G \in \mathcal{L}(Z,X)$ such that $A + GH$ is exponentially stable. We now state an important result in Control Theory,whose proof for infinite dimensional systems can be found in [5].

Theorem 2.1 *We assume that A, B is H stabilizable and that A, H is detectable. Then there exists one and only one operator $\Gamma \in \mathcal{L}(\mathcal{X},\mathcal{X})$ with $\Gamma = \Gamma^* \geq 0$ satisfying $A - BB^*\Gamma$ is exponentially stable and*

$$(2.\ 3) \qquad\qquad\qquad \Gamma A + A^*\Gamma - \Gamma BB^*\Gamma + H^*H = 0$$

The interpretation of the operator Γ is important. Consider the functional

$$K_h(v) = \int_0^\infty (|Hx|^2 + |v|^2)\, dt$$

where x is the solution of (2. 2). Then one has

$$(\Gamma h, h) = \min K_h(v).$$

The minimum in $K_h(v)$ is attained for $u = -B^*\Gamma y$ where y is the solution of

$$y' = (A - BB^*\Gamma)y \quad y(0) = h.$$

Remark 2.1 *In equation 2. 3, one must interpret the operator*

$$\Gamma A + A^* \Gamma$$

according to the general theory of algebraic Riccati equations(see [5] for details). It is sufficient to notice that for any pair h, k in $D(A)$ the bilinear form

$$< (\Gamma A + A^*\Gamma)h, k > = (Ah, \Gamma k) + (\Gamma h, Ak)$$

makes sense.

Note also that if $h \in D(A)$ then $\Gamma h \in D(A^*)$.

3 γ^2 robustness property with full observation

3.1 Setting of the Result

We state the result due to [1].

Theorem 3.1 *We assume that A, B is stabilizable and that A, H is detectable. Then the γ^2 robustness property with full observation holds for the equation (2. 1) and the cost function $K_0(v, w)$ iff there exists a $P \in \mathcal{L}(\mathcal{X}, \mathcal{X})$ with $P = P^* \geq 0$ satisfying*

$$(3.\ 1) \qquad\qquad PA + A^*P - P(BB^* - \frac{1}{\gamma^2}DD^*)P + H^*H = 0$$

and $A - (BB^* - \frac{1}{\gamma^2}DD^*)P$ is exponentially stable.

The operator P has an interesting interpretation. Indeed let be

$$K_h(v, w) = \int_0^\infty (|Hx|^2 + |v|^2)\, dt$$

where x is the solution of

$$(3.\ 2) \qquad\qquad \begin{aligned} x' &= Ax + Bv + Dw \\ x(0) &= h. \end{aligned}$$

Let also

$$J_h(v, w) = K_h(v, w) - \gamma^2 \int_0^\infty |w|^2\, dt.$$

Then one has

$$(3.\ 3) \qquad\qquad (Ph, h) = \max_w \min_v J_h(v, w).$$

One can check the important following result

Lemma 3.1 *If P satisfies the properties stated in Theorem 3.1 then one has also $A - BB^*P$ is exponentially stable.*

♠

3.2   Sketch of the proof of Main result

Sufficiency.
We consider the solution of

$$x' = (A - BB^*P)x + Dw \quad, x(0) = 0.$$

Since $A - BB^*P$ is exponentially stable and $w \in L^2(0, \infty; W)$ the solution x belongs to $L^2(0, \infty; X)$. We apply the relation (??) with $v = -B^*Px$ and $h = 0$, letting T tend to ∞. This yields

$$\int_0^\infty |Hx|^2 \, dt + \int_0^\infty |B^*Px|^2 \, dt = \gamma^2 \int_0^\infty |w|^2 \, dt - \gamma^2 \int_0^\infty |w - \frac{1}{\gamma^2} D^*Px|^2 \, dt.$$

Therefore

$$\rho(-B^*Px, w) = \gamma^2 - \gamma^2 \frac{\int_0^\infty |w - \frac{1}{\gamma^2} D^*Px|^2 \, dt}{\int_0^\infty |w|^2 \, dt}.$$

It follows that

$$\sup_w \rho(-B^*Px, w) \leq \gamma^2 - \gamma^2 \inf_w \frac{\int_0^\infty |w - \frac{1}{\gamma^2} D^*Px|^2 \, dt}{\int_0^\infty |w|^2 \, dt}.$$

Now x can be viewed as the solution of

$$x' = \left(A - (BB^* - \frac{1}{\gamma^2} DD^*)P\right) x + D(w - \frac{1}{\gamma^2} D^*Px) \quad, x(0) = 0.$$

Since $A - (BB^* - \frac{1}{\gamma^2} DD^*)P$ is exponentially stable we have

$$\int_0^\infty |x|^2 \, dt \leq c_0 \int_0^\infty |w - \frac{1}{\gamma^2} D^*Px|^2 \, dt.$$

Hence immediately

$$\int_0^\infty |w|^2 \, dt \leq c_1 \int_0^\infty |w - \frac{1}{\gamma^2} D^*Px|^2 \, dt.$$

Therefore we can assert that

$$(3.\ 4) \qquad \inf_w \frac{\int_0^\infty |w - \frac{1}{\gamma^2} D^*Px|^2 \, dt}{\int_0^\infty |w|^2 \, dt} > 0.$$

We deduce

$$\sup_w \rho(-B^*Px, w) \leq \gamma^2.$$

This is stronger than

$$\sup_w \inf_v \rho(v, w) < \gamma^2$$

which proves the γ^2 robustness property.

Necessity.
We consider a control problem where the system is defined by (3. 2) in which w is given, the control being v. We minimize the cost $K_h(v, w)$. Note that the assumptions of Theorem2.1 are satisfied, thus the equation

$$(3.\ 5) \qquad \Gamma A + A^*\Gamma - \Gamma BB^*\Gamma + H^*H = 0.$$

has a unique solution $\Gamma = \Gamma^* \geq 0$, such that the operator $A - BB^*\Gamma$ is exponentially stable. Consider next the linear equation

$$(3.\ 6) \qquad r' + (A^* - \Gamma BB^*)r + \Gamma Dw = 0$$

where the solution $r(.)$ belongs to $L^2(0, \infty; X)$. Note that in (3. 6) the initial condition is given at ∞ and not at 0. The equation (3. 6) has a unique solution. The optimal feedback is then descibed by

$$\hat{u} = -B^*(\Gamma \hat{x} + r)$$

where $\hat{x}$ the optimal state is the solution of

$$\frac{d}{dt}\hat{x} = (A - BB^*\Gamma)\hat{x} - BB^*r + Dw \ ; \hat{x}(0) = h.$$

We can also express the value of $J_h(\hat{u}, w)$ as follows

$$(3.\ 7) \qquad \begin{aligned} J_h(\hat{u}, w) &= -\int_0^\infty |B^*r|^2\, dt \\ &-\gamma^2 \int_0^\infty |w|^2\, dt + 2\int_0^\infty (r, Dw)\, dt + 2(h, r(0)) + (\Gamma h, h). \end{aligned}$$

The calculations leading to the expression (3. 7) are standard and not detailed here. We now look at the problem of maximizing the expression $J_h(\hat{u}, w)$ for $w \in L^2(0, \infty; W)$. Note that although we have an LQ problem, the concavity is not a priori verified. This is where we use the assumption of γ^2 robustness property. Set $\Phi_h(w) = J_h(\hat{u}, w)$, then we note the relation

$$\Phi_h(w) = \Phi_0(w) + 2(h, r(0)) + (\Gamma h, h).$$

By the assumption of γ^2 robustness there exists a positive $\delta < \gamma$ such that $\sup_w \inf_v \rho(v, w) \leq \gamma^2 - \delta^2$. It follows that

$$J_0(\hat{u}, w) \leq -\delta^2 \int_0^\infty |w|^2\, dt$$

for any w. Therefore, we can assert that

$$(3.\ 8) \qquad \Phi_0(w) \leq -\delta^2 \int_0^\infty |w|^2\, dt.$$

Note also the relation

$$\Phi_h(\theta w_1 + (1 - \theta)w_2) = \theta\Phi_h(w_1) + (1 - \theta)\Phi_h(w_2) - 2\theta(1 - \theta)\Phi_0(w_1 - w_2).$$

This relation shows immediately that $\Phi_h(w)$ is strictly concave. From (3. 8) it is a coercive functional. Therefore we can apply to the control problem (3. 6) and cost function $\Phi_h(w)$ the standard theory of existence, uniqueness of an optimal control. Moreover, the theory of necessary and sufficient conditions of optimality hold. Calling $\hat{w}$ and $\hat{r}$ the optimal control and state, we have the relations

$$(3.\ 9) \qquad \begin{aligned} \frac{d}{dt}\hat{r} + (A^* - \Gamma BB^*)\hat{r} + \Gamma D\hat{w} &= 0 \\ \frac{d}{dt}p = (A - BB^*\Gamma)p - BB^*\hat{r} + D\hat{w} \end{aligned}$$

$$(3.\ 10) \qquad \begin{aligned} p(0) &= h \\ \gamma^2\hat{w} - D^*\hat{r} - D^*\Gamma p &= 0 \end{aligned}$$

The classical decoupling argument applies to the system (3. 9) and (3. 10), therefore there exists $\Sigma = \Sigma^*$, such that $\hat{r} = \Sigma p$. The operator Σ is a solution of the Riccati equation

$$\begin{aligned} \Sigma\left(A - (BB^* - \frac{1}{\gamma^2}DD^*)\Gamma\right) &+ \left(A^* - \Gamma(BB^* - \frac{1}{\gamma^2}DD^*)\right)\Sigma \\ &- \Sigma(BB^* - \frac{1}{\gamma^2}DD^*)\Sigma + \frac{\Gamma DD^*\Gamma}{\gamma^2} = 0. \end{aligned}$$

Set $P = \Gamma + \Sigma$, then P is self adjoint and is a solution of (3. 1), as easily seen. Note that p is the solution of

$$\frac{d}{dt}p = \left(A - (BB^* - \frac{1}{\gamma^2}DD^*)P\right)p$$
$$p(0) = h.$$

Since $p \in L^2(0, \infty; X)$, we deduce that $A - (BB^* - \frac{1}{\gamma^2}DD^*)P$ is exponentially stable. Moreover

$$\max \Phi_h(w) = \Phi_h(\hat{w}) = (Ph, h).$$

Therefore, we have

$$\max_w \min_v J_h(v, w) = (Ph, h).$$

It follows among other things that

$$(Ph, h) \geq \min_v J_h(v, 0) \geq 0.$$

The proof of the necessity part has been completed. This concludes the end of the proof.

4 γ^2 robustness property with partial observation

4.1 Presentation of the problem

Consider again the system

$$\begin{aligned} x' &= Ax + Bv + Dw \\ x(0) &= 0. \end{aligned}$$
(4. 1)

and the cost function

$$K_0(v, w) = \int_0^\infty (|Hx|^2 + |v|^2)\, dt.$$

In the present situation, the controller has access only to a partial observation described as follows

(4. 2) $$y = Cx + \eta$$

where $C \in \mathcal{L}(X; Y)$,Y being a new Hilbert space, called the space of observations.Next η is another perturbation, modelling the measurement error. The controller can use only a causal functional on the observation y.A natural class of controllers is the following one

$$v = Lp$$

$$p' = (A + M)p + Ny \;, p(0) = 0.$$

In other words the controller is characterized by three maps $L \in \mathcal{L}(X; U), M \in \mathcal{L}(X; X), N \in \mathcal{L}(Y; X)$. We call it a *feedback controller*. We have not combined $A + M$ into a single operator since A is unbounded, whereas M is bounded.If we use such a controller in the equation 4. 1, then we get a coupled system as follows

$$\begin{aligned} x' &= Ax + BLp + Dw \\ p' &= (A + M)p + NCx + N\eta \\ x(0) &= 0 \\ p(0) &= 0. \end{aligned}$$
(4. 3)

The coupled system introduces an operator

$$\mathcal{A} = \begin{pmatrix} A & BL \\ NC & A + M \end{pmatrix}$$

The operator $\mathcal{A}$ is the operator related to the feedback controller L, M, N. We shall consider only controllers whose corresponding operator is exponentially stable. For such a controller the cost

$$K_0(Lp, w) = \int_0^\infty (|Hx|^2 + |Lp|^2)\, dt$$

is finite. We can then define the ratio corresponding to the feedback controller L, M, N

$$\rho(L, M, N) = \sup_{w, \eta} \frac{K_0(Lp, w)}{\displaystyle\int_0^\infty (|w|^2 + |\eta|^2)\, dt}.$$

We state the

Definition 4.1 *We say that the γ^2 robustness property(with partial observation) holds for the equation (4. 1), the observation (4. 2) and the cost function $K_0(v, w)$ if there exists a feedback controller L, M, N such that the corresponding operator $\mathcal{A}$ is exponentially stable and if one has*

$$\rho(L, M, N) < \gamma^2.$$

Our objective is naturally to give a necessary and sufficient condition of γ^2 robustness property(with partial observation) which leads to a computable feedback.We shall see that this property is equivalent to 3 systems and corresponding costs enjoying the γ^2 robustness property (with full observation). From section 3 corresponding Riccati equations can be introduced. In fact, the solution of one of them is expressible in terms of the others. Moreover one of the system and cost is (4. 1) with cost $K_0(v, w)$. Therefore γ^2 robustness property(with partial observation) implies γ^2 robustness property (with full observation).

4.2   Statement of the main result

We state the result due to [2].

Theorem 4.1 *We assume that A, D is stabilizable and that A, H is detectable. Then the γ^2 robustness property with partial observation holds for the equation (4. 1), the observation (4. 2) and the cost function $K_0(v, w)$ iff there exist solutions of the Riccati equations*

$$\text{(4. 4)} \qquad \begin{aligned} &PA + A^*P - P(BB^* - \tfrac{1}{\gamma^2}DD^*)P + H^*H \quad = \quad 0 \\ &P\,symmetric, \geq 0 \\ &A - (BB^* - \tfrac{1}{\gamma^2}DD^*)P\,is\ exponentially\ stable \end{aligned}$$

$$\text{(4. 5)} \qquad \begin{aligned} &\Sigma A^* + A\Sigma - \Sigma(C^*C - \tfrac{1}{\gamma^2}H^*H)\Sigma + DD^* \quad = \quad 0 \\ &\Sigma\,symmetric, \geq 0 \\ &A^* - (C^*C - \tfrac{1}{\gamma^2}H^*H)\Sigma\,is\ exponentially\ stable \end{aligned}$$

$$\text{(4. 6)} \qquad I - \frac{1}{\gamma^2}P\Sigma\,is\ invertible\ ;\ \Sigma\left(I - \frac{1}{\gamma^2}P\Sigma\right)^{-1} \geq 0$$

If the conditions (4. 4), (4. 5), (4. 6) hold, then the feedback controller

$$L = -B^*P\ ,\ M = -(BB^* - \frac{1}{\gamma^2}DD^*)P - \Pi C^*C\ ,\ N = \Pi C^*$$

where

$$\Pi = \Sigma\left(I - \frac{1}{\gamma^2}P\Sigma\right)^{-1}$$

satisfies the conditions of the definition 4.1.

Remark 4.1 *The Riccati equation (4. 4) is the same as the one characterizing the γ^2 robustness property with full observation. Note that the pair A, B is stabilizable, as a consequence of the fact that A is exponentially stable. From formula (3. 3) the solution is unique. A similar property holds for (4. 5) since it characterizes the γ^2 robustness property with full observation of a different system and cost. This will be made precise in the proof.*

4.3 Duality Considerations

We shall give here some norm equalities which will be instrumental in the proof of Theorem 4.1 and are interesting in themselves. Consider Hilbert spaces Ξ, Φ, Ψ, identified with their duals. Consider operators $\mathcal{A}, \mathcal{B}, \mathcal{C}$, which are linear, $\mathcal{A}$ is unbounded in Ξ and is the infinitesimal generator of a C_0 semigroup in Ξ. Next

$$\mathcal{B} \in \mathcal{L}(\Phi; \Xi) \ , \mathcal{C} \in \mathcal{L}(\Xi; \Psi).$$

To the triple $\mathcal{A}, \mathcal{B}, \mathcal{C}$ we associate a linear system

$$(4.\ 7) \qquad\qquad \xi' = \mathcal{A}\xi + \mathcal{B}\phi \ ; \xi(0) = 0$$

and a corresponding observation $\mathcal{C}\xi$. We define next a "dual" triple $\mathcal{A}^*$, $\mathcal{C}^*$, $\mathcal{B}^*$ to which corresponds the system

$$(4.\ 8) \qquad\qquad \zeta' = \mathcal{A}^*\zeta + \mathcal{C}^*\psi \ ; \zeta(0) = 0$$

and the corresponding observation $\mathcal{B}^*\zeta$. In equations (4. 7) and (4. 8) the quantities ϕ and ψ are inputs(or controls). We state the following

Proposition 4.1 *Assume that $\mathcal{A}$ is exponentially stable. Then one has the relation*

$$\sup_{\phi} \frac{\displaystyle\int_0^{\infty} |\mathcal{C}\xi|^2\, dt}{\displaystyle\int_0^{\infty} |\phi|^2\, dt} = \sup_{\psi} \frac{\displaystyle\int_0^{\infty} |\mathcal{B}^*\zeta|^2\, dt}{\displaystyle\int_0^{\infty} |\psi|^2\, dt}.$$

We begin with some preliminary results.

Lemma 4.1 *Let $\phi \in L^2(-\infty; +\infty; \Phi)$, then there exists one and only one function*

$$\xi \in L^2(-\infty; +\infty; \Xi) \cap C^0(-\infty; +\infty; \Xi)$$

solution of the equation

$$(4.\ 9) \qquad\qquad \xi' = \mathcal{A}\xi + \mathcal{B}\phi$$

Let

$$\alpha = \sup_{\phi} \frac{\displaystyle\int_0^{\infty} |\mathcal{C}\xi|^2\, dt}{\displaystyle\int_0^{\infty} |\phi|^2\, dt}$$

where we refer to the system (4. 7). Similarly, set

$$\beta = \sup_{\phi} \frac{\displaystyle\int_{-\infty}^{\infty} |\mathcal{C}\xi|^2\, dt}{\displaystyle\int_{-\infty}^{\infty} |\phi|^2\, dt}$$

referring now to the system (4. 9). Then we have

Lemma 4.2 *The two numbers α and β are equal.*

We can now proceed with the

Proof of Proposition 4.1 In a way similar to (4. 9) we consider the dual system on $-\infty, +\infty$

$$(4.\ 10)\qquad\qquad \zeta' = A^*\zeta + C^*\psi$$

where $\psi \in L^2(-\infty; +\infty; \Psi)$ and the solution ζ belongs to $L^2(-\infty; +\infty; \Xi) \cap C^0(-\infty; +\infty; \Xi)$. Define a linear map from $L^2(-\infty; +\infty; \Psi)$ to $L^2(-\infty; +\infty; \Phi)$ by setting

$$\Upsilon(\psi) = B^*\zeta.$$

In view of Lemma 4.1 the desired result will be demonstrated if we prove

$$(4.\ 11)\qquad\qquad \|\Theta\| = \|\Upsilon\|$$

To any ψ associate $\bar{\psi}$ by setting $\bar{\psi}(t) = \psi(-t)$. The key point is to verify that

$$(4.\ 12)\qquad\qquad \Upsilon(\psi)(t) = \Theta^*(\bar{\psi})(-t).$$

This property implies the result (4. 11). Now (4. 12) is easily deduced from the explicit formula (??) and the corresponding one for ζ the solution of (4. 10). Details are left to the reader. The proof has been completed. ♠

5 Proof of Theorem 4.1

5.1 Necessary Conditions

Proof of (4. 4)

In fact, the assumptions of Theorem 3.1 are satisfied, since the property of γ^2 robustness with full observation holds and that the pair A, B is stabilizable(see Remark 4.1). Therefore There exists a unique P solution of (4. 4).

Proof of (4. 5)

We shall use the duality considerations of paragraph 4.3, see Proposition 4.1. Let $\Xi = X \times X$, $\Phi = W \times Y$, $\Psi = Z \times U$. Let next

$$\mathcal{A} = \begin{pmatrix} A & BL \\ NC & A+M \end{pmatrix}$$

and

$$\mathcal{B} = \begin{pmatrix} D & 0 \\ 0 & N \end{pmatrix} \quad \mathcal{C} = \begin{pmatrix} H & 0 \\ 0 & L \end{pmatrix}.$$

Setting $\xi = (x, p)$, $\phi = (w, \eta)$, then we see immediately that

$$\rho(L, M, N) = \alpha$$

where the number α has been defined in the proof of Lemma 4.2. We can then make use of Proposition 4.1. The dual system of (4. 3) is (apply (4. 8) and set $\zeta = m, q$, $\psi = \lambda, \mu$)

$$(5.\ 1)\qquad \begin{aligned} m' &= A^*m + C^*N^*q + H^*\lambda \\ q' &= L^*B^*m + (A^* + M^*)q + L^*\mu \\ m(0) &= 0 \\ q(0) &= 0. \end{aligned}$$

Using then Proposition 4.1 we can assert that

$$\rho(L, M, N) = \sup_{\lambda, \mu} \frac{\displaystyle\int_0^\infty (|D^*m|^2 + |N^*q|^2)\, dt}{\displaystyle\int_0^\infty (|\lambda|^2 + |\mu|^2)\, dt}.$$

Therefore from the assumption, we have the property

$$\sup_{\lambda,\mu} \frac{\displaystyle\int_0^\infty (|D^*m|^2 + |N^*q|^2)\,dt}{\displaystyle\int_0^\infty (|\lambda|^2 + |\mu|^2)\,dt} < \gamma^2.$$

In particular, restricting to $\mu = 0$, we have

$$\begin{aligned}
m' &= A^*m + C^*N^*q + H^*\lambda \\
q' &= L^*B^*m + (A^* + M^*)q \\
m(0) &= 0 \\
q(0) &= 0.
\end{aligned}$$

(5. 2)

and

(5. 3)

$$\sup_{\lambda} \frac{\displaystyle\int_0^\infty (|D^*m|^2 + |N^*q|^2)\,dt}{\displaystyle\int_0^\infty |\lambda|^2\,dt} < \gamma^2.$$

Consider the dynamic system

(5. 4)

$$\begin{aligned}
m' &= A^*m + C^*v + H^*\lambda \\
m(0) &= 0
\end{aligned}$$

where v is the control and λ is the perturbation. We first note that the pair A^*, C^* is stabilizable, as a consequence of the fact that the operator $\mathcal{A}$, hence its dual $\mathcal{A}^*$ is exponentially stable. Consider also the cost function

$$\mathcal{K}_0(v,\lambda) = \int_0^\infty (|D^*m|^2 + |v|^2)\,dt.$$

From the assumptions A^*, D^* is detectable. Moreover, from (5. 3) we can assert that the γ^2 robustness property holds for the system (5. 4) and the cost function $\mathcal{K}_0(v,\lambda)$. Therefore, relying on Theorem 3.1we obtain the existence and uniqueness of the solution Σ of (4. 5).

Proof of (4. 6)

The proof will be decomposed in several steps. We begin with
-a:The matrix operator

$$A_P = \begin{pmatrix} A + \frac{1}{\gamma^{*2}}DD^*P & BL \\ NC & A + M \end{pmatrix}$$

is exponentially stable. For that purpose consider the dynamic system

(5. 5)

$$\begin{aligned}
\bar{x}' &= \left(A + \frac{1}{\gamma^{*2}}DD^*P\right)\bar{x} + BL\bar{p} \\
\bar{p}' &= (A + M)\bar{p} + NC\bar{x} \\
\bar{x}(0) &= h \\
\bar{p}(0) &= k.
\end{aligned}$$

Then we must prove that

(5. 6)

$$\bar{x} \in L^2(-\infty; +\infty; X) \quad ; \bar{p} \in L^2(-\infty; +\infty; X)$$

Consider then the system

(5. 7)

$$\begin{aligned}
x' &= Ax + BLp + Dw \\
p' &= (A + M)p + NCx \\
x(0) &= h \\
p(0) &= k.
\end{aligned}$$

where $w \in L^2(-\infty; +\infty; W)$ and the system corresponding to $w = 0$

$$
\begin{aligned}
x_1' &= Ax_1 + BLp_1 \\
p_1' &= (A + M)p_1 + NCx_1 \\
x_1(0) &= h \\
p_1(0) &= k.
\end{aligned}
$$
(5. 8)

Considering the differences $x - x_1$ and $p - p_1$ we can make use of the assumption

$$
\rho(L, M, N) < \gamma^2
$$

to assert that there exists a number $\delta < \gamma$ such that

$$
\int_0^\infty (|H(x - x_1)|^2 + |L(p - p_1)|^2)\, dt - \gamma^2 \int_0^\infty |w|^2\, dt \le
$$
$$
-\delta^2 \int_0^\infty |w|^2\, dt
$$
(5. 9)

Furthermore, since $\mathcal{A}$ is exponentially stable

$$
\int_0^\infty (|x_1|^2 + |p_1|^2)\, dt \le C(|h|^2 + |k|^2)
$$

where C is a constant independant of h, k. Combining this estimate and (5. 9), it is easy to deduce the following

$$
\int_0^\infty (|Hx|^2 + |Lp|^2)\, dt - \gamma^2 \int_0^\infty |w|^2\, dt \le
$$
$$
-\delta_0^2 \int_0^\infty |w|^2\, dt + C_0(|h|^2 + |k|^2)
$$
(5. 10)

where $\delta_0 < \delta$ and C_0 is an appropriate constant. Let us set

$$
\mathcal{J}_{h,k}(w) = \int_0^\infty (|Hx|^2 + |Lp|^2)\, dt - \gamma^2 \int_0^\infty |w|^2\, dt.
$$

Note that in the functional $\mathcal{J}_{h,k}(w)$ the triple L, M, N is fixed and the control is w. The estimate (5. 10) shows easily that the functional $\mathcal{J}_{h,k}(w)$ is strictly concave and tends to $-\infty$ as $w \to \infty$. Therefore for any pair h, k there exists an optimal $\hat{w}_{h,k}$ which maximizes $\mathcal{J}_{h,k}(w)$ with respect to w. We denote by $\hat{x}_{h,k}, \hat{p}_{h,k}$ the corresponding optimal state. Let T be arbitrary and set $\bar{x}_T = \bar{x}(T), \bar{p}_T = \bar{p}(T)$ where $\bar{x}, \bar{p}$ is the solution of (5. 5). We now define

$$
\tilde{x}_T(t) = \left|
\begin{array}{ll}
\bar{x}(t) & \text{if } t \le T \\
\hat{x}_{\bar{x}_T, \bar{p}_T}(t - T) & \text{if } t > T
\end{array}
\right.
\qquad
\tilde{p}_T(t) = \left|
\begin{array}{ll}
\bar{p}(t) & \text{if } t \le T \\
\hat{p}_{\bar{x}_T, \bar{p}_T}(t - T) & \text{if } t > T
\end{array}
\right.
$$

and

$$
\tilde{w}_T(t) = \left|
\begin{array}{ll}
\frac{1}{\gamma^2} D^* P \bar{x}(t) & \text{if } t < T \\
\hat{w}_{\bar{x}_T, \bar{p}_T}(t - T) & \text{if } t > T
\end{array}
\right.
$$

By construction $\tilde{x}_T, \tilde{p}_T$ is the solution of (5. 7) corresponding to $\tilde{w}_T$. Therefore from (5. 10) we can write the estimate

$$
\int_0^\infty (|H\tilde{x}_T|^2 + |L\tilde{p}_T|^2)\, dt - \gamma^2 \int_0^\infty |\tilde{w}_T|^2\, dt \le
$$
$$
-\delta_0^2 \int_0^\infty |\tilde{w}_T|^2\, dt + C_0(|h|^2 + |k|^2)
$$
(5. 11)

Now we have

$$\int_T^\infty \left(|H\tilde{x}_T|^2 + |L\tilde{p}_T|^2 - \gamma^2|\tilde{w}_T|^2 \right) dt = \int_0^\infty \left(|H\hat{x}_{\bar{x}_T,\bar{p}_T}|^2 + |L\hat{p}_{\bar{x}_T,\bar{p}_T}|^2 - \gamma^2|\hat{w}_{\bar{x}_T,\bar{p}_T}|^2 \right) dt$$

which is by construction

$$\max_w \left[\int_0^\infty \left(|H x_{\bar{x}_T,\bar{p}_T}|^2 + |L p_{\bar{x}_T,\bar{p}_T}|^2 - \gamma^2|w|^2 \right) dt \right]$$

where we have denoted by $x_{\bar{x}_T,\bar{p}_T}$, $p_{\bar{x}_T,\bar{p}_T}$ the solution of (5. 7) corresponding to initial conditions $h = \bar{x}_T, k = \bar{p}_T$.Clearly this quantity is larger or equal to

$$\max_w \min_v J_{\bar{x}_T}(v,w) = (P\bar{x}_T, \bar{x}_T).$$

Therefore we have proved that

$$\int_T^\infty \left(|H\tilde{x}_T|^2 + |L\tilde{p}_T|^2 - \gamma^2|\tilde{w}_T|^2 \right) dt \geq (P\bar{x}_T, \bar{x}_T)$$

But from the Riccati equation (4. 4) and the first equation (5. 5) we have

$$(P\bar{x}_T, \bar{x}_T) = (Ph, h) + \int_0^T \left(|B^*P\bar{x} + L\bar{p}|^2 + \frac{1}{\gamma^2}|D^*P\bar{x}|^2 - |L\bar{p}|^2 - |H\bar{x}|^2 \right) dt.$$

Combining the two last relations we deduce

$$(5.\ 12) \qquad \int_0^\infty \left(|H\tilde{x}_T|^2 + |L\tilde{p}_T|^2 - \gamma^2|\tilde{w}_T|^2 \right) dt \geq (Ph, h) + \int_0^T |B^*P\bar{x} + L\bar{p}|^2 \, dt$$

Finally from (5. 11) and (5. 12) we get the estimate

$$(Ph, h) + \int_0^T |B^*P\bar{x} + L\bar{p}|^2 \, dt \leq -{\delta_0}^2 \int_0^\infty |\tilde{w}_T|^2 \, dt + C_0(|h|^2 + |k|^2)$$

Recalling the definition of $\tilde{w}_T$, we deduce in particular

$$\delta_0^2 \int_0^T |\frac{1}{\gamma^2} D^*P\bar{x}|^2 \, dt \leq C_0(|h|^2 + |k|^2).$$

Since T is arbitrary, we have proved that $D^*P\bar{x} \in L^2(-\infty;+\infty;W)$. Since $\mathcal{A}$ is exponentially stable, this suffices (see (5. 5) to prove (5.2).

-b:Consider the system

$$(5.\ 13) \qquad \begin{aligned} x' &= \left(A + \tfrac{1}{\gamma^2}DD^*P \right) x + BLp + Dw \\ p' &= (A + M)p + NCx + N\eta \\ x(0) &= 0 \\ p(0) &= 0. \end{aligned}$$

then we can prove

$$(5.\ 14) \qquad \sup_{w,\eta} \frac{\displaystyle\int_0^\infty |B^*Px + Lp|^2 \, dt}{\displaystyle\int_0^\infty (|w|^2 + |\eta|^2) \, dt} < \gamma^2.$$

This is obtained by computing $\frac{d}{dt}(Px, x)$ and integrating between 0 and T, then letting $T \to \infty$,and using the basic assumption $\rho(L, M, N) < \gamma^2$.

-c:duality considerations We again use the duality considerations of paragraph 4.3, see Proposition 4.1. Let $\Xi = X \times X$, $\Phi = W \times Y$, and this time $\Psi = U$. Let then

$$\mathcal{A} = \begin{pmatrix} A + \frac{1}{\gamma^{*2}}DD^*P & BL \\ NC & A + M \end{pmatrix}$$

and

$$\mathcal{B} = \begin{pmatrix} D & 0 \\ 0 & N \end{pmatrix} \quad \mathcal{C} = \begin{pmatrix} B^*P & L \end{pmatrix}.$$

The dual system is (apply (4. 8) and set $\zeta = m, q$, $\psi = \mu$)

$$(5.\ 15) \qquad \begin{aligned} m' &= (A^* + \tfrac{1}{\gamma^2}PDD^*)m + C^*N^*q + PB\mu \\ q' &= L^*B^*m + (A^* + M^*)q + L^*\mu \\ m(0) &= 0 \\ q(0) &= 0. \end{aligned}$$

Using Proposition 4. 1 and (5. 14) we can assert that

$$(5.\ 16) \qquad \sup_{\mu} \frac{\displaystyle\int_0^\infty (|D^*m|^2 + |N^*q|^2)\, dt}{\displaystyle\int_0^\infty |\mu|^2\, dt} < \gamma^2.$$

Consider the dynamic system

$$(5.\ 17) \qquad \begin{aligned} m' &= (A^* + \tfrac{1}{\gamma^{*2}}PDD^*)m + C^*v + PB\mu \\ m(0) &= 0 \end{aligned}$$

where v is the control and μ is the perturbation. We observe that the pair $A^* + \frac{1}{\gamma^{*2}}PDD^*, C^*$ is stabilizable, as a consequence of the fact that the operator $\mathcal{A}_P$, hence its dual $\mathcal{A}_P^*$ is exponentially stable (see part -a of the present proof and beware of the fact $\mathcal{A}_P$ has been designated now by $\mathcal{A}$ by consistency with the generic notation used when dealing with duality considerations). Consider also the cost function

$$\mathcal{K}_0(v,\mu) = \int_0^\infty (|D^*m|^2 + |v|^2)\, dt.$$

From the assumption that A^*, D^* is detectable it follows that the pair $A^* + \frac{1}{\gamma^2}PDD^*, D^*$ is detectable. Using (5. 16) we can assert that the γ^2 robustness property holds for the system (5. 17) and the cost function $\mathcal{K}_0(v,\mu)$. Therefore, we may rely on Theorem 3.1 to obtain the existence and uniqueness of a self adjoint operator $\Pi \in \mathcal{L}(X, X) \geq 0$, solution of the Riccati equation

$$(5.\ 18) \qquad \Pi(A^* + \frac{1}{\gamma^2}PDD^*) + (A + \frac{1}{\gamma^2}DD^*P)\Pi - \Pi(C^*C - \frac{1}{\gamma^2}PBB^*P)\Pi + DD^* = 0$$

and $A^* + \frac{1}{\gamma^2}PDD^* - (C^*C - \frac{1}{\gamma^2}PBB^*P)\Pi$ is exponentially stable. Note also that as in Lemma 3.1 we have also $A^* + \frac{1}{\gamma^2}PDD^* - C^*C\Pi$ is exponentially stable.

- d:algebraic manipulations We check here that

$$(5.\ 19) \qquad \Sigma = \Pi(I - \frac{1}{\gamma^2}P\Sigma)$$

If (5. 19) is true then clearly (using also the symmetry of Σ)

$$(I - \frac{1}{\gamma^2}P\Sigma)(I + \frac{1}{\gamma^2}P\Pi) = (I + \frac{1}{\gamma^2}P\Pi)(I - \frac{1}{\gamma^2}P\Sigma) = I$$

which proves that $I - \frac{1}{\gamma^2} P\Sigma$ is invertible .Moreover

$$\Pi = \Sigma(I - \frac{1}{\gamma^2} P\Sigma)^{-1} \geq 0$$

and the proof of (4. 6) will then be complete. To prove (5. 19) we proceed by algebraic manipulations combining the Riccati equations of P, Σ, Π namely (4. 4),(4. 5) and (5. 18). To simplify notation write

$$\Lambda = -\Sigma + \Pi(I - \frac{1}{\gamma^2} P\Sigma)$$

$$A_1 = A^* - (C^*C - \frac{1}{\gamma^2} H^*H)\Sigma$$

$$A_2 = A^* + \frac{1}{\gamma^2} PDD^* - (C^*C - \frac{1}{\gamma^2} PBB^*P)\Pi$$

Then we can check after an easy calculus

(5. 20) $$\qquad\qquad\qquad \Lambda A_1 + A_2^* \Lambda = 0.$$

Note that A_1, A_2 are exponentially stable.Then the relation (5. 20) implies $\Lambda = 0$ and (5. 19) has been proven.

5.2   Sufficient Conditions

So we assume that (4. 4),(4. 5),(4. 6) hold. We set

$$\Pi = \Sigma(I - \frac{1}{\gamma^2} P\Sigma)^{-1}.$$

Note that Π is symmetric. This follows from the relation

$$\Sigma(I - \frac{1}{\gamma^2} P\Sigma) = (I - \frac{1}{\gamma^2} \Sigma P)\Sigma.$$

By assumption we know that $\Pi \geq 0$.Moreover using the notation A_1, A_2, Λ as above, we have the relation (5. 20) since $\Lambda = 0$. After using the equations of Σ and P we deduce easily that the left hand side of (5. 18) multiplied to the right by $I - \frac{1}{\gamma^2} P\Sigma$ vanishes. From the invertibility of $I - \frac{1}{\gamma^2} P\Sigma$ we deduce that Π is a solution of the left hand side of (5. 18). Details to make precise this formal calculation are left to the reader. Note also that

$$A_2 = (I - \frac{1}{\gamma^2} P\Sigma)A_1(I - \frac{1}{\gamma^2} P\Sigma)^{-1}.$$

Consider the equation

$$x' = A_2 x \quad ; x(0) = h.$$

Then setting

$$x_1 = (I - \frac{1}{\gamma^2} P\Sigma)^{-1} x$$

we have

$$x_1' = A_1 x_1 \quad ; x_1(0) = (I - \frac{1}{\gamma^2} P\Sigma)^{-1} h.$$

Since we know that A_1 is exponentially stable we deduce that $x_1 \in L^2(-\infty; +\infty; X)$ hence also $x \in L^2(-\infty; +\infty; X)$.Therefore we get that A_2 is exponentially stable. Hence the operator Π satisfies all the properties stated in the part -c of the proof of (4. 6) in the necessary conditions, paragraph 5.1.

We define next L, M, N as in the statement of Theorem 4.1, definition of the feedback controller and we shall prove that this feedback controller satisfies the conditions of γ^2 robustness property(with partial observation), as stated in Definition 4.1. We associate to the triple L, M, N the matrix operator $\mathcal{A}$ as in the proof of (4. 4)in paragraph 5.1.With the present choice of L, M, N it amounts to

$$\mathcal{A} = \begin{pmatrix} A & -BB^*P \\ \Pi C^*C & A - (BB^* - \frac{1}{\gamma^2}DD^*)P - \Pi C^*C \end{pmatrix}$$

We prove that

(5. 21) $\qquad\qquad\qquad\qquad \mathcal{A}$ is exponentially stable.

We decompose the proof in several steps.

-a:The matrix operator

$$\mathcal{A}_P = \begin{pmatrix} A + \frac{1}{\gamma^{*2}}DD^*P & -BB^*P \\ \Pi C^*C & A - (BB^* - \frac{1}{\gamma^2}DD^*)P - \Pi C^*C \end{pmatrix}$$

is exponentially stable. Consider indeed the dynamic system

$$(5.\ 22) \qquad \begin{aligned} x' &= (A + \tfrac{1}{\gamma^{*2}}DD^*P)x - BB^*Pp \\ p' &= \left(A - (BB^* - \tfrac{1}{\gamma^2}DD^*)P - \Pi C^*C\right)p + \Pi C^*Cx \\ x(0) &= h \\ p(0) &= k. \end{aligned}$$

Setting

$$\xi = x - p$$

we see that ξ is the solution of

$$(5.\ 23) \qquad \begin{aligned} \xi' &= (A + \tfrac{1}{\gamma^2}DD^*P - \Pi C^*C)\xi \\ \xi(0) &= h - k. \end{aligned}$$

Since $A + \frac{1}{\gamma^2}DD^*P - \Pi C^*C$ is exponentially stable we get that $\xi \in L^2(-\infty; +\infty; X)$. Next x appears as the solution of

$$(5.\ 24) \qquad \begin{aligned} x' &= (A + \tfrac{1}{\gamma^2}DD^*P - BB^*P)x + BB^*P\xi \\ x(0) &= h. \end{aligned}$$

Since $A + \frac{1}{\gamma^2}DD^*P - BB^*P$ is exponentially stable we deduce that $x \in L^2(-\infty; +\infty; X)$ and thus also $p \in L^2(-\infty; +\infty; X)$. This completes the proof of -a.

- b:consider the system

$$(5.\ 25) \qquad \begin{aligned} x' &= (A + \tfrac{1}{\gamma^{*2}}DD^*P)x - BB^*Pp + Dw \\ p' &= \left(A - (BB^* - \tfrac{1}{\gamma^2}DD^*)P - \Pi C^*C\right)p + \Pi C^*Cx + \Pi C^*\eta \\ x(0) &= h \\ p(0) &= k \end{aligned}$$

then one has the estimate

$$\begin{aligned} &\int_0^\infty (|Hx|^2 + |B^*Pp|^2)\,dt \quad - \quad \gamma^2 \int_0^\infty \left(|w + \tfrac{1}{\gamma^{*2}}D^*Px|^2 + |\eta|^2\right) dt \\ (5.\ 26) \qquad\qquad\qquad &< \quad -\delta_0^2 \int_0^\infty (|w|^2 + |\eta|^2)\,dt + C_0(|h|^2 + |k|^2) \end{aligned}$$

where δ_0 and C_0 are appropriate positive constants. To prove the estimate (5. 26) we shall exploit the Riccati equation (5. 18) whose solution is Π. Consider the system

$$(5.\ 27) \qquad \begin{aligned} m' &= (A^* + \tfrac{1}{\gamma^{*2}} PDD^*)m + C^*v + PB\mu \\ m(0) &= 0 \end{aligned}$$

where v is the control and μ is the perturbation. The pair $A^* + \tfrac{1}{\gamma^{*2}} PDD^*, C^*$ is stabilizable. Consider the cost function

$$\mathcal{K}_0(v,\mu) = \int_0^\infty (|D^*m|^2 + |v|^2)\, dt.$$

The pair $A^* + \tfrac{1}{\gamma^2} PDD^*, D^*$ is detectable. The existence of Π implies that the γ^2 robustness property with full observation holds for the system (5. 27) and the cost function $\mathcal{K}_0(v,\mu)$. In fact consider

$$(5.\ 28) \qquad \begin{aligned} m' &= (A^* + \tfrac{1}{\gamma^{*2}} PDD^* - C^*C\Pi)m + PB\mu \\ m(0) &= 0 \end{aligned}$$

then one has the property

$$\sup_{\mu} \frac{\displaystyle\int_0^\infty (|D^*m|^2 + |C\Pi m|^2)\, dt}{\displaystyle\int_0^\infty |\mu|^2\, dt} < \gamma^2.$$

Using duality considerations we introduce the dual system

$$(5.\ 29) \qquad \begin{aligned} \xi' &= (A + \tfrac{1}{\gamma^2} DD^*P - \Pi C^*C)\xi + Dw + \Pi C^*\eta \\ \xi(0) &= 0. \end{aligned}$$

Then we can assert the estimate

$$(5.\ 30) \qquad \sup_{w,\eta} \frac{\displaystyle\int_0^\infty |B^*P\xi|^2\, dt}{\displaystyle\int_0^\infty (|w|^2 + |\eta|^2)\, dt} < \gamma^2.$$

In particular, we can find $\delta < \gamma$ such that

$$(5.\ 31) \qquad \frac{\displaystyle\int_0^\infty |B^*P\xi|^2\, dt}{\displaystyle\int_0^\infty (|w|^2 + |\eta|^2)\, dt} \le \gamma^2 - \delta^2.$$

Consider now the system (5. 25) for initial values $h = 0; k = 0$. We denote by x_0, p_0 the corresponding solution. We see that the solution ξ of (5. 29) is equal to $x_0 - p_0$. We shall also use the following relation

$$\int_0^\infty \left(|Hx_0|^2 + |B^*Pp_0|^2 - |B^*P\xi|^2 + \gamma^2(|w|^2 - |w + \tfrac{1}{\gamma^2} D^*Px_0|^2) \right)\, dt = 0$$

which is obtained by computing $\tfrac{d}{dt}(Px_0, x_0)$ and integrating between 0 and ∞. In this equality, we make use of the estimate (5. 31) to obtain

$$(5.\ 32) \quad \int_0^\infty \left(|Hx_0|^2 + |B^*Pp_0|^2 - \gamma^2(|\eta|^2 + |w + \tfrac{1}{\gamma}{}^2 D^*Px_0|^2) \right)\, dt \le -\delta^2 \int_0^\infty (|w|^2 + |\eta|^2)\, dt.$$

Introduce $x_1 = x - x_0, p_1 = p - p_0$ which depend only on h, k and not on w, η. We replace in (5. 26) x by $x_0 + x_1$ and p by $p_0 + p_1$. Using inequalities like $|Hx|^2 \le (1+\epsilon)|Hx_0|^2 + (1+\tfrac{1}{\epsilon})|Hx_1|^2$ where ϵ

is arbitrarily small and making use of (5. 32) we easily deduce the desired estimate (5. 26).
- c:Proof of (5. 21) Consider the system

$$(5. 33) \qquad \begin{aligned} \bar{x}' &= A\bar{x} - BB^*P\bar{p} \\ \bar{p}' &= \left(A - (BB^* - \tfrac{1}{\gamma^2}DD^*)P - \Pi C^*C\right)\bar{p} + \Pi C^*C\bar{x} \\ \bar{x}(0) &= h \\ \bar{p}(0) &= k. \end{aligned}$$

then we must prove
$$(5. 34) \qquad \bar{x}, \bar{p} \in L^2(-\infty; +\infty; X).$$

This is done in a way similar to that of .
To complete the proof of the γ^2 robustness property(with partial observation) for the triple L, M, N consider the system

$$(5. 35) \qquad \begin{aligned} x' &= Ax - BB^*Pp + Dw \\ p' &= \left(A - (BB^* - \tfrac{1}{\gamma^2}DD^*)P - \Pi C^*C\right)p + \Pi C^*Cx + \Pi C^*\eta \\ x(0) &= 0 \\ p(0) &= 0 \end{aligned}$$

we must prove

$$(5. 36) \qquad \sup_{w,\eta} \frac{\displaystyle\int_0^\infty (|Hx|^2 + |B^*Px|^2)\, dt}{\displaystyle\int_0^\infty (|w|^2 + |\eta|^2)\, dt} < \gamma^2.$$

Apply the estimate (5. 26) to (5. 35) with $w = w - \tfrac{1}{\gamma^2}D^*Px$ and $h = 0, k = 0$ to obtain

$$(5. 37) \qquad \begin{aligned} \int_0^\infty &\left(|Hx|^2 + |B^*Pp|^2 - \gamma^2(|w|^2 + |\eta|^2)\right) dt \leq \\ &-\delta_0^2 \int_0^\infty \left(|w - \tfrac{1}{\gamma^2}D^*Px|^2 + |\eta|^2\right) dt \end{aligned}$$

hence

$$\sup_{w,\eta} \frac{\displaystyle\int_0^\infty (|Hx|^2 + |B^*Px|^2)\, dt}{\displaystyle\int_0^\infty (|w|^2 + |\eta|^2)\, dt} \leq$$

$$\gamma^2 - \delta_0^2 \inf_{w,\eta} \frac{\displaystyle\int_0^\infty \left(|w - \tfrac{1}{\gamma^2}D^*Px|^2 + |\eta|^2\right) dt}{\displaystyle\int_0^\infty (|w|^2 + |\eta|^2)\, dt}$$

Using the exponential stability of $\mathcal{A}_P$ we can check as done previously that

$$\inf_{w,\eta} \frac{\displaystyle\int_0^\infty \left(|w - \tfrac{1}{\gamma^2}D^*Px|^2 + |\eta|^2\right) dt}{\displaystyle\int_0^\infty (|w|^2 + |\eta|^2)\, dt} > 0.$$

Therefore (5. 36) has been proven.
The proof of Theorem 4.1 has been completed.

References

[1] Bert van Keulen, Marc Peters and Ruth Curtain, H_∞ Control with state feedback :The infinite dimensional case, W-9015, University of Gronigen.

[2] Bert van Keulen, H_∞ Control with measurement feedback for linear infinite-dimensional systems,W-9103,University of Gronigen.

[3] Richard Datko ,Extending a theorem of A.M. Liapunov to Hilbert space,*Journal of Mathematical analysis and applications*,vol.32,pp. 610-616,1970.

[4] Tamer Başar, Pierre Bernhard, H_∞ *Optimal Control and Related Minimax Design Problems. A Dynamic Game Approach*, Birkhäuser, Boston, 1991.

[5] A. Bensoussan, G. Da Prato, M. Delfour, S. Mitter, *Infinite Dimensional System Theory*, Birkhäuser, Boston, to be published.

The dynamic programming method

Optimality and Characteristics of Hamilton-Jacobi-Bellman Equations

Nathalie Caroff & Hélène Frankowska

Abstract. We study the Bolza problem arising in nonlinear optimal control and investigate under what circumstances the necessary conditions for optimality of Pontryagin's type are also sufficient. This leads to the question when shocks do not occur in the method of characteristics applied to the associated Hamilton-Jacobi-Bellman equation. In this case the value function is its (unique) continuously differentiable solution and can be obtained from the canonical equations. In optimal control this corresponds to the case when the optimal trajectory of the Bolza problem is unique for every initial state and the optimal feedback is an upper semicontinuous set-valued map with convex, compact images.

1   Introduction

This paper is concerned with the Hamilton-Jacobi equation

$$(1) \qquad -\frac{\partial V}{\partial t} + H\left(x, -\frac{\partial V}{\partial x}\right) = 0, \quad V(T,\cdot) = \varphi(\cdot)$$

associated to the Bolza type problem in optimal control:

$$(2) \qquad \text{minimize} \int_{t_0}^{T} L(x(t), u(t))dt + \varphi(x(T))$$

over solution-control pairs (x, u) of control system

$$(3) \qquad \begin{cases} x'(t) &= f(x(t)) + g(x(t))u(t), \quad u(t) \in U \\ x(t_0) &= x_0 \end{cases}$$

where U is a finite dimensional space and

$$H(x, p) = \sup_{u \in U} \langle p, f(x) + g(x)u \rangle - L(x, u)$$

In general H is not differentiable, but here we shall restrict our attention only to problems with smooth Hamiltonians.

The characteristics of the Hamilton-Jacobi-Bellman equation (1) are solutions to the *Hamiltonian system*

$$(4) \qquad \begin{cases} x'(t) &= \dfrac{\partial H}{\partial p}(x(t), p(t)), \quad x(T) = x_T \\[2mm] -p'(t) &= \dfrac{\partial H}{\partial x}(x(t), p(t)), \quad p(T) = -\nabla\varphi(x_T) \end{cases}$$

Such system is also called "canonical equations" or "equations of the extremals" in optimal control theory, since the Pontryagin maximum principle claims that if $x : [t_0, T] \to \mathbf{R}^n$ is optimal for problem (2), (3), then there exists $p : [t_0, T] \to \mathbf{R}^n$ such that (x, p) solves (4) with $x_T = x(T)$. This is not however a sufficient condition for optimality because it may happen that to a given $x_0 \in \mathbf{R}^n$ corresponds a solution (x, p) of (4) with $x(t_0) = x_0$ and x is not optimal. If such is the case and the optimal solution to (2), (3) does exist, then by the maximum principle, we can find another solution (x_1, p_1) of (4) with $x_1(t_0) = x_0$ and $p_1(t_0) \neq p(t_0)$.

The situation when there are two solutions (x_i, p_i), $i = 1, 2$ of (4) satisfying $x_i(t_0) = x_0$ and $p_1(t_0) \neq p_2(t_0)$ is called *shock* arising in the method of characteristics.

If shocks never occur on the time interval $[0, T]$, then the solution of (1) can be constructed by considering all trajectories (x, p) of (4) and setting

$$V(t_0, x(t_0)) = \varphi(x(T)) + \int_{t_0}^{T} L(x(t), u(t))dt$$

where $u(t) \in U$ is such that

$$H(x(t), p(t)) = \langle p(t), f(x(t)) + g(x(t))u(t) \rangle - L(x(t), u(t)) \quad \text{a.e. in } [t_0, T]$$

Then, by [3], V is continuously differentiable,

$$\frac{\partial V}{\partial x}(t, x(t)) = -p(t) \quad \& \quad \frac{\partial V}{\partial t}(t, x(t)) = H(x(t), p(t))$$

Furthermore V is the so called value function of our optimal control problem. In summary if we can guarantee that on some time interval $[t_0, T]$ there is no shocks, then the value function would be the continuously differentiable on $[t_0, T] \times \mathbf{R}^n$ solution to (1).

It is well known that (unfortunately) shocks do happen. This is the very reason why the value function is nonsmooth and why one should not expect to have smooth solutions. Also it was shown in [4] and [3] that the value function is not regularly differentiable at some point (t_0, x_0) if and only if the optimal trajectory of the control problem (2), (3) is not unique.

Thus if we provide conditions that guarantee the absence of shocks in the same time we get the useful information about uniqueness of optimal solutions. Furthermore, under the same assumptions as in [3] we get the optimal feedback low on $[t_0, T] \times \mathbf{R}^n$:

$$U(t, x) = \{u \in U \mid H(x, -\frac{\partial V}{\partial x}(t, x)) = \langle -\frac{\partial V}{\partial x}(t, x), f(x) + g(x)u \rangle - L(x, u)\}$$

with the set-valued map $U(\cdot)$ being upper semicontinuous with convex compact images. In this case there exists also exactly one solution of

$$x' = f(x) + g(x)u(t, x), \ u(t, x) \in U(t, x), \quad x(t_0) = x_0$$

and it is optimal for problem (2), (3).

It was proved in [3] that the shocks would not occur till time t_0 if for every (x, p) solving (4) on $[t_0, T]$ the matrix Riccati equation

$$(5) \quad \begin{cases} P' \; + \; \dfrac{\partial^2 H}{\partial p \partial x}(x(t), p(t))P \; + \; P\dfrac{\partial^2 H}{\partial x \partial p}(x(t), p(t)) + \\[2mm] \qquad + \; P\dfrac{\partial^2 H}{\partial p^2}(x(t), p(t))P \; + \; \dfrac{\partial^2 H}{\partial x^2}(x(t), p(t)) \; = \; 0 \\[2mm] P(T) \; = \; -\varphi''(x(T)) \end{cases}$$

has a solution on $[t_0, T]$.

In this paper we provide some sufficient conditions for global solvability of the above Riccati equation for all (x, p) verifying (4).

In Section 2 we recall some results from [3]. Section 3 is devoted to few useful informations about the matrix Riccati equations. In particular (5) is reduced to a much simpler equation

$$S' + S^2 + D(t) = 0, \quad S(T) = S_T$$

where $D(t)$, S_T are defined from the coefficients of (5) and which is much simpler to investigate. In Section 4 we provide some applications to the optimal control problem mentioned above.

2 Matrix Riccati Equations and Shocks

In this section we recall some results concerning differentiability of the value function and shocks of the Hamilton-Jacobi-Bellman equation (1).

Consider the Bolza problem in the nonlinear optimal control setting:

$$(P) \qquad \min \int_{t_0}^{T} L(x(t), u(t))dt \; + \; \varphi(x(T))$$

over solution-control pairs (x, u) of control system

$$(6) \qquad \begin{cases} x'(t) \; = \; f(x(t)) + g(x(t))u(t), \quad u(\cdot) \in L^1(t_0, T; \mathbf{R}^m) \\ x(t_0) \; = \; x_0 \end{cases}$$

where $t_0 \in [0, T]$, $x_0 \in \mathbf{R}^n$, $f : \mathbf{R}^n \mapsto \mathbf{R}^n$, $g : \mathbf{R}^n \mapsto L(\mathbf{R}^m, \mathbf{R}^n)$, $L : \mathbf{R}^n \times \mathbf{R}^m \mapsto \mathbf{R}$, $\varphi : \mathbf{R}^n \mapsto \mathbf{R}$.

We associate to these data the *Hamiltonian* H defined on $\mathbf{R}^n \times \mathbf{R}^n$ by

$$H(x, p) = \sup_{u} \langle p, f(x) + g(x)u \rangle - L(x, u)$$

If H is differentiable, then the *Hamiltonian system*

$$(7) \qquad \begin{cases} x'(t) \; = \; \dfrac{\partial H}{\partial p}(x(t), p(t)) \\[4mm] -p'(t) \; = \; \dfrac{\partial H}{\partial x}(x(t), p(t)) \end{cases}$$

is called *complete* if for all $t_0 \in [0,T]$, x_0, $p_0 \in \mathbf{R}^n$ it has a solution (x,p) defined on $[0,T]$ and satisfying $x(t_0) = x_0$, $p(t_0) = p_0$.

We impose the following assumptions:

$H_1)$ f and g are differentiable, locally Lipschitz and have linear growth:

$$\exists\, M \geq 0,\ \forall\, x \in \mathbf{R}^n,\ \ \|f(x)\| + \|g(x)\| \leq M(\|x\| + 1)$$

$H_2)$ $\varphi \in C^1$, $\liminf_{\|x\| \to \infty} \varphi(x) = +\infty$,

$H_3)$ $L(x,\cdot)$ is continuous, convex, $\exists\, c > 0$, $\forall\, (x,u) \in \mathbf{R}^n \times \mathbf{R}^m$, $L(x,u) \geq c\,\|u\|^2$. Furthermore for all $r > 0$, there exists $k_r \geq 0$ such that

$$\forall\, u \in \mathbf{R}^m,\ \ L(\cdot,u)\ \text{is differentiable and}\ \ k_r - \text{Lipschitz on}\ \ B_r(0)$$

$H_4)$ The Hamiltonian H is differentiable, its gradient $\nabla H(\cdot,\cdot)$ is locally Lipschitz and the Hamiltonian system (7) is complete.

We denote by $x(\cdot; t_0, x_0, u)$ the solution to (6) starting at time t_0 from the initial state x_0 and corresponding to the control $u(\cdot)$.

The value function associated to this problem is given by

$$V(t_0, x_0)\ =\ \inf_{u \in L^1(t_0,T)} \int_{t_0}^{T} L(x(t; t_0, x_0, u), u(t))dt\ +\ \varphi(x(T; t_0, x_0, u))$$

where (t_0, x_0) range over $[0,T] \times \mathbf{R}^n$. It is well known that whenever V is differentiable, it satisfies the Hamilton-Jacobi-Bellman equation (1). The following result was proved in [3]:

Theorem 2.1 *Assume that* $H_1) - H_4)$ *hold true. Then the following three statements are equivalent:*

i) The value function V is continuously differentiable
ii) $\forall\, (t_0, x_0) \in [0,T] \times \mathbf{R}^n$ the optimal trajectory to problem (P) is unique
iii) The system (4) does not exhibit shocks on $[0,T]$.

Furthermore, if one of the above (equivalent) statements holds true, then any solution (x,p) to (4) satisfies:
for all $t \in [0,T]$, $p(t) = -\frac{\partial V}{\partial x}(t, x(t))$ and x restricted to $[t_0, T]$ is optimal for problem (P) with $x_0 = x(t_0)$.

The above implies that whenever shocks do not occur on $[t_0, T]$, then the Pontryagin's necessary conditions for optimality of a solution $\bar{x}(\cdot)$ to (6): there exists $p : [t_0, T] \to \mathbf{R}^n$ such that $(\bar{x}, p)$ solves (4) on $[t_0, T]$ with $x_T = \bar{x}(T)$ are also sufficient.

It was observed in [3] that if φ, H are twice continuously differentiable and H'' is locally Lipschitz, then $V(t,\cdot) \in C^2$ for all $t \in [0,T]$ if and only if for every (x,p) solving (4) on $[0,T]$ the equation (5) has a solution on $[0,T]$. Since (5) describes the evolution of the tangent space to the set $\text{Graph}(-\frac{\partial V}{\partial x}(t,\cdot))$ at $(x(t), p(t))$ in the sense that $\text{Graph}(P(t))$ is tangent to this set at $(x(t), p(t))$, $-\frac{\partial^2 V}{\partial x^2}(t, x(t))$ solves the Riccati differential equation (5) on $[0,T]$.

3   Properties of Solutions to Riccati Equations

By the classical theory of Riccati equations (5) if for all $(x,p) \in \mathbf{R}^n \times \mathbf{R}^n$, $\frac{\partial^2 H}{\partial x^2}(x,p) \leq 0$ and $\varphi'' \geq 0$ (i.e. φ is convex), then the solution $P(\cdot)$ to (5) exists on $[0,T]$ for every choice of continuous $(x(\cdot), p(\cdot))$.

We recall next two comparison properties for solutions of Riccati equations. Results of a similar nature can be found in [2], [7], [5].

Proposition 3.1 *Let* $A, E_i, D_i : [0,T] \mapsto L(\mathbf{R}^n, \mathbf{R}^n)$, $i = 1,2$ *be integrable. We assume that* $E_i(t)$, $D_i(t)$ *are self-adjoint for almost every* $t \in [0,T]$ *and*

$$D_1(t) \leq D_2(t), \quad E_1(t) \leq E_2(t) \ \text{a.e. in} \ [0,T]$$

Consider self-adjoint operators $P_{iT} \in L(\mathbf{R}^n, \mathbf{R}^n)$ *such that* $P_{1T} \leq P_{2T}$ *and solutions* $P_i(\cdot) : [t_0, T] \mapsto L(\mathbf{R}^n, \mathbf{R}^n)$ *to the equations*

$$P' + A(t)^* P + PA(t) + PE_i(t)P + D_i(t) = 0, \quad P_i(T) = P_{iT}$$

Then $P_1 \leq P_2$ *on* $[t_0, T]$.

The result we provide below can be proved also in more general situations. However it is sufficient for our purposes.

Proposition 3.2 *Let* $E_i, D_i : [0,T] \mapsto L(\mathbf{R}^n, \mathbf{R}^n)$, $i = 1,2$ *be integrable. We assume that* $E_i(t)$, $D_i(t)$ *are self-adjoint for almost every* $t \in [0,T]$, $E_1 \in L^\infty(0,T)$ *and*

$$D_1(t) \leq D_2(t), \quad 0 \leq E_1(t) \leq E_2(t) \ \text{a.e. in} \ [0,T]$$

Consider self-adjoint operators $P_{iT} \in L(\mathbf{R}^n, \mathbf{R}^n)$ *such that* $P_{1T} \leq P_{2T}$ *and solutions* $P_i(\cdot) : [t_i, T] \mapsto L(\mathbf{R}^n, \mathbf{R}^n)$ *to the equations*

$$P' + PE_i(t)P + D_i(t) = 0, \quad P_i(T) = P_{iT}$$

Then the solution P_1 *can be extended at least to* $[t_2, T]$.

Corollary 3.3 *Let* $E, D : [0,T] \mapsto L(\mathbf{R}^n, \mathbf{R}^n)$ *be integrable. Assume that* $E(t)$, $D(t)$ *are self-adjoint,* $E(t) \geq 0$ *for almost every* $t \in [0,T]$ *and* $E \in L^\infty(0,T)$. *Consider a self-adjoint operator* $P_T \in L(\mathbf{R}^n, \mathbf{R}^n)$ *and assume that there exists an absolutely continuous* $P : [t_0, T] \mapsto L(\mathbf{R}^n, \mathbf{R}^n)$ *such that for every* $t \in [t_0, T]$, $P(t)$ *is self-adjoint and*

$$P'(t) + P(t)E(t)P(t) + D(t) \leq 0 \ \text{a.e. in} \ [t_0, T], \quad P_T \leq P(T)$$

Then the solution $\overline{P}$ *to the equation*

$$P' + PE(t)P + D(t) = 0, \quad P(T) = P_T$$

is defined at least on $[t_0, T]$ *and* $\overline{P} \leq P$ *on* $[t_0, T]$.

Our next aim is to associate to the Riccati equation

$$(8) \qquad P' + A(t)^{*}P + PA(t) + PE(t)P + D(t) = 0, \quad P(T) = P_T$$

a new equation

$$(9) \qquad S' + S^2 + C(t) = 0, \quad S(T) = S_T$$

in such way that the existence of the solution to (9) on $[t_0, T]$ implies that of (8).

Theorem 3.4 *Consider an absolutely continuous* $A : [0,T] \mapsto L(\mathbf{R}^n, \mathbf{R}^n)$, *an integrable* $D : [0,T] \mapsto L(\mathbf{R}^n, \mathbf{R}^n)$, $E, P_T \in L(\mathbf{R}^n, \mathbf{R}^n)$ *and* $t_0 \in [0,T]$. *We assume that*

$$EA(t)^{*} = A(t)E \quad \text{for almost every } t \in [0,T]$$

Then the solution to

$$(10) \qquad P' + A(t)^{*}P + PA(t) + PEP + D(t) = 0, \quad P(T) = P_T$$

exists on $[t_0, T]$ *if and only if so does the solution to*

$$(11) \qquad S' + S^2 + ED(t) - A'(t) - A(t)^2 = 0, \quad S(T) = A(T) + EP_T$$

Proof — Let P solves (10) on $[t_0, T]$. Set $S(t) = A(t) + EP(t)$. Then differentiating this expression we get

$$S'(t) = A'(t) - E\left(A(t)^{*}P(t) + P(t)A(t) + P(t)E(t)P(t) + D(t)\right)$$

$$= -(A(t) + EP(t))^2 + A(t)^2 + A'(t) - ED(t)$$

Thus S solves (11). Conversely let

$$t_1 = \inf_{t \in [0,T]} \{ \text{ The solution } P \text{ to } (10) \text{ is defined on } [t,T] \}$$

and S solves (11) on $[t_0, T]$. It is enough to prove that if $t_0 \leq t_1$, then P can be extended by continuity to t_1, that is it can happen only if $t_0 = t_1 = 0$. So let $t_0 \leq t_1$. From the first part of the proof and uniqueness of solution we know that for every $t \in]t_1, T]$, $S(t) = A(t) + EP(t)$. Hence $\sup_{t \in [t_1, T]} \|EP(t)\| < \infty$. Integrating (10) we deduce that for all $x \in \mathbf{R}^n$ with $\|x\| \leq 1$

$$\|P(t)x\| \leq \|P_T\| + \int_t^T \|D(s)\| \, ds + \int_t^T \left(2\|A(t)\| + \|EP(t)\|\right)\|P(t)\| \, dt$$

Since x is an arbitrary element of the unit ball we proved that for some $c \geq 0$

$$\|P(t)\| \leq c + \int_t^T c\,\|P(t)\| \, dt$$

This and the Gronwall lemma imply that $\sup_{t \in]t_1, T]} \|P(t)\| < \infty$. Thus P can be extended on $[t_1, T]$. ∎

Corollary 3.5 *Under all the assumptions of Theorem 3.4, suppose P_T is self-adjoint and that for almost every $t \in [0,T]$, $A'(t) + A(t)^2 - ED(t)$ is self-adjoint. If there exists $a \in \mathbf{R}$ such that*

$$\text{for a.e. } t \in [0,T], \ A'(t) + A(t)^2 - ED(t) \geq a^2 I \quad \& \quad A(T) + EP_T \leq aI$$

then the solution to the Riccati equation (10) is defined on $[0,T]$.

Proof — By Theorem 3.4 it is enough to show that the problem (11) has a solution on $[0,T]$. For all $t \in [0,T]$, set $S(t) = aI$. Then

$$S'(t) + S(t)^2 + ED(t) - A'(t) - A(t)^2 \ \leq \ 0, \ \ S(T) = aI$$

Corollary 3.3 ends the proof. ∎

The result we provide below is related to [7, Chapter 7].

Theorem 3.6 *Consider $E : [0,T] \mapsto L(\mathbf{R}^n, \mathbf{R}^n)$ such that for some $\omega > 0$ and a.e. $t \in [0,T]$, $E(t) \geq \omega I$ and is self-adjoint. We assume that the square root of $E(t)$, denoted by $C(t)$, is absolutely continuous. Let $A : [0,T] \mapsto L(\mathbf{R}^n, \mathbf{R}^n)$ be absolutely continuous, $D : [0,T] \mapsto L(\mathbf{R}^n, \mathbf{R}^n)$ be integrable and $P_T \in L(\mathbf{R}^n, \mathbf{R}^n)$. Then the solution to (8) exists on $[t_0, T]$ if and only if so does the solution to*

$$S' + S^2 + \mathcal{D}(t) = 0, \ S(T) = \frac{1}{2}(A_1(T) + A_1(T)^\star) + C(T)^\star P_T C(T)$$

where

$$A_1(t) = C(t)^{-1}A(t)C(t) - C(t)^{-1}C'(t), \ D_1(t) = C(t)^\star D(t)C(t)$$

$$\mathcal{D}(t) \ = \ \frac{1}{4}X(t)^\star \left(A_1(t)A_1(t)^\star - A_1(t)^2 - A_1(t)^{\star^2} \right) X(t) +$$

$$+ \ X(t)^\star (D_1(t) - \tfrac{1}{2}A_1'(t) - \tfrac{1}{2}A_1'(t)^\star - \tfrac{3}{4}A_1(t)^\star A_1(t))X(t)$$

and $X(\cdot)$ denotes the solution to

$$X' \ = \ \frac{1}{2}(A_1(t) - A_1(t)^\star)X, \ \ X(T) = Id$$

Proof — ⋅ Let P solves (8) on $[t_0, T]$. Set $R(t) = C(t)^\star P(t)C(t)$. Differentiating this relation we obtain

$$R'(t) = C'(t)^\star P(t)C(t) + C(t)^\star P(t)C'(t) -$$

$$- C(t)^\star \left(A(t)^\star P(t) + P(t)A(t) + P(t)E(t)P(t) + D(t) \right) C(t)$$

$$= C'(t)^\star C(t)^{\star^{-1}} R(t) + R(t)C(t)^{-1}C'(t) - C(t)^\star A(t)^\star C(t)^{\star^{-1}} R(t) -$$

$$- R(t)C(t)^{-1}A(t)C(t) - R(t)^2 - C(t)^\star D(t)C(t)$$

We conclude that R solves

$$(12) \qquad R' + A_1(t)^\star R + R A_1(t) + R^2 + D_1(t) = 0, \quad R(T) = C(T)^\star P_T C(T)$$

Conversely, consider a solution R to (12). Then $P(t) := C(t)^{\star^{-1}} R(t) C(t)^{-1}$ solves (8). Set

$$\mathcal{A}(t) \;=\; \frac{1}{2}(A_1(t) + A_1(t)^\star) \;\;\&\;\; \overline{\mathcal{A}}(t) \;=\; \frac{1}{2}(A_1(t) - A_1(t)^\star)$$

and observe that $\overline{\mathcal{A}}(t)^\star = -\overline{\mathcal{A}}(t)$. Therefore $X(t)^\star X(t) = Id$. We rewrite the equation (12) in the following form

$$R' + (\mathcal{A}(t) - \overline{\mathcal{A}}(t))R + R(\mathcal{A}(t) + \overline{\mathcal{A}}(t)) + R^2 + D_1(t) = 0, \quad R(T) = C(T)^\star P_T C(T)$$

Set $S(t) = X(t)^\star(\mathcal{A}(t) + R(t))X(t)$. Then, differentiating this equality, we obtain

$$S'(t) = X(t)^\star \overline{\mathcal{A}}(t)^\star (\mathcal{A}(t) + R(t))X(t) + X(t)^\star(\mathcal{A}(t) + R(t))\overline{\mathcal{A}}(t)X(t) +$$

$$X(t)^\star(\mathcal{A}'(t) - (\mathcal{A}(t) - \overline{\mathcal{A}}(t))R(t) - R(t)(\mathcal{A}(t) + \overline{\mathcal{A}}(t)) - R(t)^2 - D_1(t))X(t)$$

$$= -X(t)^\star \overline{\mathcal{A}}(t)(\mathcal{A}(t) + R(t))X(t) + X(t)^\star(\mathcal{A}(t) + R(t))\overline{\mathcal{A}}(t)X(t) +$$

$$X(t)^\star(\mathcal{A}'(t) - (\mathcal{A}(t) + R(t))^2 + \mathcal{A}(t)^2 + \overline{\mathcal{A}}(t)R(t) - R(t)\overline{\mathcal{A}}(t) - D_1(t))X(t)$$

$$= -S(t)^2 + X(t)^\star \left(\mathcal{A}'(t) - D_1(t) - \overline{\mathcal{A}}(t)\mathcal{A}(t) + \mathcal{A}(t)\overline{\mathcal{A}}(t) + \mathcal{A}(t)^2 \right) X(t)$$

and the result follows.

4   Applications to the Bolza Problem

We apply the previous results to the problem treated in Section 2.

4.1 Linear with Respect to Controls Systems. We consider the problem

$$(13) \qquad \text{minimize} \int_{t_0}^{T} \left(l(x(t)) + \frac{1}{2}\langle Ru, u\rangle \right) dt \;+\; \varphi(x(T))$$

over solution-control pairs (x, u) of control system

$$(14) \qquad \begin{cases} x'(t) &=& f(x(t)) + Bu(t), \quad u \in L^1(t_0, T; \mathbf{R}^m) \\ x(t_0) &=& x_0 \end{cases}$$

where $t_0 \in [0, T]$, $x_0 \in \mathbf{R}^n$, $f = (f_1, ..., f_n) : \mathbf{R}^n \mapsto \mathbf{R}^n$, $l : \mathbf{R}^n \mapsto \mathbf{R}$, $\varphi : \mathbf{R}^n \mapsto \mathbf{R}$, $B \in L(\mathbf{R}^m, \mathbf{R}^n)$ and $R \in L(\mathbf{R}^n, \mathbf{R}^n)$ is a self-adjoint operator such that for some $\omega > 0$ and all $u \in \mathbf{R}^m$, $\langle Ru, u\rangle \geq \omega \|u\|^2$.

We impose the following assumptions:

H_1) $\exists\, M \geq 0,\; \forall\, x \in \mathbf{R}^n,\; \|f(x)\| \leq M(\|x\| + 1)$

$H_2)$ $\liminf_{\|x\|\to\infty} \varphi(x) = +\infty$

$H_3)$ $f, l, \varphi \in C^2$ and their second derivatives are locally Lipschitz

$H_4)$ The Hamiltonian system (7) is complete

$H_5)$ $\forall\, x \in \mathbf{R}^n$, $f'(x)$ is self-adjoint

$H_6)$ $\forall\, x \in \mathbf{R}^n$, $f'(x)BR^{-1}B^* = BR^{-1}B^*f'(x)$, $l''(x)BR^{-1}B^* = BR^{-1}B^*l''(x)$

$H_7)$ For every $j = 1,...,n$ and $x \in \mathbf{R}^n$, $f_j''(x)BR^{-1}B^* = BR^{-1}B^*f_j''(x)$.

We recall that $H_5)$ is equivalent to the assumption that f is gradient of a smooth function and $H_6), H_7)$ are satisfied for instance when $U = \mathbf{R}^n$ and $B = R = Id$.

Theorem 4.1 *Assume that there exists $a \in \mathbf{R}$ such that for every $x \in \mathbf{R}^n$*

$$BR^{-1}B^*l''(x) + \left(\frac{1}{2}\|f\|^2\right)''(x) \geq a^2 I \quad \& \quad f'(x) - BR^{-1}B^*\varphi''(x) \leq aI$$

Then

a) the value function is continuously differentiable, $V_{xx}(t,\cdot)$ is continuous

b) the optimal control problem (13), (14) has a unique optimal control for any initial condition $(t_0, x_0) \in [0,T] \times \mathbf{R}^n$

c) for every solution (x,p) to the system

$$(15) \quad \begin{cases} x' = f(x) + BR^{-1}B^*p, & x(T) = x_T \\[2mm] -p' = f'(x)p - \nabla l(x), & p(T) = -\nabla\varphi(x_T) \end{cases}$$

and $t_0 \in [0,T]$, $x(\cdot)$ restricted to $[t_0, T]$ is optimal for (13), (14) with $x_0 = x(t_0)$ and $p(t) = -\frac{\partial V}{\partial x}(t, x(t))$.

d) The map $t \mapsto -\frac{\partial^2 V}{\partial x^2}(t, x(t))$ solves the Riccati equation

$$P' + f'(x(t))P + Pf'(x(t)) + PBR^{-1}B^*P - \left(\frac{1}{2}\|f\|^2\right)''(x(t)) = 0, \quad P(T) = -\varphi''(x(T))$$

Furthermore the optimal feedback low $u : [0,T] \mapsto \mathbf{R}^n$ is given by

$$\forall\, (t,x) \in [0,T] \times \mathbf{R}^n, \quad u(t,x) = -R^{-1}B^*\frac{\partial V}{\partial x}(t,x)$$

Corollary 4.2 *Let us assume that $U = \mathbf{R}^n$, $R = B = Id$, the map $x \mapsto l(x) + \frac{1}{2}\|f(x)\|^2$ is convex and for every $x \in \mathbf{R}^n$, $f'(x) - \varphi''(x) \leq 0$. Then all the conclusions of Theorem 4.1 hold true.*

We observe first that the Hamiltonian corresponding to this problem is given by

$$\forall\, x, p \in \mathbf{R}^n, \quad H(x,p) = \langle p, f(x) \rangle - l(x) + \frac{1}{2}\langle BR^{-1}B^*p, p \rangle$$

Thus, by $H_5)$,

$$\frac{\partial H}{\partial x}(x,p) = f'(x)p - l'(x) \quad \& \quad \frac{\partial H}{\partial p}(x,p) = f(x) + BR^{-1}B^*p$$

$$\frac{\partial^2 H}{\partial x \partial p}(x,p) = \frac{\partial^2 H}{\partial p \partial x}(x,p) = f'(x) \quad \& \quad \frac{\partial^2 H}{\partial p^2}(x,p) = BR^{-1}B^\star$$

$$\frac{\partial^2 H}{\partial x^2}(x,p) = (\psi_{r,s})_{r,s=1,\ldots,n} - l''(x), \quad \text{where } \psi_{r,s} = \sum_{i=1}^n \frac{\partial^2 f_r}{\partial x_i \partial x_s}(x)p_i$$

and

$$BR^{-1}B^\star \frac{\partial^2 H}{\partial x^2}(x,p) = (c_{ij})_{i,j=1,\ldots,n}$$

where

$$c_{ij} = \sum_{s=1}^n p_s \sum_{k=1}^n \frac{\partial^2 f_j}{\partial x_s \partial x_k}(x)e_{ik}, \quad BR^{-1}B^\star = (e_{ij})_{i,j=1,\ldots n}$$

Proof of Theorem 4.1 — From Section 2 we know that if for every solution (x,p) of (15) the matrix Riccati equation (5) has a solution on $[0,T]$, then all the conclusions of our theorem hold true. Set

$$A(t) = f'(x(t)) \quad \& \quad D(t) = \frac{\partial^2 H}{\partial x^2}(x(t),p(t))$$

Differentiating A we get

$$A'(t) = (\varphi_{ij}(t))_{i,j=1,\ldots n}$$

where

$$\varphi_{ij}(t) = \sum_{k=1}^n \frac{\partial^2 f_i}{\partial x_k \partial x_j}(x(t))f_k(x(t)) + \left\langle \nabla \frac{\partial f_i}{\partial x_j}(x(t)), BR^{-1}B^\star p(t) \right\rangle$$

Denote by γ_{is}^j the elements of the matrix $f_j''(x)BR^{-1}B^\star$. Then, by H_7), $\gamma_{is}^j = \gamma_{si}^j$. On the other hand, by H_5), for all $p = (p_1,\ldots,p_n)$

$$\left\langle \nabla \frac{\partial f_i}{\partial x_j}(x(t)), BR^{-1}B^\star p \right\rangle = \sum_{k=1}^n \sum_{s=1}^n \frac{\partial^2 f_i}{\partial x_k \partial x_j}(x)e_{ks}p_s =$$

$$= \sum_{s=1}^n p_s \sum_{k=1}^n \frac{\partial^2 f_j}{\partial x_k \partial x_i}(x)e_{ks} = \sum_{s=1}^n \gamma_{is}^j p_s = \sum_{s=1}^n \gamma_{si}^j p_s = \sum_{s=1}^n p_s \sum_{k=1}^n \frac{\partial^2 f_j}{\partial x_s \partial x_k}(x)e_{ki}$$

Using that $e_{ki} = e_{ik}$ for all $k,\ i = 1,\ldots,n$, we obtain

$$BR^{-1}B^\star D(t) - A'(t) - A(t)^2 = -BR^{-1}B^\star l''(x) - \left(\sum_{k=1}^n \frac{\partial^2 f_i}{\partial x_k \partial x_j}(x(t))f_k(x(t)) \right)_{i,j=1,\ldots,n} -$$

$$-f'(x(t))^2 = -BR^{-1}B^\star l''(x(t)) - \left(\tfrac{1}{2}\|f(\cdot)\|^2 \right)''(x(t)) \le -a^2 I$$

Corollary 3.5 ends the proof.

4.2 Linear Convex Bolza Problem. We consider the problem

$$(16) \qquad \text{minimize} \int_{t_0}^T \left(l(x(t)) + \frac{1}{2}\langle Ru, u \rangle \right) dt + \varphi(x(T))$$

over solution-control pairs (x, u) of control system

$$(17) \qquad \begin{cases} x' &= Ax + Bu(t), \quad u \in L^1(t_0, T; \mathbf{R}^m) \\ x(t_0) &= x_0 \end{cases}$$

where $t_0 \in [0, T]$, $x_0 \in \mathbf{R}^n$, $l : \mathbf{R}^n \mapsto \mathbf{R}$, $\varphi : \mathbf{R}^n \mapsto \mathbf{R}$, $A \in L(\mathbf{R}^n, \mathbf{R}^n)$, $B \in L(\mathbf{R}^m, \mathbf{R}^n)$ and $R \in L(\mathbf{R}^n, \mathbf{R}^n)$ is a self-adjoint operator such that for some $\omega > 0$ and all $u \in \mathbf{R}^n$, $\langle Ru, u \rangle \geq \omega \|u\|^2$. We assume that $H_2)$, $H_3)$ from the previous subsection hold true and that l, φ are convex. Then for every $(x, p) \in \mathbf{R}^n \times \mathbf{R}^n$,

$$\frac{\partial^2 H}{\partial x \partial p}(x, p) = A, \qquad \frac{\partial^2 H}{\partial x^2}(x, p) = -l''(x), \qquad \frac{\partial^2 H}{\partial p^2}(x, p) = BR^{-1}B^\star$$

Since $l''(x) \geq 0$, $\varphi''(x) \geq 0$ for every $x \in \mathbf{R}^n$, from the classical theory of Riccati equations (see for instance [2] or [7]), the solution to (5) is defined on $[0, T]$ for every choice of continuous $(x(\cdot), p(\cdot))$. Hence all the conclusions of Theorem 4.1 hold true.

4.3 Local Regularity of the Value Function. In the general case we do not have existence of solutions to the matrix Riccati equations (5) for all the extremals (x, p). When we have some a priori bounds on the data, then we can estimate the interval of time $[t_0, T]$ during which there is no shocks and so the value function is continuously differentiable on $[t_0, T] \times \mathbf{R}^n$. We provide here a result of a similar nature than that in [1].

We consider the problem

$$(P) \qquad \text{minimize} \int_{t_0}^{T} \left(l(x(t)) + \frac{1}{2} \langle Ru, u \rangle \right) dt \ + \ \varphi(x(T))$$

over solution-control pairs (x, u) of the control system

$$(18) \qquad \begin{cases} x'(t) &= f(x(t)) + g(x(t))u(t), \quad u(\cdot) \in L^1(t_0, T) \\ x(t_0) &= x_0 \end{cases}$$

where $t_0 \in [0, T]$, $x_0 \in \mathbf{R}^n$, $f : \mathbf{R}^n \mapsto \mathbf{R}^n$, $g : \mathbf{R}^n \mapsto L(\mathbf{R}^m, \mathbf{R}^n)$, $l : \mathbf{R}^n \mapsto \mathbf{R}$, $\varphi : \mathbf{R}^n \mapsto \mathbf{R}$ are continuously differentiable and $R \in L(\mathbf{R}^n, \mathbf{R}^n)$ is a self-adjoint operator such that for some $\omega > 0$ and all $u \in \mathbf{R}^n$, $\langle Ru, u \rangle \geq \omega \|u\|^2$.

We assume that $f, g, l, \varphi, f', g', f'', g''$ are bounded and f'', g'' are locally Lipschitz. Then the Hamiltonian H is given by

$$H(x, p) = \langle p, f(x) \rangle + \frac{1}{2} \left\langle R^{-1}g(x)^\star p, g(x)^\star p \right\rangle$$

and

$$\frac{\partial H}{\partial x}(x, p) = f'(x)^\star p + (g^\star(\cdot)p)'(x) \quad \& \quad \frac{\partial H}{\partial p}(x, p) = f(x) + g(x)R^{-1}g(x)^\star p$$

Thus for any solution (x, p) to (4) we have: $p(\cdot)$ solves the system

$$-p' = f'(x(t))^\star p + (g^\star(\cdot)p)'(x(t)), \quad p(T) = -\nabla\varphi(x(T))$$

Hence the norms of the co-states $p(\cdot)$ are bounded by a constant independent of x_T. On the other hand every solution x of (4) verifies $x'(t) = f(x) + g(x)R^{-1}g(x)^*p(t)$. Thus there exists $c > 0$ such that every solution (x,p) of (4) satisfies

$$\|x'(\cdot)\|_\infty + \|p(\cdot)\|_\infty + \|p'(\cdot)\|_\infty \leq c$$

By Theorem 3.6 and our assumptions we may reduce (5) to the Riccati equations

$$S' + S^2 + Q_{(x(\cdot),p(\cdot))}(t) = 0, \quad S(T) = S_{(x(\cdot),p(\cdot))}$$

with $Q_{(x(\cdot),p(\cdot))}(t)$ and $S_{(x(\cdot),p(\cdot))}$ self-adjoint and such that

$$\forall\, t \in [0,T], \quad Q_{(x(\cdot),p(\cdot))}(t) \leq \lambda I, \quad S_{(x(\cdot),p(\cdot))} \leq \lambda I$$

where λ is independent from the solution (x,p) of (4). Setting $S(t) = \lambda I + (T - t)\gamma I$ and choosing γ large enough we prove that for some $t_0 \in [0,T[$ and

$$\forall\, t \in [t_0,T], \quad S'(t) + S^2(t) + Q_{(x(\cdot),p(\cdot))}(t) \leq 0$$

for all (x,p) solving (4). Corollary 3.3 completes the proof.

References

[1] BARBU V. & DA PRATO G. *Hamilton-Jacobi equations and synthesis of nonlinear control processes in Hilbert space*, J. Diff. Eqs., 48, 350-372

[2] BENSOUSSAN A., DA PRATO G., DELFOUR M.C. & MITTER S.K. (1993) REPRESENTATION AND CONTROL OF INFINITE DIMENSIONAL SYSTEMS, Birkhäuser, Boston, Basel, Berlin

[3] BYRNES CH. & FRANKOWSKA H. (1992) *Uniqueness of optimal trajectories and the nonexistence of shocks for Hamilton-Jacobi-Bellman and Riccati Partial Differential Equations*, Preprint

[4] CANNARSA P. & FRANKOWSKA H. (1991) *Some characterizations of optimal trajectories in control theory*, SIAM J. Control and Optimiz.

[5] ESCHENBURG J.-H., HEINTZE E. (1990) *Comparison theory for Riccati equations*, Manuscripta Math., 68, 209-214

[6] RICCATI, Count J.F. (1724) *Animadversationes in aequationes differentiales secundi gradus*, Actorum Eruditorum quae Lipsiae Publicantur, Supplementa 8, 66-73

[7] REID W.T. (1972) RICCATI DIFFERENTIAL EQUATIONS, Academic Press

Authors address: CEREMADE
Université de Paris-Dauphine
75775 Paris Cx 16, FRANCE

International Series of Numerical Mathematics, Vol. 107, © 1992 Birkhäuser Verlag Basel

VERIFICATION THEOREMS OF DYNAMIC PROGRAMMING TYPE
IN OPTIMAL CONTROL

Stefan Mirică

Abstract. The aim of this paper is to present a short
survey of the existing results and to announce several new
verification theorems for optimal control problems with semicontinuous value functions. In our approach, a selection of
admissible trajectories is given and a verification theorem
contains sufficient conditions in the form of differential
inequalities and regularity properties of the corresponding
value function that insure the optimality of all the admissible trajectories in the given selection.

Many simple examples and some recent results (e.g. Boboc
1988, Frankowska 1989, Mirică 1990) show that the most general
regularity property of the value function (as an infimum-type
marginal function) of an optimal control problem is lower
semicontinuity (rather than upper semicontinuity) and much
more restrictive hypotheses imply its continuity, lipschitzia-
nity, etc..

On the other hand, in view of the fact that the Dynamic
Programming approach is best suited for the "synthesis (feedback) problem" (aimed at finding an optimal trajectory with
respect to every initial point in a given subset of the phase
space), we consider value functions associated to given selections of admissible trajectories. This setting is justified
also by the fact that, in practical applications, the value

function is obtained in this way using certain selections of extremal trajectories ("fields of extremals" in classical Calculus of Variations, "regular synthesis" in Boltyansky's approach, "generalized hamiltonian flows" introduced by Mirică 1985,1990, etc.).

We consider the optimal control problem of minimizing a (cost) functional of the form:

$$(1) \quad C(x(\cdot))=g(t_1,x(t_1)), \quad x(\cdot) \in \mathcal{X}_a(t_0,x_0), \quad (t_0,x_0) \in E_0 \subset \mathbb{R} \times \mathbb{R}^m$$

over a specified class, $\mathcal{X}_a(t_0,x_0)$, of admissible trajectories that are absolutely continuous solutions of the differential inclusion:

$$(2) \quad x' \in F(t,x), \quad x(t_0)=x_0$$

and satisfy phase space and terminal constraints of the form:

$$(3) \quad (t,x(t)) \in E_0 \quad \forall \quad t \in [t_0,t_1)$$

$$(4) \quad (t_1,x(t_1)) \in E_1 \subset \mathbb{R} \times \mathbb{R}^m$$

As usual in Optimal Control Theory, the class $\mathcal{X}_a$ of admissible trajectories is one of the following sets of solutions of the differential inclusion in (2): the set $\mathcal{X}_{pc}$ of piecewise continuously differentiable solutions, the set $\mathcal{X}_r$ of regular ones (for which the derivatives, $x'(\cdot)$, are reguated having a countable number of discontinuities), and, for any $p \in [1,\infty]$, the class $\mathcal{X}_p$ of solutions $x(\cdot)$ for which $x'(.)$ is p-integrable; the set $\mathcal{X}_c$ of classical (continuously differentiable) solutions is usually avoided since the problem (1)-(4) may not have a solution in this class.

In case (2) is a parametrized differential inclusion ("standard control system") of the form:

(2)' $x' \in f(t,x,U)$, $x(t_o)=x_o$

the set $\mathcal{X}_a(t_o,x_o)$ of admissible trajectories may be generated by a corresponding set, $\mathcal{U}_a(t_o,x_o)$, of (piecewise continuous, regulated, etc.) admissible controls $u(\cdot): [t_o,t_1] \to U$ in
the sense that the corresponding admissible trajectory, $x(\cdot)$,
is a solution of the differential equation:

(2)'' $x'=f(t,x,u(t))$, $x(t_o)=x_o$

As it is well known, the Dynamic Programming Method consists in using the value function defined by:

(5) $W(s,y)=\begin{cases} g(s,y) & \text{if } (s,y) \in E_1 \\ \inf\{C(x(\cdot)); \ x(\cdot) \in \mathcal{X}_a(s,y)\} & \text{if } (s,y) \in E_o \end{cases}$

to obtain necessary and/or sufficient optimality conditions.

In our setting, a selection of admissible trajectories,
$A(s,y)=x_{s,y}(\cdot) \in \mathcal{X}_a(s,y)$, $(s,y) \in E_o$, is assumed to be given and
the problem consists in finding (verifiable) conditions on the
associated value function defined by:

(6) $W_A(s,y)=\begin{cases} g(s,y) & \text{if } (s,y) \in E_1 \\ C(x_{s,y}(\cdot))=g(t_1(s,y),x_{s,y}(t_1(s,y))) & \text{if } (s,y) \in E_o \end{cases}$

that imply the optimality of every admissible trajectory, $x_{s,y}(\cdot)$,
in the selection (i.e. such that $W_A(.,.)$ coincides with the
value function in (5)).

The starting point for obtaining verification theorems in
this setting is the following simple result whose proof is
straightforward (see Cesari 1983, Fleming and Rishel 1975):

PROPOSITION 1. A selection $A(.,.)$, of admissible trajectories, is optimal iff the corresponding value function in (6)
has the following monotonicity property: for any $(s,y) \in E_o$,
$x(\cdot) \in \mathcal{X}_a(s,y)$, the function $w_x(t)=W_A(t,x(t))$, $t \in [s,t_1]$, is

nondecreasing on the interval $[s, t_1]$.

Thus, our problem is reduced to that of finding sufficient conditions under which a given function, $W_A(.,.)$, is nondecreasing along a given set of solutions of a differential inclusion (Mirică 1989, 1990, 1992b, etc.).

The simplest case is that in which the function $W_A(.,.)$
is differentiable (see Cesari 1983, Fleming et al. 1975, etc.).

THEOREM 2 (Elementary verification theorem). Let $A(.,.)$
be a selection of admissible trajectories such that the corresponding value function in (6) has the following properties:

(i) $W_A(.,.)$ is upper semicontinuous at the points in E_1:

$$(7) \quad \lim_{E_\bullet \ni (s,y) \to (t,x)} \sup W_A(s,y) \leqslant W_A(t,x) = g(t,x) \quad \forall \ (t,x) \subsetneq E_1$$

(ii) The restriction $W_\bullet(.,.)$ defined by:

$$(8) \quad W_\bullet(.,.) = W_A(.,.)\big| E_\bullet$$

is (Fréchet) differentiable and satisfies the inequality:

$$(9) \quad DW_\bullet(s,y).(1,v) \geqslant 0 \quad \forall \ v \in F(s,y), \ (s,y) \in E_\bullet$$

(iii) Either $\mathcal{X}_a \subset \mathcal{X}_r$ (i.e. any admissible trajectory is
at least regular) or $W_\bullet(.,.)$ is locally Lipschitz.

Then the selection $A(.,.)$ is optimal (i.e. $x_{s,y}(\cdot) \in \mathcal{X}_a(s,y)$
is optimal for any $(s,y) \in E_\bullet$).

REMARK 3. Theorem 2 is already a slight refinement of the
usual elementary verification theorem (Boltyansky 1971, Cesari
1983, etc.) in which $W_A(.,.)$ is of class C^1 on the set $E=$
$=E_\bullet \cup E_1$; simple examples show that the value function is often
discontinuous at the terminal points in E_1.

Further refinements of Theorem 2 may be obtained using
concepts and results from Nonsmooth Analysis: tangent cones to
arbitrary subsets of $\mathbb{R}^n$ and corresponding generalized deivatives.

From the multitude of tangent cones in the literature,
among the most suitable for our problem seem to be the contingent (Bouligand-Severi) cones to a subset $X \subset \mathbb{R}^n$ at $x \in X$:

$$(10) \quad K_x^+ X = \left\{ v \in \mathbb{R}^n; \; \exists \, (\theta_m, v_m) \to (0+, v): \left\{ x + \theta_m v_m \right\} \subset X \right\}, \quad K_x^- X = -K_x^+ X$$

and the quasitangent ("intermediate") cones:

$$(11) \quad Q_x^+ X = \left\{ v \in \mathbb{R}^n; \; \exists \, a(\cdot) : [0, \theta_0) \to X: \; a'(0) = v \right\}, \quad Q_x^- X = -Q_x^+ X$$

As it is well known, the classical (Fréchet) differentiability is generalized by the "contingent differentiability" of
a mapping $h(\cdot) : X \subset \mathbb{R}^n \to \mathbb{R}^m$ assuming the existence of the contingent directional derivative:

$$(12) \quad h_K^+(x; v) = \lim_{\substack{(\theta, u) \to (0+, v) \\ x + \theta \cdot u \in X}} (h(x + \theta \cdot u) - h(x))/\theta, \quad \forall \; v \in K_x^+ X$$

It is easy to see that Theorem 2 remains valid if hypothesis (ii) is replaced by the following one:

(ii)' The restriction $W_0(\cdot, \cdot)$ in (8) is contingent
differentiable and satisfies the inequality:

$$(9)' \quad (W_0)_K^+((s,y); (1,v)) \ge 0 \quad \forall \; v \in F_Q^+(s,y), \; (s,y) \in E_0$$
$$F_Q^+(s,y) = \left\{ v \in F(s,y); \; (1,v) \in Q_{(s,y)}^+ E_0 \right\}$$

A different type of refinement of Theorem 2 and also of
the results based on the concept of "regular synthesis" (see
Boltyansky 1971, Cesari 1983, Fleming et al. 1975, etc.) has
been obtained by Miricǎ (1985, 1986) in the case the function
$W_0(\cdot, \cdot)$ in (8) is "weakly C^1-stratified" in the sense that
there exists a locally finite partition, $\mathcal{S}$, of E_0, into
differentiable manifolds such that for any $S \in \mathcal{S}$, the restric-
tion $W_S(\cdot, \cdot) = W_0(\cdot, \cdot) | S$ is of class C^1; in this case the tan-
gent space $T_{(s,y)} E_0$ of E_0 at $(s,y) \in E_0$ is taken as the

tangent space in the sense of Differential Geometry, $T_{(s,y)}S$ if $(s,y) \in S \in \mathcal{J}$ and the derivative of $W_{\bullet}(\cdot,\cdot)$ is defined by: $DW_{\bullet}(s,y)=DW_S(s,y)$ if $(s,y) \in S \in \mathcal{J}$.

THEOREM 4 (Mirică 1985, 1986). If hypothesis (i) in Theorem 2 is verified and hypotheses (ii), (iii) are replaced, respectively, by:

(ii)'' The restriction $W_{\bullet}(\cdot,\cdot)$ in (8) is weakly C^1-stratified and satisfies:

$$(9)'' \quad \begin{aligned} & DW_{\bullet}(s,y).(1,v) \geqslant 0 \quad (\forall) \ v \in F_T(s,y), \ (s,y) \in E_{\bullet} \\ & F_T(s,y)=\{v \in F(s,y); \ (1,v) \in T_{(s,y)}E_{\bullet}\} \end{aligned}$$

(iii)' Either $W_{\bullet}(\cdot,\cdot)$ is continuous and $\mathcal{X}_a \subset \mathcal{X}_r$ or $W_{\bullet}(\cdot,\cdot)$ is locally Lipschitz,

then the selection $A(\cdot,\cdot)$, of admissible trajectories, is optimal.

Other refinements of Theorem 2 may be obtained using the upper contingent directional derivatives of a function $h(\cdot):X \subset \mathbb{R}^n \longrightarrow \overline{R}=[-\infty,+\infty]$ at a point $x \in \operatorname{dom} h(\cdot)= x \in X; \ h(x) \in \overline{R}\}$:

$$(13) \quad \begin{aligned} & \overline{D}_K^+ h(x;v)= \lim_{\substack{(\theta,u) \to (0+,v) \\ x+\theta.u \in X}} \sup \ (h(x+\theta.u)-h(x))/\theta \ , \ v \in K_X^+ X \\[2ex] & \overline{D}_K^- h(x;v)= \lim_{\substack{(\theta,u) \to (0-,v) \\ x+\theta.u \in X}} \sup \ (h(x+\theta.u)-h(x))/\theta \ , \ v \in K_X^- X \end{aligned}$$

The easiest case is that of problems whose value functions are locally Lipschitz on $E_{\bullet}$; the proof of the following result is very similar to that of Theorem 2:

THEOREM 5 (Mirică 1985, 1990). If hypothesis (i) in Theorem 2 is satisfied and hypothesis (ii) is replaced by:

(ii)''' The restriction $W_{\bullet}(\cdot,\cdot)$ in (8) is locally Lipschitz and satisfies: there exists a null subset $J \subset \mathbb{R}$ such that:

$$(9)''' \quad \begin{aligned} & \max\{\overline{D}_K^+ W_{\bullet}((s,y);(1,v)),\overline{D}_K^- W_{\bullet}((s,y);(1,v))\} \geqslant 0 \ \forall \ v \in F_Q(s,y) \\ & F_Q(s,y)=F_Q^+(s,y) \cap F_Q^-(s,y), \quad \forall \ (s,y) \in E_{\bullet} \ , \ s \notin J \end{aligned}$$

then the selection $A(.,.)$ is optimal.

Further refinements of Theorem 5 have been obtained by
Mirică (1985,1990) using the monotonicity results in Mirică(1989)
for problems defined by continuously parametrized differential
inclusions of the form in (2)' with semicontinuous value
functions; a similar result has been obtained by Frankowska(1989)
(Theorem 3.1) for problems defined by locally Lipschitz compact-
valued differential inclusions whose value functions are upper
semicontinuous:

THEOREM 6 (Frankowska 1989). Let $E_0=[0,1)\times\mathbb{R}^m$, $E_1=\{1\}\times\mathbb{R}^n$,
let $F(.,.)$ be a locally Lipschitz (with respect to both varia-
bles) compact-valued multifunction from $E=[0,1]\times\mathbb{R}^n$ to $\mathbb{R}^n$ and
let $A(.,.)$ be a selection of admissible trajectories such that
the value function $W_A(.,.)$ in (6) is upper semicontinuous
and satisfies:

(14) $\overline{D}_K^+ W_A((s,y);(1,v)) \geqslant 0$ $\forall$ $v \in$ co $F(s,y)$, $(s,y) \in E \cap$ dom $W_A(.,.)$

Then the selection $A(.,.)$ is optimal.

Simple examples show that the value function may not have
the upper semicontinuity property in (7):

EXAMPLE 7 (Cesari 1983, Example 3.7.6). Using either Cauchy's
Method of Characteristics (e.g. Mirică 1987,1990) or the usual
necessary optimality conditions for the problem of minimizing
$C(x(\cdot))=x_2(1)$ subject to:

(15) $\begin{cases} x_1'=u(t)\in\mathbb{R}, & x_1(t_0)=x_1^\bullet\in\mathbb{R}, \ x_1(1)=0 \\ x_2'=t(u(t))^2, & x_2(t_0)=x_2^\bullet\in\mathbb{R}, \ t_0<1 \end{cases}$

one may find the following selection of admissible controls:
$u_{s,y}(t)=y_1/t\bullet\log s$, $t \in [s,1]$ $\forall$ $(s,y) \in E_0=(0,1)\times\mathbb{R}^2$, for which

the value function in (6) is given by:

$$(16) \quad W_A(s,y) = \begin{cases} y_2 & \text{is} \quad (s,y) \in E_1 = \{1\} \times (\{0\} \times \mathbb{R}), \ y = (y_1, y_2) \\ y_2 - (y_1)^2/\log s & \text{if} \quad (s,y) \in E_o = (0,1) \times \mathbb{R}^2 \end{cases}$$

Though the function $W_A(.,.)$ in (16) is C^1-stratified by the stratification $\mathcal{J} = \{E_o, E_1\}$ and $W_o(.,.)$ defined as in (8) is of class C^1 on E_o, neither of the Theorems 2, 4, 5, 6 may be used to prove the optimality of the above admissible controls (even if we restrict ourselves to the class $\mathcal{U}_{o\omega}$ of measurable bounded admissible controls) since $W_A(.,.)$ does not have the upper semicontinuity property in (7); besides, the multifunction $F(.,.)$ defining the control system in (15) is not locally Lipschitz with respect to the first variable and has unbounded values.

More realistic verification theorems are readily available from the refinements obtained by Mirică (1992b) of the existing results concerning locally invariant sets and nondecreasing functions with respect to solutions of differential inclusions.

In what follows $F(.,.)$ is a multifunction from an open subset $D \subset \mathbb{R} \times \mathbb{R}^n$ to $\mathbb{R}^n$ and we shall use the following notations:

$$(17) \quad \begin{aligned} \hat{F}(t-,x) &= \bigcap_{\mu(J)=0} \ \bigcap_{\theta, r>0} \overline{co} \ F((t-\theta, t] \setminus J, B_r(x)), \quad (t,x) \in D \\ \hat{F}(t+,x) &= \bigcap_{\mu(J)=0} \ \bigcap_{\theta, r>0} \overline{co} \ F([t, t+\theta) \setminus J, B_r(x)) \end{aligned}$$

where $\mu(\cdot)$ is the Lebesgue measure and $B_r(x) = \{y \in \mathbb{R}^n; \|y-x\| < r\}$.

We note that according to the results in Mirică (1992a), if $F(.,.)$ is upper hemicontinuous then one has:

$$(18) \quad \hat{F}(t-,x) = \hat{F}(t+,x) = \overline{co} \ F(t,x) \quad \forall \ (t,x) \in D$$

From the several possible variants of verification theorems that may be obtained directly from the monotonicity results in

Mirică (1992b), we mention only the following:

THEOREM 8. Let $A(.,.)$ be a selection of admissible trajectories of the problem (1)-(4) such that the following hypotheses are satisfied:

(a) The subset $E=E_0 \cup E_1 \subset D = \mathrm{Int}(D)$ is locally closed, the value function $W_A(.,.)$ in (6) is lower semicontinuous and the following conditions hold:

$$(19) \quad (1,\hat{F}(s-,y)) \subset K^-_{(s,y)} E \quad \forall\ (s,y) \in E$$

$$(20) \quad \bar{D}^-_K W_A((s,y);(1,v)) \ngtr 0 \quad \forall\ v \in \hat{F}(s-,y),\ (s,y) \in E \cap \mathrm{dom}\ W_A(.,.)$$

(b) The multifunction $F(.,.)$ has one of the following properties:

(b$_1$) $F(.,.)$ is locally Lipschitz (with respect to both variables);

(b$_2$) $F(.,.)$ is locally Lipscitz only with respect to the second variable and either the multifunction $E(t)=\{x \in \mathbb{R}^n;$ $(t,x) \in E\}$ is locally absolutely continuous on $\mathrm{pr}_1 E$ or $\mathcal{X}_a \subset \mathcal{X}_{aa}$ (i.e. any admissible trajectory is locally Lipschitz; in particular, $F(.,.)$ is locally bounded).

Then the selection $A(.,.)$ is optimal.

We note that in the case of unbounded controls (i.e. $\mathcal{X}_a \not\subset \mathcal{X}_{aa}$) the requirement that $E(\cdot)$ is absolutely continuous (in particular, locally Lipschitz) may be replaced by a differential inequality of $W_A(.,.)$ on the set of unbounded directions of $F(.,.)$ introduced in Mirică(1992a).

A similar result, extending Theorem 6 may be obtained in the case the value function $W_A(.,.)$ is upper semicontinuous, in which case conditions (19) and (20) should be replaced, respectively by:

$$(19)' \quad (1,\hat{F}(s+,y)) \subset K^+_{(s,y)} E, \quad \forall \ (s,y) \in E$$

$$(20)' \quad \bar{D}^+_K W_A((s,y);(1,v)) \geqslant 0 \quad \forall \ v \in \hat{F}(s+,y), \ (s,y) \in E$$

We note that according to some examples in Mirică(1989,1990), if $W_A(.,.)$ is lower semicontinuous then conditions (19), (20) may not be replaced by (19)', (20)' which are suitable only for upper semicontinuous value functions.

REFERENCES

Bebec, C.(1988), The lower semicontinuity of the value function in optimal control, Studii Cerc.Mat.,40, 119-124.

Boltyansky V.G.(1971), Mathematical Methods in Optimal Control, Holt& Rinehart, New York.

Cesari L., Optimization - Theory and Applications, Springer, New York, 1983.

Clarke F.H.(1983), Optimization and Nonsmooth Analysis, Wiley, New York.

Fleming W.H. and Rishel R.W.(1975), Deterministic and Stochastic Optimal Control, Springer, New York.

Frankowska H.(1989), Optimal Trajectories Associated with a Solution of the Contingent Hamilton-Jacobi Equation, Appl.Math.Optim., 19, 291-311.

Mirică St.(1985), Sufficient Optimality Conditions of Dynamic Programming Type in Control Theory, Proc. 24th C.D.C., Ft.Lauderdale, Florida, 626-630.

Mirică St.(1986), Sufficient Optimality Conditions for Stratified Optimal Control Problems, SIAM J.Control Opt.,24,675-691.

Mirică St.(1987), Generalized Solutions by Cauchy's Method of Characteristics, Rend.Sem.Mat.Univ.Padova,77,317-350.

Mirică St.(1989), Invariant Sets and Monotone Functions with
    respect to Solutions of Ordinary Differential Equations,
    Revue Roum.Math.Pures Appl.,34, 419-453.
Mirică St.(1990), Optimal Control. Sufficient Conditions and
    Synthesis, Ed.Stiinţifică, Bucureşti (Roumanian).
Mirică St.(1992a),Tangent and Contingent Directions to Trajec-
    tories of Differential Inclusions, Revue Roum.Math.Pures
    Appl. (to appear).
Mirică St.(1992b), Invariance and Monotony with Respect to
    Solutions of Differential Inclusions, submitted.

Author's address:

Stefan Mirică
Faculty of Mathematics
University of Bucharest
Academiei 14
70109 Bucharest, Romania.

International Series of Numerical Mathematics, Vol. 107, © 1992 Birkhäuser Verlag Basel

Isaacs' equations for value-functions of differential games

Hélène Frankowska* & Marc Quincampoix **

May 7, 1992

⋆ CEREMADE	⋆⋆ Département de Mathématiques
Université Paris-Dauphine	Université François Rabelais
Place de Lattre de Tassigny	Parc de Grandmont
75775 Paris cedex 16	37200 Tours

Abstract: We study value functions of a differential game with payoff which depends on the state at a given end time. We consider differential games with feedback strategies and with nonanticipating strategies. We prove that value-functions are solutions to some Hamilton-Jacobi-Isaacs equations in the viscosity and contingent sense. For these two notions of strategies, with some regularity assumptions, we prove that value-functions are the unique solution of Isaacs' equations.

1 Introduction

Let us consider the following differential game:

$$(1) \qquad \begin{cases} i) & x'(t) = f(t, x(t), u(t), v(t)), \ t \in [t_0, T] \\ ii) & u(t) \in U \quad v(t) \in V \end{cases}$$

The two players act on the state $x(\cdot)$ by choosing controls, u for the first player and v for the second one. The goal of the first player is to maximize at the given end time T the payoff $g(x(T))$, the second player wants to minimize it.

Let us recall that the game with the following payoff:

$$(2) \qquad g(x(T)) + \int_0^T L(t, x(t), u(t), v(t)) dt$$

may be reduced to the above one. In fact, by the simple change of variable $z := (x, y)$, we obtain the new game

$$(3) \quad \begin{cases} i) & (x'(t), y'(t)) = (f(t, x(t), u(t), v(t)), L(t, x(t), u(t), v(t))) \\ ii) & u(t) \in U \\ iii) & v(t) \in V \end{cases}$$

with the payoff $G(x(T), y(T)) := g(x(T)) + y(T)$ which is equal to (2).

Since Isaacs (cf [14]), it is well-known that the value-function satisfies a partial differential equation (the Isaacs' equation) when the game is regular enough. The solutions of this equation have been studied by Isaacs himself in C^1 case (see [14]), lipschitz solutions have been studied for example in [15], and in [9], [5], [17] for viscosity solutions. However such regularity is not always the case (see also for instance [12], [4], [6]... for control systems).

We introduce two notions of strategies and we prove that the associated value-functions are solutions to Isaacs' equation without any assumptions concerning the regularity of g. In this paper, we mainly state results (see [13] for more detailed proofs).

2   Feedback strategies of differential games

Consider a function $f : [0, T] \times I\!\!R^n \times U \times V \mapsto I\!\!R^n$ where U and V are two complete separable metric spaces. Let us denote by $t \mapsto x(t, t_0, x_0, u(\cdot), v(\cdot))$ the solution to (1) corresponding to controls $u(\cdot)$, $v(\cdot)$, starting from x_0 at time t_0 (i.e. such that $x(t_0) = x_0$). We shall need the following assumptions:

$$(4) \quad \begin{cases} i) & f \text{ is continuous} \\[2mm] ii) & \forall\, (t,x) \in [0,T] \times \mathbb{R}^n,\ \forall\, (u,v) \in U \times V \text{ the sets} \\ & f(t,x,u,V),\ f(t,x,U,v) \text{ are compact and convex} \\[2mm] iii) & \forall\, (t,x,u,v) \in [0,T] \times \mathbb{R}^n \times U \times V, \\ & \text{functions } f(t,x,\cdot,v) \text{ and } f(t,x,u,\cdot) \text{ are} \\ & l - \text{Lipschitz, where } l > 0 \\[2mm] iv) & \forall\, R > 0,\ \exists\, c_R \in L^1(0,T) \text{ such that for almost all} \\ & t \in [0,T] \text{ and for all } (u,v) \in U \times V,\ f(t,\cdot,u,v) \\ & \text{is } c_R(t) - \text{Lipschitz on } B(0,R). \\[2mm] v) & \exists\, k \in L^1(0,T) \text{ such that for almost all } t \in [0,T] \\ & \sup_{u \in U} \sup_{v \in V} \|f(t,x,u,v)\| \le k(t)(1 + \|x\|) \end{cases}$$

We call *feedback strategy* for the first player any function $\varphi : [0,T] \times X \mapsto U$ such that $(t,x) \mapsto f(t,x,\varphi(t,x),V)$ is upper semicontinuous with respect to (t,x). We denote by Φ the set of feedback strategies for the first player. In a similar way we can define feedback strategies for the second player and Ψ the set of such strategies. We denote by $\mathcal{U}$ (respectively $\mathcal{V}$) the set of measurable functions $[0,T] \mapsto U$ (respectively $[0,T] \mapsto V$).

We assume furthermore the following crucial condition which allows to define the value function of the game:

There exists a pair of feedback strategies $(\varphi^\star, \psi^\star) \in \Phi \times \Psi$ such that for any measurable control $u(\cdot)$, there exists an unique solution to

$$x'(t) = f(t, x(t), u(t), \psi^\star(t, x(t)))$$

such that $x(t_0) = x_0$ and we denote by $x(\cdot, t_0, x_0, u(\cdot), \psi^\star(\cdot, \cdot))$ this solution. In a similar way, we assume also the existence and unicity of $x(\cdot, t_0, x_0, \varphi^\star(\cdot, \cdot), v(\cdot))$ and $x(\cdot, t_0, x_0, \phi^\star(\cdot, \cdot), \psi^\star(\cdot, \cdot))$.

$$(5) \quad \begin{cases} \forall\, (t_0, x_0),\ \forall\, (u(\cdot), v(\cdot)) \in \mathcal{U} \times \mathcal{V} \\ g(x(T, t_0, x_0, u(\cdot), \psi^\star(\cdot, \cdot))) \le g(x(T, t_0, x_0, \phi^\star(\cdot, \cdot), \psi^\star(\cdot, \cdot))) \\ \le g(x(T, t_0, x_0, \phi^\star(\cdot, \cdot), v(\cdot))) \end{cases}$$

Definition 2.1 *If (5) is satisfied, we call*

$$W(t_0, x_0) := g(x(T, t_0, x_0, \phi^\star(\cdot, \cdot), \psi^\star(\cdot, \cdot)))$$

the value function[1] of the differential game with feedback strategy.

3   Contingent solutions to Isaacs' equations

Consider the following contingent[2] inequalities:

$$(6)\quad \begin{cases} \Theta(T,\cdot) = g(\cdot) \ \text{ and } \ \forall\,(t,x) \in Dom(\Theta) \\ i)\ \text{if}\ t \in [0,T[,\ \text{then} \\ \sup_{u\in U}\inf_{v\in V} D_\uparrow\Theta(t,x)(1,f(t,x,u,v)) \le 0 \\ ii)\ \text{if}\ t \in [0,T[,\ \text{then} \\ \sup_{u\in U}\inf_{v\in V} D_\downarrow\Theta(t,x)(1,f(t,x,u,v)) \ge 0 \end{cases}$$

$$(7)\quad \begin{cases} \Theta(T,\cdot) = g(\cdot) \ \text{ and } \ \forall\,(t,x) \in Dom(\Theta) \\ i)\ \text{if}\ t \in [0,T[,\ \text{then} \\ \inf_{v\in V}\sup_{u\in U} D_\uparrow\Theta(t,x)(1,f(t,x,u,v)) \le 0 \\ ii)\ \text{if}\ t \in [0,T[,\ \text{then} \\ \inf_{v\in V}\sup_{u\in U} D_\downarrow\Theta(t,x)(1,f(t,x,u,v)) \ge 0 \end{cases}$$

A such $\Theta : [0,T] \times {I\!\!R}^n \mapsto \overline{I\!\!R}$ is called contingent solution to the Isaacs' equation. We can prove *without any assumptions on g* the following:

Proposition 3.1 *Assume that (4) and (5) hold true, then the value function satisfies (6)i) and (7)ii).*

Proof — Let us prove (6)i). Fix $\bar{u} \in U$ and $h > 0$. Define x_h a solution of $x'(t) = f(t,x(t),\bar{u},\psi^\star)\ t \in [t_0,t_0+h]$ such that $x(t_0) = x_0$. Let us introduce

$$u_h(t) := \begin{cases} \bar{u} \ \ \text{if} \ \ t \in [t_0,t_0+h] \\ \varphi^\star(t,x(t,t_0+h,x_h(t_0+h),\varphi^\star(\cdot,\cdot),\psi^\star(\cdot,\cdot)))) \\ \text{if} \ \ t > t_0+h \end{cases}$$

---

[1]This definition is very related to the one of Pierre Bernhard (see [7]).

[2]Recall the definition of the contingent epiderivative of $\Theta : R^n \mapsto \overline{R}$ at $x_0 \in Dom(\Theta)$,

$$D_\uparrow\Theta(x_0)(u) = \liminf_{h\longrightarrow 0+,\, v\longrightarrow u} \frac{\Theta(x_0+hv) - \Theta(x_0)}{h}$$

or equivalently $EpiD_\uparrow\Theta(x_0) = T_{Epi\Theta}(x_0,\Theta(x_0))$, where Epi states for the epigraph. In a similar way for the contingent hypoderivative of Θ at $x_0 \in Dom(\Theta)$ is defined by $D_\downarrow\Theta(x)(u) := -D_\uparrow(-\Theta)(x)(u)$, and the contingent derivative of Θ at $x_0 \in Dom(\Theta)$ is defined by:

$$GraphD\Theta(x_0) = T_{Graph\Theta}(x_0,\Theta(x_0)).$$

Then by the very definition of W,

$W(t_0 + h, x_h(t_0 + h)) = g(x(T, t_0 + h, x(t_0 + h), \phi^\star(\cdot, \cdot), \psi^\star(\cdot, \cdot)))$ which is equal to $g(x(T, t_0, x_0, u_h(\cdot), \psi^\star(\cdot, \cdot)))$ and by (5), $g(x(T, t_0, x_0, u_h(\cdot), \psi^\star(\cdot, \cdot))) \leq g(x(T, t_0, x_0, \phi^\star(\cdot, \cdot), \psi^\star(\cdot, \cdot))) = W(t_0, x_0)$. Hence

$$\liminf_{h \longrightarrow 0^+} \frac{W(t_0 + h, x_h(t_0 + h)) - W(t_0, x_0))}{h} \leq 0$$

Thanks to our assumptions and by the Mean Value Theorem, there exists $h_n \longrightarrow 0^+$ such that:

$$\frac{x_{h_n}(t_0 + h_n) - x_0}{h_n} \longrightarrow w \in f(t_0, x_0, \overline{u}, V)$$

This yields $\inf_{v \in V} D_\uparrow W(t_0, x_0)(1, f(t, x, \overline{u}, v)) \leq 0$ and proves (6)i. $\quad\square$

We can obtain other contingent inequalities with suitable regularity assumptions concerning strategies

Proposition 3.2 *Assume that (4) and (5) hold true. If $(t, x) \mapsto f(t, x, \varphi^\star(t, x), V)$ is continuous at (t_0, x_0), then the value function W satisfies (6)ii. If $(t, x) \mapsto f(t, x, U, \psi^\star(t, x))$ is continuous at (t_0, x_0), then the value function W satisfies (7)i.*

Now we shall state an unicity result[3]:

Theorem 3.3 *Assume that (4), (5) hold true and that W is continuous.*

- *If $(t, x) \mapsto f(t, x, \varphi^\star(t, x), V)$ is continuous, then any lower semicontinuous (l.s.c.) function Θ satisfying (6)i is larger or equal than W*

- *If $(t, x) \mapsto f(t, x, U, \psi^\star(t, x))$ is continuous, then any upper semicontinuous (u.s.c.) function Θ satisfying (7)ii is lower or equal than W.*

Corollary 3.4 *When (4), (5) hold true, and W is continuous and when $(t, x) \mapsto f(t, x, U, \psi^\star(t, x))$ and $(t, x) \mapsto f(t, x, \varphi^\star(t, x))$ are continuous, the value function W is the unique continuous solution to Hamilton-Jacobi-Isaacs inequalities (6) and (7).*

---

[3]cf the proof in [13]

4   Viscosity solutions to Isaacs' equation

We first define the lower and upper Hamiltonians of the differential (1):

$$H_-(t,x,p) := \max_{v \in V} \min_{u \in U} < p, f(t,x,u,v) >$$
$$H_+(t,x,p) := \min_{u \in U} \max_{v \in V} < p, f(t,x,u,v) >,$$

Consider two Hamilton-Jacobi equations:

$$(8) \qquad \begin{cases} -\frac{\partial \Theta}{\partial t}(t,x) + H_+(t,x,-\frac{\partial \Theta}{\partial x}(t,x)) = 0 \\ \Theta(T,\cdot) = g(\cdot) \end{cases}$$

$$(9) \qquad \begin{cases} -\frac{\partial \Theta}{\partial t}(t,x) + H_-(t,x,-\frac{\partial \Theta}{\partial x}(t,x)) = 0 \\ \Theta(T,\cdot) = g(\cdot) \end{cases}$$

In this section, we give some results concerning viscosity solutions to Hamilton-Jacobi-Isaacs equations. First, we recall the definition of viscosity solution by using sub and super differentials[4]:

Definition 4.1 *Consider $H : [0,T] \times \mathbb{R}^n \times \mathbb{R}^n \mapsto \mathbb{R}$ Let us recall that the function $\Theta : [0,T] \times \mathbb{R}^n \mapsto \overline{\mathbb{R}}$ is a viscosity supersolution to the following Hamilton-Jacobi equation $-\frac{\partial \Theta}{\partial t}(t,x) + H(t,x,-\frac{\partial \Theta}{\partial x}(t,x)) = 0$ if and only if:*
$$\forall (t,x) \in Dom(\Theta), \ \forall (p_t,p_x) \in \partial_- \Theta(t,x), \ -p_t + H(t,x,-p_x) \geq 0$$
The function Θ is a viscosity subsolution if and only if:
$$\forall (t,x) \in Dom(\Theta), \ \forall (p_t,p_x) \in \partial_+ \Theta(t,x), \ -p_t + H(t,x,-p_x) \leq 0$$
A function Θ is a viscosity solution if it is a supersolution and a subsolution.

We can prove *without any assumptions* concerning g the following existence result:

Proposition 4.2 *Assume that (4) holds true, then the value-function W is a supersolution to (8) and a subsolution to (9).*

But when the value-function W is continuous, we have the more precise

Proposition 4.3 *Assume (4). Then if W is continuous and $(t,x) \mapsto f(t,x,U,\psi^\star(t,x))$ is continuous, then W is a viscosity solution to (8).*
If W is continuous and $(t,x) \mapsto f(t,x,\varphi^\star(t,x),V)$ is continuous, then W is a viscosity solution to (9).

---

[4]Recall the definition of the subdifferential of $\phi : R^n \mapsto \overline{R}$, at $x_0 \in Dom(\phi)$
$\partial_- \phi(x_0) := \{ p \in R^n \mid \liminf_{x \to x_0} \frac{\phi(x) - \phi(x_0) - <p, x - x_0>}{\|x - x_0\|} \geq 0 \}$ and the super differential
of ϕ at x_0 is given by: $\partial_+ \phi(x_0) := -\partial_-(-\phi)(x_0)$.

Theorem 4.4 *Let assumptions of Corollary 3.4 hold true. Let $\Theta : [0,T] \times \mathbb{R}^n \mapsto \mathbb{R}$ be continuous. Then Θ is the value function of the game if and only if it is a viscosity supersolution to (8) and a viscosity subsolution to (9).*

Corollary 4.5 *Let us assume (4), (5) and let $\Theta : [0,T] \times \mathbb{R}^n \mapsto \mathbb{R}$ be a continuous function. If we assume the following Isaacs' condition:*

$$(10) \qquad \forall\, (t,x,p), \;\; H_-(t,x,p) = H_+(t,x,p),$$

then the value function is the unique *viscosity solution to (9) (or equivalently(8)).*

These results follow from results of the previous section and from the following section.

5 Comparison between viscosity and contingent solutions to Hamilton Jacobi Isaacs equations

Proposition 5.1 *Consider $\Theta : [0,T] \times \mathbb{R}^n \mapsto \overline{\mathbb{R}}$ verifying (6) (respectively (7)). Then Θ is a viscosity solution to (8) (respectively to (9)).*

This result is a consequence of the following

Lemma 5.2 *Consider $\Theta : [0,T] \times \mathbb{R}^n \mapsto \overline{\mathbb{R}}$.*

- *If Θ satisfies (6)i), then it is a supersolution of (8).*

- *If Θ satisfies (6)ii), then it is a subsolution of (8).*

- *If Θ satisfies (7)i), then it is a supersolution of (9).*

- *If Θ satisfies (7)ii), then it is a subsolution of (9).*

Proof — Let us prove the first statement. If Θ satisfying (6)i) then[5]:

$$(11) \qquad \begin{cases} \forall\, (p_t, p_x) \in \partial_- \Theta(t,x), \; \forall\, (u,v) \in U \times V, \\ D_\uparrow \Theta(t,x)(1, f(t,x,u,v)) \geq p_t + \, < p_x, f(t,x,u,v) > \end{cases}$$

---

[5]Let us recall (see [2] chapter 6 for instance) that we have the following equivalent definition for the subdifferential of a function ϕ

$$\partial_- \phi(x_0) = \{ p \,|\, \forall\, q \in \mathbb{R}^n, \, D_\uparrow \phi(x_0)(q) \geq < p,q > \}$$

Then by taking the "supinf" of this inequality, we prove that Θ is a super-solution to (8). The proofs of the other statements are similar. $\quad\square$

When value functions are continuous, the notions of contingent and viscosity solutions of Isaacs' equations are equivalent.

Theorem 5.3 *Let* $\Theta : [0,T] \times \mathbb{R}^n \mapsto \mathbb{R}$ *be a continuous function and let (4) hold true. Then* Θ *satisfies the contingent inequalities (6) (respectively (7)) if and only if it is a viscosity solution to the Hamilton-Jacobi-Isaacs equation (8) (respectively (9)).*

Lemma 5.4 *If (4) holds true.*

- *Any l.s.c. function* Θ *is a supersolution of (8) if and only if it satisfies (6)i).*

- *Any l.s.c. function* Θ *is a supersolution of (9) if and only if it satisfies (7)i).*

- *Any u.s.c. function* Θ *is a subsolution of (8) if and only if it satisfies (6)ii).*

- *Any u.s.c. function* Θ *is a subsolution of (9) if and only if it satisfies (7)ii).*

Proof of Lemma — We already know, thanks to Proposition 5.1 and Lemma 5.2, that contingent solutions are viscosity solutions. Let us prove the converse implication.

Assume that Θ is a supersolution to (8), i.e.:

$$(12) \qquad \forall \, (p_t, p_x) \in \partial_- \Theta(t,x), \ \sup_{u \in U} \inf_{v \in V} p_t + \; < p_x, f(t,x,u,v) > \leq 0$$

Hence, for any $u \in U$, $\inf_{v \in V} p_t + \; < p_x, f(t,x,u,v) > \leq 0$. But we know, (cf [12]) that $(p_t, p_x) \in \partial_- \Theta(t,x)$ if and only if $(p_t, p_x, -1)$ belongs to the normal cone $(T_{Epi\Theta}(t,x,\Theta(t,x)))^-$. We claim that

$$(13) \quad \forall \, u \in U, \ \{1\} \times f(t,x,u,V) \times \{0\} \cap co(T_{Epi\Theta}(t,x,\Theta(t,x))) \neq \emptyset$$

Assume for a moment that is false, then, by the separation theorem we should have:

$$(14) \quad \begin{cases} \exists \, (p_t, p_x, q) \in (T_{Epi\Theta}(t,x,\Theta(t,x)))^-, \ \exists \, u \in U \text{ such that} \\ \forall \, v \in V, \ p_t + \; < p_x, f(t,x,u,v) > \; > 0 \end{cases}$$

This is a contradiction with (12). So

$$\begin{cases} \forall\,(t,x,y) \in Epi(\Theta), \text{ for all } u \in U \\ \{1\} \times f(t,x,u,V) \times \{0\} \cap co(T_{Epi\Theta}(t,x,y)) \neq \emptyset \end{cases}$$

and we can deduce from[6] Theorem 3.2.4 in [1], that $\{1\} \times f(t,x,u,V) \times \{0\} \cap T_{Epi\Theta}(t,x,\Theta(t,x)) \neq \emptyset$, for any $(t,x) \in Dom\Theta$. This implies the following contingent equation:

$$\forall\,(t,x) \in Dom(\Theta), \forall\,u \in U, \inf_{v \in V} D_\uparrow \Theta(t,x)(1, f(t,x,u,v)) \leq 0$$

Let us prove the third statement. Assume that Θ satisfies

$$(15) \qquad \forall\,(p_t, p_x) \in \partial_+ \Theta(t,x), \sup_{u \in U} \inf_{v \in V} p_t + < p_x, f(t,x,u,v) > \geq 0$$

We claim that

$$(16) \quad \exists\,u \in U, \{1\} \times f(t,x,u,V) \times \{0\} \subset co(T_{Hypo\Theta}(t,x,\Theta(t,x)))$$

If (16) is not satisfied, by the separation Theorem

$$(17) \qquad \begin{cases} \forall\,u \in U, \exists\,v \in V \text{ such that} \\ \exists\,(p_t, p_x, q) \in (T_{Hypo\Theta}(t,x,\Theta(t,x)))^-, \\ p_t + < p_x, f(t,x,u,v) > < 0 \end{cases}$$

This is a contradiction with (15). Then, thanks to (4), and since (cf [2] p.130),

$$\liminf_{(t',x',y') \mapsto (t,x,\Theta(t,x))} co(T_{Hypo\Theta}(t',x',y')) \subset T_{Hypo\Theta}(t,x\Theta(t,x))$$

we can deduce that $\{1\} \times f(t,x,u,V) \times \{0\} \subset T_{Hypo\Theta}(t,x,\Theta(t,x))$ and consequently (7)ii) holds true. The proofs are similar for the other statements. $\square$

---

[6] Let us recall a duality result in viability theory (due to Ushakov see for instance Theorem 3.2.4 in [1]). Consider a closed set $K \subset R^n$ and let F be u.s.c set-valued map with compact convex values. Then the following two statements are equivalent:

$$\begin{aligned} i) &\quad \forall\,x \in K, F(x) \cap T_K(x) \neq \emptyset \\ ii) &\quad \forall\,x \in K, F(x) \cap co(T_K(x)) \neq \emptyset \end{aligned}$$

where co is the closed convex hull.

6   Nonanticipating strategies

We shall define value-functions for a concept of strategy studied by Elliot-Kalton (see also [9]). We denote by $\mathcal{U}(t)$ (respectively by $\mathcal{V}(t)$) the set of measurable functions $u : [t, T] \mapsto U$ (respectively $v : [t, T] \mapsto V$).

Firstly let us recall the definition of *nonanticipating strategies*.

Definition 6.1 *We call* nonanticipating strategy for the first player *any function* $\alpha : \mathcal{V}(t) \mapsto \mathcal{U}(t)$ *such that*

$$\forall\, t \in [0, T],\ \forall\, (v, \bar{v}) \in \mathcal{V}(t),\ \forall\, s \in [0, T],$$
$$v \equiv \bar{v} \text{ a. e. in } [t, s] \ \Rightarrow\ \alpha(v) \equiv \alpha(\bar{v}) \text{ a. e. in } [t, s]$$

and we denote by $\Gamma(t)$ *the set of such nonanticipating strategies.*

We call nonanticipating strategy for the sevond player *any function* $\beta : \mathcal{U}(t) \mapsto \mathcal{V}(t)$ *such that*

$$\forall\, t \in [0, T],\ \forall\, (u, \bar{u}) \in \mathcal{U}(t),\ \forall\, s \in [0, T],$$
$$u \equiv \bar{u} \text{ a. e. in } [t, s] \ \Rightarrow\ \beta(u) \equiv \beta(\bar{u}) \text{ a. e. in } [t, s]$$

and we denote by $\Delta(t)$ *the set of such nonanticipating strategies.*

This notion of strategies enables us to define the two value-functions:

Definition 6.2 *Consider the upper value-function of the game:*

$$\Phi(t_0, x_0) := \inf_{\beta \in \Delta(t_0)} \sup_{u(\cdot) \in \mathcal{U}(t_0)} g(x(T, t_0, x_0, u(\cdot), \beta(u)))$$

and the lower value-function:

$$\Psi(t_0, x_0) := \sup_{\alpha \in \Gamma(t_0)} \inf_{v(\cdot) \in \mathcal{V}(t_0)} g(x(T, t_0, x_0, \alpha(v), v(\cdot)))$$

Proposition 6.3 *Assume that (4) holds true. If g is continuous, then Ψ and Φ are continuous.*

Proof —   We shall prove that Ψ is continuous[7] at some t_1, x_1. Consider $\varepsilon > 0$, , x_2 and $0 \le t_1 \le t_2 \le T$. By the very definition of the value-function Ψ, there exists $\alpha \in \Gamma(t_1)$ such that

$$(18) \qquad \Psi(t_1, x_1) \le \inf_{v(\cdot) \in \mathcal{V}(t_1)} g(x(T, t_1, x_1, \alpha(v(\cdot)), v(\cdot))) + \varepsilon$$

---

[7]It's easy to extend the proof when g is uniformly continuous and then the value-functions are uniformly continuous too.

Fix $\bar{v} \in V$. For any $v(\cdot) \in \mathcal{V}(t_2)$, we define $\underline{v}(s) =$

$$
\begin{cases}
\bar{v} & \text{if } s \in [t_1, t_2] \\
v(s) & \text{if } s \in [t_2, T]
\end{cases}
$$

and for any α we define $\underline{\alpha}(v) = \alpha(\underline{v})$.

Hence, there exists $v(\cdot) \in \mathcal{V}(t_2)$ such that $\Psi(t_2, x_2) \geq g(x(T, t_2, x_2, \underline{\alpha}(v), v))) - \varepsilon$ and according to (18), we have $\Psi(t_1, x_1) \leq g(x(T, t_1, x_1, \alpha(v), v)) + \varepsilon$. On the other hand, from Gronwall's Lemma, there exists some $R > 0$ such that

$$
\|x(T, t_1, x_1, \underline{\alpha}(v)(\cdot), v(\cdot)) - x(T, t_2, x_2, \alpha(v)(\cdot), v(\cdot))\| \leq R(\|x_1 - x_2\| + (t_1 - t_2))
$$

Since g is continuous, there exists $\delta > 0$ such that for any $(t_2, x_2) \in R([0, 1] \times B)$ we have

$$
|g(x(T, t_1, x_1, \underline{\alpha}(v)(\cdot), v(\cdot))) - g(x(T, t_2, x_2, \alpha(v)(\cdot), v(\cdot)))| \leq \varepsilon
$$

Hence $\Psi(t_1, x_1) - \Psi(t_2, x_2) \leq 3\varepsilon$. On the other hand for every α: $\Psi(t_1, x_1) \geq \inf_{v(\cdot) \in \mathcal{V}(t_0)} g(x(T, t_1, x_1, \alpha(v), v(\cdot))) \geq \inf_{v(\cdot) \in \mathcal{V}(t_0)} g(x(T, t_2, x_2, \alpha(v), v(\cdot))) - \varepsilon$ Hence $\Psi(t_1, x_1) \geq \Psi(t_2, x_2) - \varepsilon$. We have similar result when $t_2 < t_1$ and for the value-function Φ. $\square$

7 Solutions to Isaacs equations with nonanticipating strategies

Proposition 7.1 *If (4) holds true, then Φ satisfies (6)i and Ψ satisfies (7)ii.*

Proof — Fix $\bar{u} \in U$. Consider $\beta_h \in \Delta(t_0)$ such that

$$
\sup_{u \in \mathcal{U}(t_0)} g(x(T, t_0, x_0, u(\cdot), \beta_h(u))) \leq \Phi(t_0, x_0) + h^2
$$

Let define $\mathcal{U}_h(t_0)$ the subset of measurable controls $u(\cdot) \in \mathcal{U}(t_0)$ such that $u(s) = \bar{u}$ for almost every $s \in [t_0, t_0 + h]$. then

$$
(19) \qquad \sup_{u \in \mathcal{U}_h(t_0)} g(x(T, t_0, x_0, u(\cdot), \beta_h(u))) \leq \Phi(t_0, x_0) + h^2
$$

By the very definition of β_h, there exists some $v(\cdot) \in \mathcal{V}(t_0)$ such that for any $u(\cdot) \in \mathcal{U}_h$, $v(s) = \beta_h(u)(s)$ for almost every $s \in [t_0, t_0 + h]$.

Let $x_h(\cdot)$ denote the solution to $x'(t) = f(t, x(t), \bar{u}, v(t))$ on $[t_0, t_0 + h]$ such that $x_h(t_0) = x_0$. From (19), we deduce

$$\sup_{u \in \mathcal{U}_h(t_0)} g(x(T, t_0 + h, x_h(t_0 + h), u(\cdot), \beta_h(u))) \leq \Phi(t_0, x_0) + h^2$$

Define $\beta \in \Delta(t_0)$ such that for any $u(\cdot) \in \mathcal{U}(t_0)$ we have $\beta(u) := \beta_h(\underline{u})$ with

$$\underline{u}(s) := \begin{cases} \bar{u} & \text{if } s \in [t_0, t_0 + h] \\ u(s) & \text{if } s > t_0 + h \end{cases}$$

Hence $\sup_{u \in \mathcal{U}(t_0)} g(x(T, t_0, x_0, u(\cdot), \beta(u))) \leq \Phi(t_0, x_0) + h^2$ and therefore $\inf_{\beta \in \Delta(t_0)} \sup_{u \in \mathcal{U}(t_0)} g(x(T, t_0, x_0, u(\cdot), \beta_h(u))) \leq \Phi(t_0, x_0) + h^2$. This proves the following inequality

$$\Phi(t_0 + h, x_h(t_0 + h)) \leq \Phi(t_0, x_0) + h^2.$$

On the other hand, there exists a sequence $h_i \longrightarrow 0$ and $\bar{v} \in V$ such that

$$\frac{x_{h_i}(t_0 + h_i) - x_0}{h_i} \longrightarrow f(t_0, x_0, \bar{u}, \bar{v})$$

this yields $D_{\uparrow}\Phi(t_0, x_0)(1, f(t_0, x_0, \bar{u}, \bar{v})) \leq 0$ and consequently (6)i). The proof is similar for the second statement. $\square$

Proposition 7.2 *If g is continuous, then Φ satisfies (6)ii) and Ψ satisfies (7)ii).*

It is possible to prove that Φ is a viscosity subsolution to (8) and thanks to results of section 5 that it is a contingent solution to (6)ii) (see [13] for the proof).

Corollary 7.3 *If g is continuous, then Φ is a viscosity solution to (8) and Ψ is a viscosity solution to (9).*

Finally we just state an existence result

Proposition 7.4 *Assume that (4) holds true and that g is uniformly continuous. If we assume the Isaacs' condition (10), then $\Phi = \Psi$ and the value-function is the unique uniformly continuous viscosity solution to the Isaacs' equation.*

The proof is based on a theorem of Crandall-Lions concerning the unicity of bounded uniformly continuous solution of Hamilton-Jacobi equations (see [16]).

References

[1] AUBIN J.-P. (1991) VIABILITY THEORY. Birkhäuser. Boston, Basel, Berlin.

[2] AUBIN J.-P. & FRANKOWSKA H. (1990) SET-VALUED ANALYSIS. Birkhäuser. Boston, Basel, Berlin.

[3] AUBIN J.-P. & FRANKOWSKA H. (to appear) *Partial differential inclusions governing feedback controls.*

[4] BARLES G. (1991) *Discontinuous viscosity solutions of first-order Hamilton-Jacobi Equations: a guided visit.* Preprint.

[5] BARRON E.N., EVANS L.C. & JENSEN R. (1984) *Viscosity solutions of Isaacs' equations and differential games with Lipschitz controls.* J. of Differential Equations, No.53, pp. 213-233.

[6] BARRON E.N. & JENSEN R. (1990) *Optimal Control and semicontinuous viscosity solutions*, Preprint, March 1990.

[7] BERNHARD P. (1979) CONTRIBUTION À L'ÉTUDE DES JEUX DIFFÉRENTIELS À SOMME NULLE ET À INFORMATION PARFAITE. Thèse de doctorat d'état, Paris VI.

[8] CRANDALL M.G. & EVANS L.C. & LIONS P.L. (1984) *Some properties of viscosity solutions of Hamilton-Jacobi Equations*, Transactions of A.M.S., 282, pp. 487-502.

[9] EVANS L.C. & SOUGANDINIS P.E. (1984) *Differential games and representation formulas for solutions of Hamilton-Jacobi-Isaacs equation.* Indiana University Mathematical J. Vol. 33, N. 5, pp. 773-797.

[10] FILIPPOV A. F. (1958) *On some problems of optimal control theory.* Vestnik Moskowskovo Universiteta, Math. No. 2, 25-32. (English translation (1962) in SIAM J. of Control, 1, 76-84).

[11] FRANKOWSKA H. (to appear) CONTROL OF NONLINEAR SYSTEMS AND DIFFERENTIAL INCLUSIONS. Birkhäuser. Boston, Basel, Berlin.

[12] FRANKOWSKA H. (to appear) *Lower semicontinuous solutions to Hamilton-Jacobi-Bellman equations* SIAM J. of Control.

[13] FRANKOWSKA H. & QUINCAMPOIX (to appear) *Value functions for differential games with two concepts of strategies.*

[14] ISAACS R. (1965) DIFFERENTIAL GAMES. Wiley.

[15] KRASSOVSKI N.N. & SUBBOTIN A.I. (1988) GAME-THEORETICAL CONTROL PROBLEMS. Springer-Verlag.

[16] LIONS P.L. (1982) GENERALIZED SOLUTIONS OF HAMILTON-JACOBI EQUATIONS, Pitman.

[17] SUBBOTIN A.I. & TARASYEV A.M. (1985) *Stability properties of the value function of a differential game and viscosity solutions of Hamilton-Jacobi equations* Probl. of Contr. and Info. Theory, Vol.15, No 6, pp. 451-463.

International Series of Numerical Mathematics, Vol. 107, © 1992 Birkhäuser Verlag Basel

OPTIMAL CONTROL
FOR ROBOT MANIPULATORS

A. KHOUKHI[1,2], Y. HAMAM[1]

ABSTRACT :

A trajectory planning method based on the dynamic model of robot manipulators is presented. First the problem of the optimal trajectography is discussed by considering several sub-problems related to the structural and modelling singularities, obstacle avoidance and technological constraints. The trajectory is then obtained through the optimization of a time-energy criterion and the identification of the conditions governing all singular configurations of a six degrees of freedom manipulator whose inverse kinematic position problem has a closed form solution. This general solution is applicable to a large class of robotic systems and takes into account constraints on the torques, non linearities and non convexities in the state space equations. The discrete augmented Lagrangian (DAL) technique is used to obtain a controller that caters for constraints on the state, as well as for system inputs, task specification and environnement modelling. The penalty coefficient asssociated with the equality and inequality constraints of the DAL is considered as a variable which is adjusted during the iterative procedure in order to improve the conditioning and the constraints satisfaction. The method was programmed on a PC-AT and some simulation results are given. The method was also used as a CAD tool for the trajectory planning of the modular assembly robot PAMIR developed at the Control Laboratory of ESIEE.

Key words : *Robotics, Dynamic Trajectory Planning, Singularity, Optimal Control, Augmented Lagrangian, CAD.*

(1) Groupe ESIEE, 2 Boulevard Blaise Pascal - BP.99 - 93160 Noisy-Le- grand Cedex - France.

(2) ENST, Département Réseaux, 46 Rue Barrault - 75634 Paris Cedex 13.

1   INTRODUCTION

As the number of industrial robots introduced into various manufactoring areas increases, requirements for their autonomy, flexibility and self-governing,and ability to work in different environments becomes more severe. Hence Automatic Trajectory Planning of a manipulator in an environment with obstacles is a fundamental problem in robotic design and control. This problem is very complex, involving several subproblems such as obstacles modeling, handling sensor information, searching for collision-free path and avoidance of singular configurations of the robot [5] [23]. In addition this task should be accomplished quite accurately in a specified time, with high speed and in cooperation or competition with other machines.

From the point of view of kinematics a manipulator is essentially a positioning device, and one must deal with all types of singular configurations with mathematical modelling and structural singular configurations. When the Jacobian of a manipulator is singular, the manipulator becomes irresolvable [5] [6] and loses one or more degrees of freedom. In this case, finite cartezian motion rate may corresponds to an infinite joint motion rate. Consequently the robot may be immovable in certain directions, even after very large joint torques are applied [6] [25]. Moreover when a robot is singular with respect to a given hand position and orientation, its inverse kinematic problem may have an infinite number of solutions. This will impair trajectory execution. For these reasons it is important to consider the singularity problem in robotic trajectory planning.

In the manual planning case [1], the trajectory is obtained as a set of generalized position which the robot must reproduce in task execution. The drawback of this planning scheme is that in a real time situation, we may have a saturation of actuators which was not predicted by the plannar and the robot fails to deliver the required speed and acceleration[6],[25].

On the other hand, robotic manipulators are essentially dynamic systems. In the case of slightly coupled mechanical configurations two degrees of freedom (DOF) and slow speeds, kinematic control produces a sufficiently good performances. However in the case of faster motion with strongly coupled joints, dynamic effects such as inertia, centrifugal, gravity forces, and viscous friction cannot be neglected, and must be taken into account to obtain good control[23] [24]. A second fact is that, the high performance of a manipulator is directly related to its speed in executing tasks as well as it's precision and reliability. Thus in many practical applications a time-optimal control with good, precise positioning is higly desirable. It is shown in [9] that the time-optimal control for a large class of robot manipulators is essentially of the bang-bang type and singular arcs in restricted regions. An algorithm which computes the time-optimal control is also given. This algorithm operates in two phases: a minimum-time ballistic phase which is essentially bang-bang followed by an adaptive feedback approach phase in order to assure proper approach and precision.In this paper we are concerned with the optimal multicriteria trajectory planning of a robot manipulator. As shown in[9] , the bang-bang control scheme has some disadvantages which make it impracticle due to dynamic effects. The first disadvantage is the fatigue of the robot due to continuous operation at the limits. The second is the sensitivity of such control scheme to frequent commutations of the robot between limits, leading to deviations exceeding the tolerance desired. The third is the non-adaptivity of that scheme to the environnement of the robot which affects its versatility.

Several attempts have been made to satisfy the above objectives, but these have fallen short of a general solution. They are based on the linearization of the dynamics in order to apply standard techniques of the linear optimal control theory. Kahn and Roth[13] linearize the dynamics constraints around the final point, and use a dynamic model in order to obtain a time optimal control. Approaches based on dynamic linearization have been cast into doubt, since it can be shown that the velocity product terms have the same segnificance relative to the acceleration terms for all steps of the movement [26]. Thus the main assumption used to justify linearization of the dynamics, namely that the velocity terms can be ignored, is fundamentally wrong. Besides, the linearised approaches which may give good results are limited to the cases of small movements, low speeds, and low performance requirements. Cases where such conditions prevail may be solved without much prejudice by simple kinematic control[27]. Unfortunatly in many interesting situations it is not the case. For example, in an assembly robot, the task may be considered as two successive steps, a large movement followed by a precise insertion final stage. The above control schemes may satisfy one or the other requirement, but not both. The shortcoming of purely kinematic approaches is that the dynamic effects are not considered, so that the constraints consist of bounds on kinematics elements (position, velocity and acceleration). These constraints are imposed by the weakest configuration of the robot. Therefore, motion in the other areas of the workspace is suboptimal[27].
Lynch [15] has proposed a control strategy based on a sequential axis minimum time control. In his scheme each axis is moved individually in minimum time using a linear dynamic equation simplification. This approach is far from being time-optimal and for an n-joint robot produces movements about n-times slower than simultaneous minimum time control of all the joints.
Kim and Shin[25] present another approach to the time-optimal trajectography. It consists of a fixed parametrized path in the generalized space: straight line segments connected at via points, and a curved transition region from one straight line segment to the next with a bounded deviation from the knot point, the intermediate points are given and it is assumed that there is a constant acceleration and deceleration to a maximum velocity along each straight line segment, and that the transition is made with a constant acceleration.

The dynamic effects are introduced into the transition points as constraints on the acceleration. The transition points are adjusted to minimise time given the acceleration constraints. Because of the heuristic nature of this method, it is possible for the solution to be far from the optimal. Also there is no guarantee that the obtained solution does not violate torque bounds in intermediat regions. Another approach is that developed by Bobrow et al [3] where the bounds on the torques (forces obtained from the motors) are used to construct a curve in the distance-velocity space that constitutes an upper limit on the performance of the arm. A set of switching points is then found, where the arm moves as close as possible to the limit curve. One actuator is always saturated and the others adjust their torques so that some constraints and limitations on the motion are not violated. A similar geometrical approach was considered by Shin and McKay [21] with the difference being a parametrization of the path in the generalized space instead of in the operation space. The only problem of this method is that the path has to be known beforehand.

In this paper, a new approach is proposed to overcome the bang-bang difficulties and to take into account the non linearities, geometrical singularities configurations of the robot, task specification, and environment constraints.

A minimum time-energy control (MTEC) is used with a weighting factor which allows the user to adapt the solution to his particular application. This approach may be viewed as a smoothing of the bang-bang trajectories; thus reducing the fatigue of the robot which would result from running it continuously at the extremes.

The multicriteria control is formulated as a non-linear programming problem under equality and inequality constraints on the state and input. As is known, the method of multipliers [16] [17] provides the solution of the constrained problem via a sequence of unconstrained problems. The main drawback of this method is that in principle, it requires an infinite sequence of unconstrained minimization problem to be solved. To overcome this difficulty, a further development was proposed by Fletcher [7] [8] and Rockafellar [14] [16] by-introducing in the augmented Lagrangian a multiplier vector continuously dependent on x. In this way a single minimization is required. A related algorithm was proposed by Mukai and Polak [17]. However this method requiers a matrix inversion at each function evaluation and this may limit their applicability. Other possibilities were proposed by Wierzbicki [18] and Dipillo and Grippo [19]. In [18] several algorithms are derived for locating directly the saddle-point of the augmented Lagrangian by simultaneous updating of the input and multiplier parameter and without resorting to matrix inversion.

In [19] a different approach obtained by adding to the augmented Lagrangian a penalty term on the first order necessary conditions.

More recently utilising the auxiliary problem idea, Cohen [11] has applied the method to solve a large class of decomposition/coordination problems in the non-differential case. Glowinski et al. have used the method as well for the numerical solution of the two dimentionnal problem in incompressible finite elasticity [13].

The algorithm proposed here is based on the discrete formulation of the augmented Lagrangian which presents many advantages in optimal control applications [11].

As remarked earlier, the augmented Lagrangian factor c has an important role in the conditioning of the convergence speed, and constraint satisfaction.
In this work a penalty coefficient readjustement is introduced in the second level of the optimization process, which is a crucial task in the conditioning of the non strongly convex constraints, such as the instantanious acceleration constraints.

The paper is organised as follows. In section 2, the robot model and associated constraints are described. In section 3, we present the non-linear programming problem using the discret augmentated Lagrangian formulation. In section 4, simulation results and numerical considerations are discussed. In section 5, by introducing disturbances to the robot model, sensitivity of MTEC to noise effects and perturbations are analysed. This is obtained by introducing disturbances to the robot model. In section 6, the application of the method as a CAD tool for the dynamic design and off-line trajectory planning is discussed. Finally we conclude by offering some remarks and perspectives.

2   ROBOT MODEL AND ASSOCIATED CONSTRAINTS

The work presented in this paper has been applied to the PAMIR robot developed at the control department of "Groupe ESIEE"[4] [9] [10], Fig.1. This robot is a pedagogic modular assembly robot designed for education and research. It's modularity (sensor, control and motorization), renders it suitable as a test bench for control strategies.

2.1 The robot model

The dynamic model is established from the Euler Lagrange equations which take the following form [10] [24]:

$$M\,(q)\,\ddot{q} + N\,(q\,,\,\dot{q}) = U + G\,(q) \qquad (1)$$

where

q is the generalized coordinates vector
M kinematic energy matrix of dimension $n\mathrm{x}n$
G is the vector of n torques due to the gravity
U is the control vector of dimension n
$N(q\,,\,\dot{q}) = B\,(q)\,\dot{q}\,\dot{q} + C\,(q)\,\dot{q}^2$

and

B is the matrix of coefficients of the coriolis torques
C is the matrix of coefficients of the centrifugal torques
$\dot{q}\,\dot{q} = (\dot{q}_1\,\dot{q}_2\,,\,\dot{q}_2\,\dot{q}_3\,\cdots\,\dot{q}_{n-1}\,\dot{q}_n)^T$

In this paper, the system controlled is of n motor-reducers and the numerical applications will be made for $n = 3$ which does not lose the generality. Each motor is subject to a torque defined by

$$R_i = \frac{q_{mi}}{q_i} = \frac{\dot{q}_{mi}}{\dot{q}_i} = \frac{\ddot{q}_{mi}}{\ddot{q}_i} = \frac{\Gamma_i}{\Gamma_{mi}}\,(R_i > 1)$$

where

R_i is the reduction ratio
q_{mi} is the ith component of the generalized coordinates on the motor side
Γ_{mi} the resisting torque on the motor side

The fundamental dynamic equation is

$$J_i\,\ddot{q}_{mi} + V_i\,\dot{q}_{mi} = K_i\,I_i - \Gamma_{mi} \qquad (2)$$

where

J_i is the moment of inertia of the motor reducer (i)
V_i is the viscous coefficient of friction of the motor (i)
K_i is the torque constant of the motor (i)

After transformation to the robot coordinate, the following equation is obtained

$$J_i\,R_i^2\,\ddot{q}_i + V_i\,R_i^2\,\dot{q}_i - \Gamma_i = K_i\,R_i\,I_i \qquad (3)$$

which may be reexpressed in the same way as for equation (1)

$$(A + JR^2)\,\ddot{q} + B\,\dot{q}\,\dot{q} + C\,\dot{q}^2 + V\,R^2\,\dot{q} - G = K\,R\,I \qquad (4)$$

where

JR^2 is a diagonal matrix of the coefficients $J_i\ R_i^2$ of dimension $n \times n$
VR^2 is a diagonal matrix of the coefficients $V_i\ R_i^2$ of dimension $n \times n$
KR is a diagonal matrix of the coefficients $K_i\ R_i$ of dimension $n \times n$
I is the vector of the motor currents of dimension n

The equation (4) gives

$$\ddot{q} = (A + JR^2)^{-1}\,(G - B\,\dot{q}\,\dot{q} - C\,\dot{q}^2 - V\,R^2\,\dot{q} + KRI) \qquad (5)$$

This leads to the following state space representation

$$\begin{bmatrix} \dot{q} \\ \ddot{q} \end{bmatrix} = \begin{bmatrix} \dot{q} \\ (A + JR^2)^{-1}\,(G - B\dot{q}\,\dot{q} - C\dot{q}^2 - V\,R\,\dot{q}) \end{bmatrix} + \begin{bmatrix} 0 \\ (A + JR^2)^{-1}\,KR \end{bmatrix} I$$

$$(6)$$

It is important to note that in equation (6), the control appears in a linear form whereas the influence of the state is non linear.

2.2   The Geometric model

Since the work zone is defined in the operation coordinates and the control in the generalized coordinates space the geometric coordinate transformation has to be used. This is given by (see Fig.1 and [4] [9] [10]).

$$Y_1 = H_1(q) = ((d_3 - y_o - d_1\ sin\ (q_2) + d_4\ sin\ (q_2 + \_3))\ cos\ (q_1) - z_o\ sin\ (q_1)$$

$$Y_2 = H_2(q) = d_2 + (d_3 - y_o - d_1)\ cos\ (q_2) + d_4\ sin\ (q_2 + q_3) \qquad (7)$$

$$Y_3 = H_3(q) = ((d_3 - y_o - d_1)\ sin\ (q_2) + d_4\ sin\ (q_2 + q_3))\ sin\ (q_1) + z_o\ cos\ (q_1)$$

2.3   The discrete model

When a robot is under computer control, the inputs u(t) are updated at each sampling instant and kept constant along this period, by digital-to-analog converters (DAC). The (DAC) are the interfaces between the digital controller and the robot. At each sampling instant, the DAC transmit piecewise constant inputs to the robot :

$$I(t) = I(KT) = I_k \quad for\ kT < T < (k+1)T \qquad (8)$$

The continuous state variables of the robot, the joints coordinates and the velocities are accessible and can be calculated from the model or measured with available instrumentation (such as encoders resolvers and gyros). The analog-to-digital convertors (ACD) sample the state variable at each sampling instant to produce the piecewise constant states :

$$q(t) \ = \ q(kT) \ = \ q_k \qquad for \ kT \ < \ T \ < \ (k+1)T$$

$$\dot{q}(t) \ = \ \dot{q}(kT) \ = \ \dot{q}_k \qquad for \ kT \ < \ T \ < \ (k+1)T \tag{9}$$

In response to constant input, the robot is thus a closed system. The number of independant integrals of the motion for a closed N degrees of freedom mechanical system is (2N-1) (see [5]).
The discrete model proposed is related to the dynamic equation of the robot and imposes constraints on the discretization scheme by considering as variable the sampling period (for more details see [4] and references there in).

We define $\dot{X}(t) \ = \ (\dot{q}(t), \ddot{q}(t))$

so the equation (6) may be written as :

$$\dot{X}(t) \ = \ F(X(t)) \ + \ G(X(t)) \, I \ + \ H(X(t))$$

Where :
$$F(X) \ = \ \begin{pmatrix} 0 & I_n \\ 0 & 0 \end{pmatrix}$$

$$G(X) \ = \ \begin{pmatrix} 0 \\ M^{-1}(q) \end{pmatrix}$$

$$H(X) \ = \ \begin{pmatrix} 0 \\ M^{-1}(q) \, N(q,\dot{q}) \end{pmatrix}$$

A similar procedure of descretization was considered in [27]. However in this paper the sampling period is taken as a control variable, and the optimization process is made with respect to it as well as to the current intensity I. Suppose that the control is piece-wise continuous,taking a constant values during a sampling length period h_j, and is calculated from the following scheme:

$$X(t+h_j) \ = \ F_d\,(X(t)) \ + \ G_d\,(X(t))\,I \ + \ H_d \tag{10}$$

with
$$F_d \ = \ e^{Fhj} \ = \ \begin{bmatrix} I_n & h_j I_n \\ 0 & I_n \end{bmatrix}$$

$$G_d \ = \ \int_t^{t+h_j} e^{F(t+h_j-1)} \, G(X(s))ds$$

$$H_d \ = \ \int_t^{t+h_j} e^{F(t+h_j-s)} \, H(X(s))ds$$

As the sampling step is optimized along the trajectory, and this is accomplished in a limited bounds while the changes of the state $X = (q, \dot{q})$ are assumed too small, so G and H may be approximated as :

$$G_d = \left[\int_t^{t+h_j} e^{F(t+h_j-s)} \, ds \right] G(X(H))$$

$$= - \left[\begin{array}{c} \frac{h_j^2}{2} I \\ h_j \, I \end{array} \right] M^{-1}(X(t)) \tag{11}$$

$$H_d = \left[\int_t^{t+h_j} e^{F(t+h-1)} \right] H(X(t))$$

$$= - \left[\begin{array}{c} \frac{h_j^2}{2} I \\ h_j \, I \end{array} \right] M^{-1}(X(t)) \, N(X(t)) \tag{12}$$

(10) - (12) lead to the following discrete equation :

$$X(t + h_j) = F_d(X(t)) + \left[\begin{array}{cc} h_j^2 & I \\ h_j & I \end{array} \right] I'$$

So the discrete representation consists of the development to second order the joints coordinates and to first order the velocity coordinates of the robot. This scheme of descretisation has been simulated and it runs well.

The state is defined by a six dimentional differential equation (6) of the robot model where $X(t) = (q_1(t), q_2(t), q_3(t), \dot{q}_1(t), \dot{q}_2(t), \dot{q}_3(t))$

A discrete representation [9] [10] may be given by:

$$\left(\begin{array}{c} X_{1k+1} \\ X_{2k+1} \end{array} \right) = \left(\begin{array}{c} X_{1k} + h_{1k} \, X_{2k} + \frac{h_k^2}{2} X_{2k} \\ X_{2k} + h_k \, X_{1k} \end{array} \right) \tag{13}$$

With $X_{2k} = M^{-1}(qk)(I_k - N(q_k))$

Note that this descretization scheme requiers the measurements of the position q and the velocity $\dot{q}$, and there is no need for measuring the joint accelerations.

2.4   Associated constraints

2.4.1   Initial and final state constraints:

The constraints on the initial and final position are

$$\begin{array}{cc} X_1(0) = X(0) & X_1(N) = X_F \\ X_2(0) = 0 & X_2(N) = 0 \end{array} \tag{14}$$

2.4.2 Initial and final velocity constraints:

It is assumed that the task starts and terminates with zero speed.

2.4.3 Control constraints:

The control parameters here are the current intensity I defined on a compact $I^{ad} \subset R^{3N}$ bounded by the saturation constraints on the actuators and which can be made state-dependant without modifying the algorithm

$$I_{min}^i \leq I_K^i \leq I_{max}^i \quad i = 1,3 \quad k = 1, N \tag{15}$$

Note that the current intensity I, may be a highly non linear function of the state, such as in electrical motors [5], and the length of commutation interval hi which is also in a compact $H^{ad} \in R^{+M}$ such that

$$hj_{min} < hj < hj_{max} \quad j = 1,3 \tag{16}$$

2.4.4 Acceleration constraints:

$$A_{min}^\ell \leq X_{2K}^\ell \leq A_{max}^\ell \quad \ell = 1,3 \quad k = 1, N \tag{17}$$

2.4.5 Work zone and environment constraints:

The work space of the manipulator is defined as the entire set of points that can be reached by the end effetor or the wrist point. Two types of singularities are distinguished.

First, structural singularities which corresponds to :

$$X_{1k} = (q_1^{(k)}, q_2^{(k)}, q_3^{(k)}) \in D_s \text{ where the robot is not resolvable with}$$

$$D_s = \left\{ (q_1, q_2, q_3) \text{ such that : a) } -d_1 + (d_3 - y_0) \sin q_2 + d_4 \sin (q_2 + q_3) = 0 \right\} \tag{18}$$

$$\text{b) } q_3 = 0 \mod [\pi]$$

This corresponds to the vacinity of the geometric and kinematic inverse. A physical interpretation of the domain D_s is that it corresponds to the structural singularity configuration of the robot. The arm is completely folded, or completely unfolded in which case the end effector can not accomplish the task.

REFERENCES

[1] M.BRADY : *"Robot motion : planning and control"*, *MIT Press, 1982.*

[2] E.BRYSON, HO : *"Applied optimal control"*, *Hemisphere Washington D.C., 1975.*

[3] J.E.BOBROW : *"Optimal control of robotic manipulators"*, *Ph D Desertation Univ. California, C.A., December 1982.*

[4] P.DURAND, F.VIETTE : *"A robot design for teaching and research"*, *Recent trends in Robotics. Proc. of the Inter. Symp. on rob. Research, nov. 1986, New Mexico, USA.*

[5] P.COIFFET : *"Les robots, modΩlisation et commande."* *Tome 1, Hermes,1981*

[6] T.LOZANO-PEREZ : *"Spatial planning : a configuration space approach"*, *IEEE Trans. on Computers. C-32:108, 1983.*

[7] R.FLETCHER : *"An exact penalty function for nonlinear programming with inequalities",in Math. Programming, Vol.5(1973), pp. 129-150.*

[8] R.FLETCHER : *"Methods related to Lagrangian functions"*, *in Numerical methods for constrained optimization, P.E.Gill, W.Murray, eds., Academic Press, New York, 1974, pp. 219-239.*

[9] A.KHOUKHI, Y.HAMAM : *"Adaptive minimum time control of robot manipulators"*, *Accepted for the 1989 IFAC conference and submitted to AUTOMATICA 1988.*

[10] D.GEORGES, Y.HAMAM : *"Planification optimale de trajectoire d'un robot manipulateur"*, *RAIRO Automatique et Productique, avril 1987, Vol.23.*

[11] G.Cohen : *"Decomposition et coordination en optimisation deterministê, These de l'Universite Paris Dauphine, 1984.*

[12] J.Y.S.LUH, C.E.CAMBELL : *"Minimum distance collision free path planning for industrial robots with prismatic joints", IEEE Trans. Aut. Cont., nov. 1984.*

[13] M.E.KAHN, B.E.ROTH : *"The near minimum-time control of open loop articulated Kinematic chains", ASME "Journal of Dynamic Systems, Measurement and Control" Sept. 1979.*

[14] T.ROCKAFELLAR : *"Penalty methods and augmented Lagrangian in nonlinear programming"*, *5th IFIP Conference on optimization techniques, Part I, R.Conti, A.Ruberti, eds., Springer-Verlag, Berlin, 1973, pp. 418-425.*

[15] P.M.LYNCH : *"Minimum-time, sequential axis operation of a cylindrical, two-axis manipulator"*, *in Proc.Joint Aut. Cont. Conf., 1981, Vol.1, paper WP-2A.*

[16] D.P.BERTSEKAS : *"Multiplier methods : a survey"*, *Automatica, 12 (1976), pp. 133-145.*

[17] E.POLAK : *"On the stabilization of locally convergent algorithms for optimization and root finding"*, *Automatica, 12 (1976), pp. 337-342.*

[18] A.P.WIERZBICKI : *"A primal-dual large scale optimization method based on augmented Lagrange functions and interaction shift prediction"*, *Ricerche di Automatica, 7 (1976), pp. 34-58.*

[19] G.Di PILLO, L.GRIPPO : *"A new class of augmented Lagrangian in non linear programming"*, *in SIAM, J.Cont. Opt., sept. 1979, Vol.17, N.5.*

[20] A.KHOUKHI, H.ZHANG, D.GEORGES, Y.HAMAM : *"Robot RACE; Modele dynamique et planification optimale de la trajectoire en temps energie, Rapport interne ESIEE-CEA/UGRA. N.*

[21] K.G.SHIN, N.D.MACKY : *"Minimum time control of robotics manipulators with geometric contraints"*, *IEEE Aut. Cont., june 1985, Vol. 30, N.6.*

[22] H.J.SUSSMAN : *"The structure of time optimal control trajectories for single input systems in the plane, SIAM, J.Cont., 8 octuber, july, 1987.*

[23] J.A.BROOKS : *"Planning collision-free motion for pick and place operations"*, *Al Lab Stanford University, 1983.*

[24] R.P.PAUL : *"Robot manipuators : mathematics, programming and control"*, *MIT Press, Mass 1981.*

[25] L.KIM,K.G.SHIN : *"Near minimum time control for robot manipulatpors"*, *IEEE Tran Aut.Cont. Juin 1985.*

[26] J.M.HOLLERBACH : *"Dynamic scaling of manipulator trajectories"*, *ASME, J.Dyn.Sys.Meas.Cont.1984. Vol 106.*

[27] M.VUKOBRATOVIC, D.STOKIC : *"Is dynamique control needed in robotics systems? and if so to what extent?"*, *The int.Jou.of Rob.Res.*

Control theory
and
Environmental problems:
Slow fast models for management of renewable ressources.

C. Lobry

Abstract: This paper is a report on work in progress within an interdisciplinary group of people supported by the "Programme Environnement " of C.N.R.S. Detailled mathematical proofs and ecological discussions will be published elsewhere. We focus on qualitative properties of two species ecological systems, one of which is exploited.

1. Introduction

The management of renewable ressources is one of the main problem of the present time. How to furnish food to an increassing population ? How to preserve future ? Is control theory of nonlinear ordinary differential equations of some help to solve these problems ? This paper have not the pretention to solve any of these three questions. I will concentrate on the question of fisheries, and more particularly on two species models of fischeries.

In the book of C. Clark(1976) one can learn very interesting applications of optimal control theory and Maximum Principle to management of fisheries. The models of fisheries considered are mainly one species models and, despite their unrealistic nature, much undersanding is gained from simple considerations. For instance system theoretist will recognize immediatly that "quota" policies are open loop control while "fishing effort" policies are feed back control Thus no surprise if the former are unstable and the later stable ! The last two chapter of the book are concerned with growth and multispecies problems. It turns out that, even with two species, the discussion is much more difficult. In the recent survey paper Munro(1992) on fischeries economics one sees that not much was gained on these questions since the 1970's. This is not surprising. Considered with the cost function, the hamiltonian system associated to a n-species model is a differential system in dimension $2(n+1)$ autonomous or not. The dynamics are essentially non linear and badly known. It is very hard to state something, even qualitative, unless in the case $n = 1$, which, thanks to structural properties (there are some !) reduces to two dimension and thus is completely tractable by phase plane analysis.

Neithertheless there is a way to open a breach: Take into account the fact that often, time scales in the dynamics, are not of the same magnitude. This was already recognised in Clark(1976) but not fully exployted. More recently , Muratory S. and S. Rinaldi(1989, 1990 a, 1990b) in a sery of interesting papers, not devoted to control but to coexistence of three species, explored systematiquely models with different time scales and showed that much qualitative understanding of the ecological phenomena is gained in this way.

On the other hand the theory of "slow-fast" differential equations in the plane and higher dimensions progressed a lot during last ten years. See Zvonkin A. and M. Shubin(1984), Diener M. and C. Lobry (Editors)(1985) and Benoit E. (Editor)(1992) for surveys on these questions). These progress where mainly obtained within the framework of Non Standard analysis. The intersting fact with non-standard analysis is that it allows us to say : "epsilon is a strictly positive fixed infinitesimal" to idealize the fact that actual epsilon is small compared to unity (say 0.1 or 0.01). This way of speaking is much appreciated by non mathematicians.

Another peculiarity of ecological systems modeled by differential equations is the fact that we usually just know the equations on a very qualitative ground. For instance it is reasonnable to assume that the growth rate of the population of Wales is a monotone increassing function of the available amount of Krill, starting from zero up to some maximum rate. But, due to the difficulty to keep populations of wales in laboratory it is difficult to say more. For this reason all the discussion in this paper will be based not on the analytical definition but on the sape of the functions defining the dynamics.

2 - The model

The amount of first species (say the Krill) is denoted by x. In absence of predator the evolution of x is given by the differential equation:

$$x'(t) = f(x(t))$$

where f have the shape given in fig 1 .

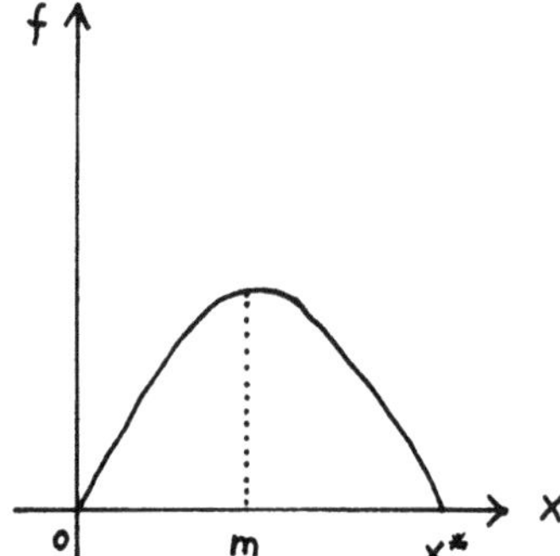

Fig. 1 graph of f

The amount of second species (say the wales) is denoted by y . It is a predator of x and the equations for the two species are:

$$x'(t) = f(x(t)) - y(t)V(x(t))$$
$$y'(t) = \varepsilon y(t) V(x(t)) - d y(t)$$

where V is the growth rate (depending on x) for y and has the shape given on the picture (fig 2), and d is a positive mortality coefficient.

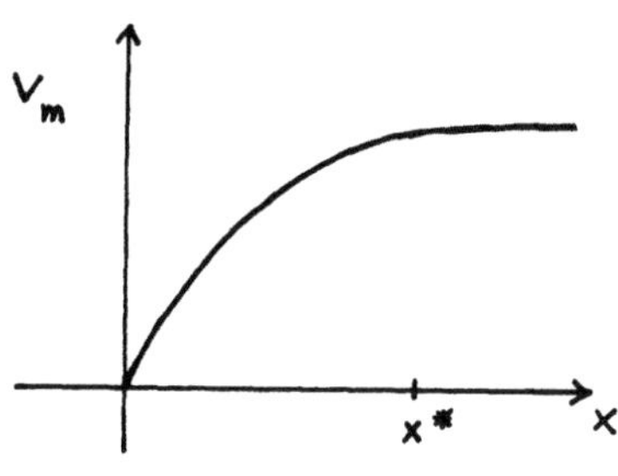

Fig 2 graph of V

The coefficient ε is a conversion coefficient of the biomass from x to y. It is smaller than one because a part of the biomass of x eaten by y is used, not to create biomass of y, but just to maintain the metabolism of y. In the model we make the idealization that ε is an infinitesimal. In the case of Krill and Wales this coefficient is taken as 0.1: May et al(1979). Now we introduce the fisching effort E in the model. We assume that the amount of fished y is proportionnal to y and to the fisching effort.

The complete model is thus:

$$x'(t) = f(x(t)) - y(t) V(x(t))$$
$$y'(t) = \varepsilon y(t) V(x(t)) - (d+E) y(t) \quad 0 \leqslant E \leqslant E_m$$

We do not claim that this model is realistic ! One one hand there are certainly other species which interact with x an y, on the other hand the response to fisching effort is far from being linear.

3 Mathematical analysis of the model.

Just by a change of unity on x and y, there is no loss in generality to suppose that $x^* = 1$ and $V_m = 1$. We have:

$$x'(t) = f(x(t)) - y(t) V(x(t))$$
$$y'(t) = \varepsilon y(t)[V(x(t)) - (d+E)/\varepsilon] \quad 0 \leqslant E$$

We consider the system for some fixed value of E.

Let us look for stationnary solutions (x_s, y_s). The nulcline $y'(t) = 0$ is a vertical line defined by :

$$v(x) = (d+E)/\varepsilon$$

which from the graph of V has a unique solution if $(d+E)/\varepsilon$ is smaller than one. (This means that mortality + catch rate must be smaller than the highest growth rate). The nulcline $x'(t) = 0$ is given by:

$$y = f(x)/v(x)$$

Corresponding to different hypothesis on V we give on fig 3-4-5 three possibilities for the nulcline $x'(t) = 0$.

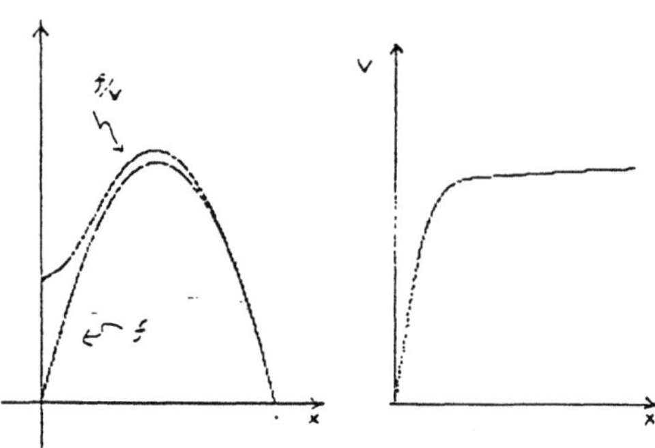

Fig 3 Hyp. 1

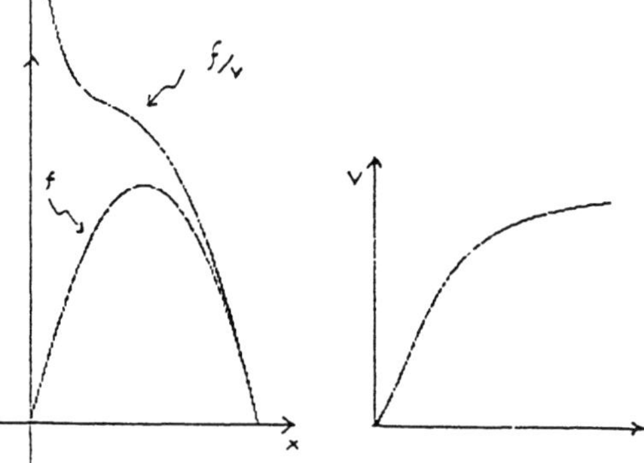

Fig 4 Hyp. 2

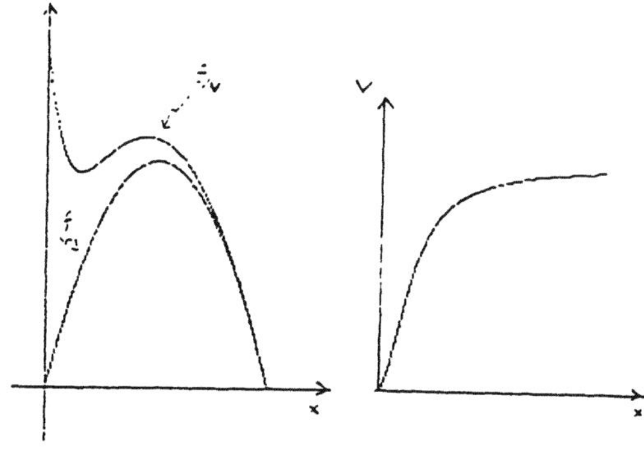

Fig. 5 Hyp. 3

Let E_m be defined by the relation:

$$V(1) = (d + E_m)/\varepsilon.$$

Provided that E is smaller than E_m there is, in the three cases, a unique steady state:

$$x_s = V^{-1}((d+E)/\varepsilon)$$
$$y_s = f(x_s)/V(x_s)$$

It is easily computed (and known in ecological litterature) that the steady state is stable if it is on the right of the hump, unstable otherwise. In the case of hyp. 1 there is no hump and the steady state is allways stable.

We compute now the yield $Y(x_s)$ corresponding to the steady state, even if it is unstable.

$$Y(x_s) = E y_s = (\varepsilon V(x_s) - d)f(x_s)/v(x_s)$$

$$V(x_s) = \varepsilon (f(x_s) - (d/\varepsilon) f(x_s)/v(x_s))$$

This formula shows that if the natural mortality is a small fraction of ε (or 0 !) the maximum yield almost coincides with the maximum of f , the growth rate of the prey . If the natural mortality is not small the maximum is obtained for sligthly greater values of x_s. It is intersting to notice that this maximum allways correspond to a stable steady state.

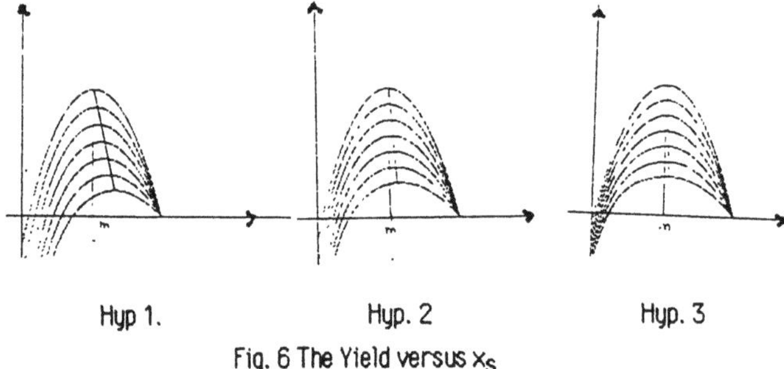

Fig. 6 The Yield versus x_S

Now we look at the case of a non stable steady state. We take advantage of the fact that ε is small. Thus if we are not in an infinitesimal neighbourhood of the nulcline $x'(t) = 0$ the trajectories are quasi horizontal and the orientations are given by direct examination of the signums. In the infinitesimal neighbourhood of the nulcline $y(t)$ is increasing if it is on the right of x_S ans decrease otherwise. Thus when the nulcline is attracting the motion is completely described.

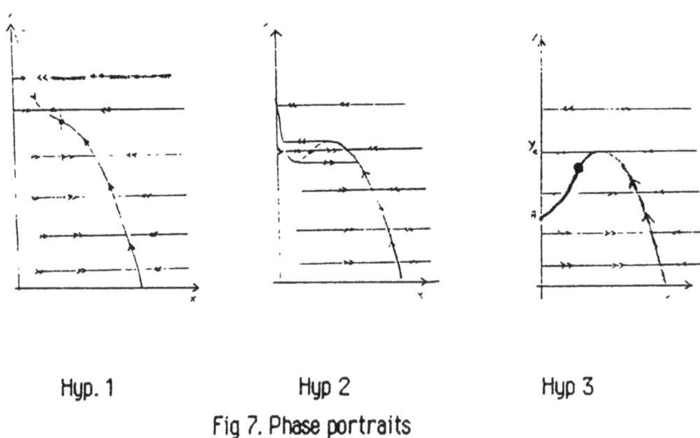

Fig 7. Phase portraits

The the case of Hip. 3 deserves more attention. What happens when $x(t)$ is infinitesimal ? Suppose that the trajectory enters the infinitesimal neigbourhood of $x = 0$ at an "entry point" y_e. From the arrows we see that it cannnot leave the vertical axe before $y(t)$ is smaller than A, by a slight modification of the techniques introduced by Benoit(1981) it is easily proved that the trajectory will actually leave at a point y_b strictly smaller (and not infintesimaly close to A) and computable from y_e , f'(0) and V'(0).

Very often people believe that the trajectory will leave at the point A. This tis he case for instance in Muratori and Rinaldi(1989, 1990a and b). It might be argued that in actual systems, due to the presence of noise the unstable branch below A is not "physic" an that in "actual world" the trajectory must leave just below the point A. This is not exactly the case. It depends of the nature of the noise. A multiplicative noise (i.e. of amplitude proportionnal to x) will have the tendency to maintain x(t) close to 0 while an additive noise has the opposite effect . See (Benoit 1992, Benoit and alii 1989).

We assume no noise at all on our system an follow on our discussion. We shall discuss later this assumption on biological grounds. For the present time we just discuss the model as a mathematical object. Thus it turns out that, in both cases hyp. 2 and hyp. 3 it exists a unique stable limit cycle as shown below.

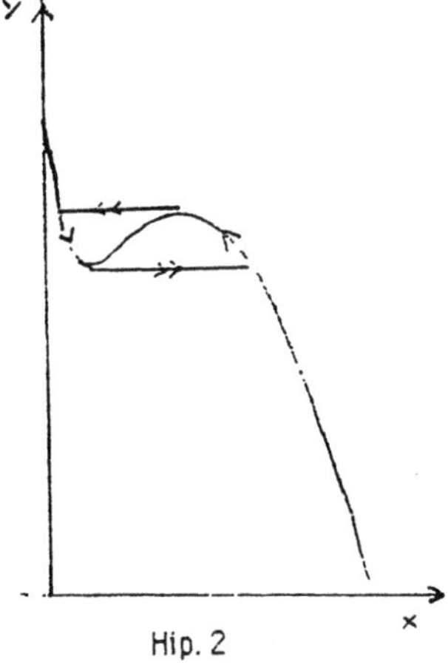

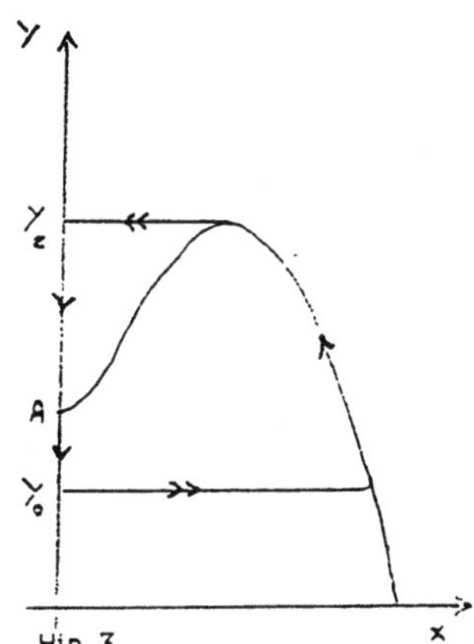

Fig 8

Now if we are interested in the resulting yield for a fishing effort corresponding to a limit cycle we see that in average y(t) is smaller than the value at the highest point on the hump of $f(x)/V(x)$ which corresponds to a steady state whose yield is smaller than the best one obtained for a greater value of x_S. On the other hand the catching effort increases when x_S moves to the rights. From these considerations it turns out that in the three hypothesis the best Yield for a constant effort E is obtained for an effort to which corresponds a stable steady state. Thus under these hypothesis the situation seams O.K. Let us look at it more closely.

4 Interpretation of the results

We will not discuss the fact that our model is over simplificated. This is allways the case when one wants to get informations from the mathematical structure of the dynamics. As a consequence the following interpretations are just indications on what could be. We also do not claim that we have explored in this paper all the possibilities of the model. With reasonnable assumpions on f (concavity) and V (monotonicity) the shape of the quotient f/V can be different.

First of all our easy discussion about the presence or absence of cycles is based on the purely mathematical hypothesis that ε is infinitesimal. To what extend the results remain true for actual values of ε. We show on the pictures below the actual phase portrait for $\varepsilon = 0.1$ and 0.2. It is a well recognised fact (Benoit and alii 1981, Muratori   Rinaldi 1990 b) that qualitative pictures on the phase portrait proved for ε infintesimal remain true on simulations for surprisingly large values of ε.

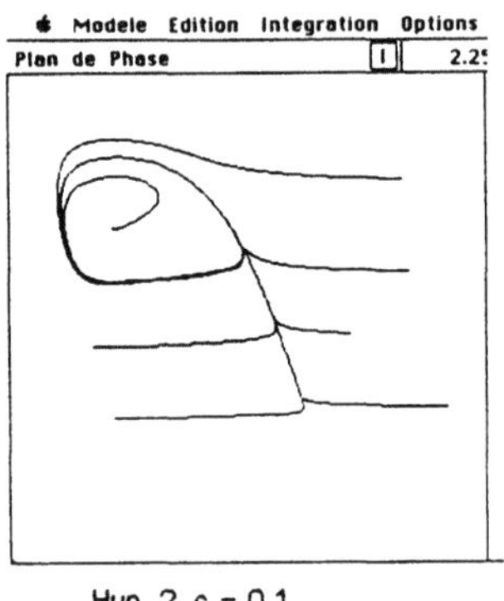

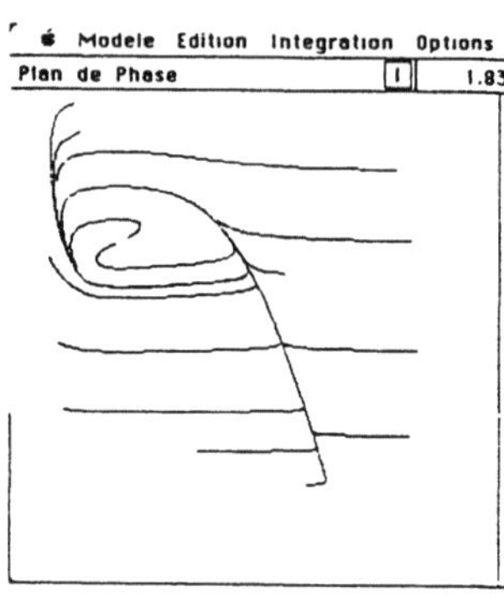

Hyp. 2 $\varepsilon = 0.1$ Hyp 2 $\varepsilon = 0.02$

Fig 9 simulations

Let us turn now to the fact that mathematicaly there is an "optimal fishing effort". Let E_0 be the optimal fishing effort and x_{s0} , y_{s0} the correspondind steady state.

When there is a hump, the optimal steady state is rather close to it. It is very important to notice that due to the form of $V(x)$ its inverse function is very sensible on E at E_0 . This means that a very small change in E around E_0 will move x_s a good deal right or left of x_{s0}.

In the case of hyp. 1 this has no other effect than a decrease of the Yield.

In the case of hip. 2 this might have the effect to make the system perform a cycle during which $y(t)$ is at a very low level during a certain duration. This might have domageable economical consequences. Notice that as soon as the trajectory is on the left of the hump, no matter the catch effort E is, the system will go to low values of y for a while.

In the case of hyp. 3 nothing is changed from hyp. 2 except that the situation is then catastrophic. The model we used, with continuous variables, is no longer valid because the values of x become very low (It can be proved, Benoit(1981) that these values are of the order of $\exp(-K/\varepsilon)$). There are some presomptions that the species x will disappear.

As a consequence of this discussion we see that it is very important to know the shape of $f(x)/V(x)$. Unfortunately this shape is very sensible to $f'(0)$ and $V'(0)$ which are certainly difficult, if not impossible, to estimate. The presence of a hump depends to some extent whether the saturation in $V(x)$ appears before the maximum growth rate for x or not. This might have ecological significance.

There is something interesting in the case of hyp 3 . Assume that x_S , y_S is a steady state and suppose that by some action on the environment the grotwh conditions for the prey are increased. This could change the shape of $f(x)/v(x)$ in order that the hump goes on the right of x_S ; the steady state is then unstable and the values of x will become excedingly low. This paradox: An enrichment of the prey drives the system to extinction, was noticed by Rosenzweig (1971) and discussed on mathematical grounds by Gilpin (1972). We have an analogous paradox if instead of changing the location of the hump we decrease the mortality rate such that x_S goes to the left of the hump. In some circonstances diminishing fishing effort may cause catastrophe !

5. Conclusion

Let us come back to the third question we addressed in the introduction. Is control theory of nonlinear ordinary differential equations of some help ?

If we assume two scales time in the equations it is possible to discuss significant features of the model, just on the basis of the shape of the nulclines. Argumentation is almost trivial and rigourous justifications can be found in mathematical literature. The same kind of argumentation will work for different models, with for instance a growth rate for y which depends on y itself. Controllability questions can be discussed by examination of the pictures. Easy computer simulations show to what extend ε has to be small. The same qualitative analysis is possible for three dimenssional systems. It has been successfully started by Muratory and Rinaldi (1989, 1990a, 1990b). But in this case the possibilities are very numerous and much remains to be done.

In conclusion my feeling is that nonlinear control theory can help, but not directly the one which was developped for engineering purposes. Many concepts and tools are to be reconsidered.

6 References

Benoit E. (1981): Relation entrée sortie, C.R.Acad. SC. Paris t. 293 pp. 293– 296.

Benoit E. (Editor)(1992): "<u>Dynamic Bifurcations</u>" Actes du colloque·de Luminy (mars 90) sur les bifurcations dynamiques. Lecture notes in mathematics 1493 .

Benoit E., Callot J. L.,Diener F and M. Diener (1981): Chasse aux canards, Collectanea Matematica, 32, 1 et 2.

Benoit E. and C. Lobry (1982): Canards de $\mathbb{R}^3$, C.R.A.S. t. 294 p. 483–488.

Benoit E., Candelpergher B. and C.Lobry (1989): Retard à la bifurcation en présence d'un bruit multiplicatif, in "<u>New trends in non linear control theory</u>", Descusse, Isidori, Fliess, Leborgne ed. Lecture notes in Control and Information Sciences no 122.

Clark C. (1976) : "<u>The optimal management of renewable resources</u>". Wiley and Sons 1976.

Diener M. and C. Lobry (Editors) (1985): "<u>Analyse non standard et représentation du réel.</u>" OPU (Alger) et CNRS (Paris).

Gilpin M. (1971): Enriched Predator – Prey Systems: Theoretical stability. Science, sept. 1972, pp 902 –904.

May R. , Beddington J., Clark W., Holt S. and Laws R. (1979): Management of Multispecies

Fischeries, Science, Vol 205, 20 July 1979 pp 267-275.

Munro R. (1992) : Mathematical Bioeconomics and the evolution of modern fischeries economics. Bull. of Math. Biology Vol 54. n° 2-3 pp 163-184

Muratory S. and S. Rinaldi (1989): Remarks on competitive coexistence. Siam J. Applied Math, Vol 49, n° 5 pp 1462-1472.

Muratory S. and S. Rinaldi (1990 a): Limit Cycles in Slow-Fast Forest-Pest Models. Working paper, IIASA, WP-90-38.

Muratory S. and S. Rinaldi (1990 b): A separation condition for the existence of limit cycles in slow-fast systems, to appear in Appl. Math. Modelling.

Rozenzweig M. (1971): Paradox of Enrichment: Destabilization of exploitation Ecosystems in Ecological Time. Science Vol 171, 29 Jan 1971pp. 385-387.

Zvonkin A. and M. Shubin (1984): Non-standard analysis and singular perturbations of differential equations, Russina Maths. Surveys 39: 69-131 (1984).

Aknowledgements

I thank E. Benoit (Ecole des Mines Sophia Antipolis), J.L. Durand (IFREMER Nantes) J.L. Gouzé (I.N.R.I.A. Sophia Antipolis), A. Sciandra and P. Nival (C.N.R.S. Villefranche sur Mer) for helpful discussions.

Author's address:

C. Lobry
8 Bd du Tzarewitch,
06 000 NICE (France)

International Series of Numerical Mathematics, Vol. 107, © 1992 Birkhäuser Verlag Basel   231

On the Riccati equations of stochastic control

Dr.Toader Morozan

Abstract. *We give sufficient conditions for the existence of bounded nonnegative solutions of Riccati equations of stochastic control. The problem of the existence of an optimal stabilizing compensator for linear control systems with state-dependent noise is discussed .*

1. Stabilizability, dectability, uniform observability and uniform controllability for linear Itô equations.

Let us consider the following stochastic linear control system

$$dx(t)=[A(t)x(t)+B(t)u(t)]dt+\sum_{j=1}^{N} G_j(t)x(t)dw_j(t) \tag{1}$$

Consider the linear system

$$dx(t)=A(t)x(t)dt+\sum_{j=1}^{N} G_j(t)x(t)dw_j(t) \tag{2}$$

with the output

$$z(t)=C(t)x(t)$$

where x,u,z are n,m,p-vectors respectively and $w_1,...,w_N$ are independent ‚standard Wiener processes.

Functions A,B,C and G_j are defined on the whole real axis $\mathbb{R}$ and are assumed to be continuous and bounded.

By $X(t,s)$, $t \geq s$ we denote the fundamental random matrix solution associated with the system (2).

Definition 1.

We say that $\{A;G_j\}$ is stable if there exist $b \geq 1$ and $a > 0$ such that $E|X(t,s)|^2 \leq be^{-a(t-s)}$ for all $t \geq s$.(E denotes expectation)

Definition 2 (see Da Prato and Ichikawa 1990)

1) We say that $\{A,B;G_j\}$ is stabilizable (the system (1) is stabilizable) if there exists F continuous and bounded on $\mathbb{R}$ such that $\{A-BF;G_j\}$ is stable

2) We say that $\{A,C;G_j\}$ is detectable if there exists L continuous and bounded on $\mathbb{R}$ such that $\{A-LC;G_j\}$ is stable.

Definition 4. (see Morozan 1993)

1) We say that $\{A,C;G_j\}$ is uniformly observable if there exists $\tau > 0$ and $\gamma > 0$ such that

$$E \int_s^{s+\tau} X^*(t,s)C^*(t)C(t)X(t,s)dt \geq \gamma I \text{ for all } s \in \mathbb{R}$$

2) We say that $\{A,B;G_j\}$ is uniformly controllable if there exists $\tau > 0$ and $\gamma > 0$ such that

$$E \int_{s-\tau}^{s} X(s,t)B(t)B^*(t)X^*(s,t)dt \geq \gamma I \text{ for all } s \in \mathbb{R}$$

In the following by H_n we denote the space of all n×n symmetric matrices. H_n is a real Hilbert space with the scalar product $\langle H_1,H_2 \rangle = tr(H_1 H_2)$; (Tr A is the trace of A)

By using the Itô formula we can prove easily that if $x(t)$ is a solution of (2) then $K_0(t)=Ex(t)x^*(t)$ is a solution of the following linear differential equation in the space H_n:

$$\frac{dK(t)}{dt} = A(t)K(t)+K(t)A^*(t)+\sum_{j=1}^{N} G_j(t)K(t)G_j^*(t) \qquad (3)$$

Let $T(t,s)$ be the evolution operator associated with equation (3).

We have

$$\frac{d}{dt}T(t,s)=U(t)T(t,s), \quad \frac{d}{dt}T^*(s,t)= -U^*(t)T^*(s,t) \qquad (4)$$

where T^* and U^* are the linear adjoint operators of T and U respectively and the linear operator $U(t)$ is given by

$$U(t)(H)=A(t)H+HA^*(t)+\sum_{j=1}^{N} G_j(t)HG_j^*(t), \quad H\in H_n$$

It is not difficult to verify (see Morozan 1993) that

$$T(t,s)(H)=EX(t,s)HX^*(t,s); \quad T^*(t,s)(H)=EX^*(t,s)HX(t,s), \qquad (5)$$

$H\in H_n$, $t\geq s$,

Thus from (5) it follows directly

Proposition 1.

(i) $\langle A,C;G_j\rangle$ is uniformly observable iff

$$\int_s^{s+\tau} T^*(t,s)(C^*(t)C(t))dt\geq\gamma I, \text{ for all } s\in\mathbb{R}$$

(ii) $\langle A,B;G_j\rangle$ is uniformly controllable iff

$$\int_{s-\tau}^{s} T(s,t)(B(t)B^*(t))dt\geq\gamma I \text{ for all } s\in\mathbb{R}$$

In the following we shall use the notation $\hat{A}(t)=A^*(-t)$, $t\in\mathbb{R}$

Proposition 2

$\langle A;G_j\rangle$ is stable iff $\langle\hat{A};\hat{G}_j\rangle$ is stable

Proof.

From (5) it follows that $\langle A;G_j\rangle$ is stable iff $\|T(t,s)\|\leq b_1 e^{-a(t-s)}$ for all $t\geq s$.

Let now $\hat{T}(t,s)$ be the evolution operator associated to $\hat{A},\hat{G}_j$. From (4) we deduce that
$$\hat{T}(t,s)=T^*(-s,-t) \qquad (6)$$
and thus the proof is finished.

Corollary 1.

$\langle A,B;G_j\rangle$ is stabilizable iff $\langle\hat{A},\hat{B};\hat{G}_j\rangle$ is detectable.
From (6) and Proposition 1 it follows directly

Corollary 2.

$\langle A,B;G_j\rangle$ is uniformly controllable iff $\langle\hat{A},\hat{B};\hat{G}_j\rangle$ is uniformly observable.

2. Bounded solutions of Riccati equations of stochastic control.

Consider the following Riccati equations with continuous and bounded coefficients:

$$P'(t) + A^*(t)P(t) + P(t)A(t) + \sum_{j=1}^{N} G_j^*(t)P(t)G_j(t) + C^*(t)C(t) -$$

$$- P(t)B(t)B^*(t)P(t) = 0 \tag{7}$$

$$S'(t) = A(t)S(t) + S(t)A^*(t) + \sum_{j=1}^{N} G_j(t)S(t)G_j^*(t) + D(t)D^*(t) -$$

$$- S(t)M^*(t)M(t)S(t) = 0 \tag{8}$$

It is known (see Da Prato,Ichikawa 1990 and Wonham 1968) that the equation (7) arises in the stochastic quadratic control problem and for $G_j \equiv 0$ the equation (8) arises in the linear filtering theory.

Definition 4.

1) A symmetric solution P of ·(7) is called stabilizing if $\{A-BF; G_j\}$ is stable, where $F(t) = B^*(t)P(t)$.
2) A symmetric solution S of (8) is called stabilizing if $\{A-LM; G_j\}$ is stable, where $L(t) = S(t)M^*(t)$.

Theorem 1.

Suppose:

1) $\{A,B; G_j\}$ is stabilizable
2) $\{A,C; G_j\}$ is either detectable or uniformly observable.
Then the Riccati equation (7) admits a unique nonnegative bounded solution $\tilde{P}(t)$. Moreover $\tilde{P}$ is a stabilizing solution.

Proof.

If $\{A,B; G_j\}$ is stabilizable and $\{A,C; G_j\}$ is detectable then from Theorem 4.1 in Da Prato and Ichikawa (1990) it follows that the equation (7) has a unique nonnegative bounded solution $\tilde{P}$ and $\tilde{P}$ is a stabilizing solution. If $\{A,B; G_j\}$ is stabilizable and $\{A,C; G_j\}$ is uniformly observable then from the main theorem in Morozan (1993) it follows that the equation (7) admits a unique nonnegative bounded solution $\tilde{P}$. Moreover $\tilde{P}$ is a stabilizing solution and there exists $\delta>0$ such that $\tilde{P}(t)\geq\delta I$ for all $t\in\mathbb{R}$.

The following result follows directly from Theorem 1, Corollary 1 and Corollary 2.

Theorem 2.

Suppose:

1) $\{A,M;G_j\}$ is detectable
2) $\{A,D;G_j\}$ is either stabilizable or uniform controllable.

Then the Riccati equation (8) admits a unique nonnegative bounded solution $\tilde{S}$. Moreover $\tilde{S}$ is a stabilizing solution.

3. Optimal stabilizing compensator for linear systems with state-dependent noise.

Let us consider the stochastic control system

$$dx(t)=[A(t)x(t)+B(t)u(t)]dt+\sum_{j=1}^{N} G_j(t)x(t)dw_j(t)+D(t)dv(t) \qquad (9)$$

with the output
$$dy(t)=M(t)x(t)dt+V(t)dv(t)$$
where $(w_1,...,w_n,v)$ are independent standard Wiener processes.
We assume the following structure properties:
$$V(t)V^*(t)=I, \quad D(t)V^*(t)=0, \quad t\in\mathbb{R}.$$
We shall stabilize (9) by using a compensator of the form

$$dx_c(t)=A_c(t)x_c(t)dt+\sum_{j=1}^{N} G_j(t)x_c(t)dw_j(t)+B_c(t)dy(t) \qquad (10)$$

$$y_c(t)=C_c(t)x_c(t)$$

The coupling of the compensator (10) to system (9) is performed by taking $u(t)=y_c(t)$.
We shall say that the compensator (10) stabilizes (9) if the system obtained after coupling (in the product space of state

variables x,x_c) is mean-square exponentially stable in the absence of additive noise.

The quality of such stabilization will be measured in a natural way by

$$\lim_{T\to\infty} E\int_{t_0}^{t_0+T} [\,|C(t)x(t)|^2+|u(t)|^2]dt \qquad (11)$$

The problem we are considering now consists in describing a stabilizing compensator of the form (10), optimal in the sense that (11) is minimal.

We remark that the above control problem is a special form of the stochastic quadratic control problem under partial observation; the control $u(t)$ depends upon the whole history of the output y up to time t, but here this dependence is determined by the structure of the compensator.In the special case $G_j\equiv 0$ the general stochastic quadratic control problem under partial observation is discussed in Da Prato and Ichikawa (1990).

We suppose that A_c,B_c and C_c are continuous and bounded on $\mathbb{R}$.
Theorem 3.

Under assumptions of Theorems 1 and 2, let $\tilde{P}$ and $\tilde{S}$ be the unique nonnegative bounded solutions of the Riccati equations (7) and (8) respectively.

Then the compensator defined by: $A_c=A-B\tilde{B}^*\tilde{P}-\tilde{S}M^*M$; $B_c=\tilde{S}M^*$; $C_c=-B^*\tilde{P}$ stabilizes system (9) and minimizes (11) in the class of stabilizing compensators of the form (10).

The proof of Theorem 3 is given in Dragan et al. (1992).

References.

1. Da Prato G., Ichikawa A.(1990), Quadratic control for linear time-varying systems, SIAM J.Control Optim., v.28, 359-381.

2. Dragan V., Halanay A., Morozan T.(1992), Optimal stabilizing compensator for linear systems with state-dependent noise, Stochastic Anal.Appl., v.10, no.3.

3. Morozan T.(1993), Stochastic uniform observability and

Riccati equations of stochastic control, Revue Roum.Math.Pures et Appl. (to appear)
4. Wonham V.M.(1968), On a matrix Riccati equation of stochastic control, SIAM J.Control Optim., V.6, 681-697.

Author's address:

Dr.Toader Morozan
Institute of Mathematics
of the Romanian Academy
P.O.Box 1-764

RO-70700 Bucharest
ROMANIA

Optimal control of nonlinear partial differential equations

International Series of Numerical Mathematics, Vol. 107, © 1992 Birkhäuser Verlag Basel   241

A boundary Pontryagin's principle for the optimal control of state-constrained elliptic systems

J.Frédéric Bonnans   Eduardo Casas

Abstract Pontryagin's principle, originally devised for the optimal control of ordinary differential equation, has recently been extended to the optimal control of semilinear elliptic systems and variational inequalities in the case of a distributed control. In this paper we show that if the control is also active at the boundary of the domain, then a boundary Hamiltonian satisfying a boundary maximum Pontryagin's principle appears in a natural way.

1 Statement and discussion of the main result

In this paper we consider the following optimal control problem. Let Ω be a bounded open subset of $\mathbb{R}^n$ with $C^{1,1}$ boundary Γ. The state equation is

$$\begin{cases} -\Delta y = F(y(x), u_d(x)), & x \in \Omega, \\ \alpha y + \dfrac{\partial y}{\partial n} = f(y(\sigma), u_b(\sigma)), & \sigma \in \Gamma. \end{cases} \tag{1}$$

Here $\alpha > 0$, F and f are continuous mapping from $\mathbb{R}^2$ into $\mathbb{R}$, differentiable with respect to y, and such that F_y and f'_y are continuous. We define $u = (u_d, u_b)$ and the value function as

$$J(y, u) := \int_\Omega L(y(x), u_d(x))dx + \int_\Gamma \ell(y(\sigma), u_b(\sigma))d\sigma. \tag{2}$$

We also assume that L and ℓ are continuous, with continuous derivative with respect to y. Let $g(x, y)$ be a mapping $\mathbb{R}^2 \to \mathbb{R}$ with continuous derivative with respect to y, and finally let K_d, K_b be two bounded subsets of $\mathbb{R}$. We define the set of controls as

$$U := \{u = (u_d, u_b) \text{ measurable} ; \ u_d(x) \in K_d, \text{ a.e. } x \in \Omega ; \ u_b(\sigma) \in K_b, \text{ a.e. } \sigma \in \Gamma\}.$$

We consider the optimal control problem

$$\min J(y, u) \text{ s.t. } (1), \ u \in U, \ g(x, y(x)) \le M, \ \forall x \in \Omega. \tag{P}_M$$

In order to obtain well-possedness of (1), we assume the following :

$$F'_y \le 0 \text{ and } f'_y \le 0. \tag{3}$$

Then the following holds :

Proposition 1 *Under the above hypotheses, for each $u \in U$, equation (1) admits a unique solution*

$$y = y_u \in Y := H^1(\Omega) \cap C(\bar{\Omega}).$$

We now define the distributed Hamiltonian (here and after we will omit the subscripts d, b on u_d, u_b when no confusion is possible)

$$H(y, u, p) := L(y, u) + pF(y, u),$$

and the boundary Hamiltonian

$$h(y, u, p) := \ell(y, u) + pf(y, u).$$

We say that $(P)_M$ is strongly stable if

$$\inf(P)_M \geq \inf(P)_{M'} + 0(M - M').$$

The hypotheses is of strong stability is generic in the following sense. We note that $M \to \inf(P)_M$ is decreasing ; let $\bar{M}$ be the smallest number such that $\inf(P)_M < \infty$ if $M > \bar{M}$.

It can be checked that $\inf(P)_M > -\infty$. Now as $\inf(P)_M$ is decreasing,is it differentiable a.e. $M \geq \bar{M}$, hence $(P)_M$ is strongly stable a.e. $M \geq \bar{M}$.

We now state our main result.

Theorem 1 *(Extension of Pontryagin's principle). Let $\bar{u}$ be a solution of $(P)_M$ with $(P)_M$ strongly stable, and $\bar{y}$ the associated state. Then there exists $\bar{p}$ in $W^{1,\sigma}(\Omega)$, for all $\sigma < n/(n-1)$, $\bar{\lambda}$ in $M(\bar{\Omega})$, such that (denoting by $\bar{\lambda}_d, \bar{\lambda}_b$ the restriction of $\bar{\lambda}$ to $M(\Omega)$ and $M(\Gamma)$) :*

$$\begin{cases} -\Delta \bar{p} & = & F_y'(\bar{y}, \bar{u})\bar{p} + L_y'(\bar{y}, \bar{u}) + g_y'(x, \bar{y}(x))\bar{\lambda}_d \ in \ \Omega, \\ \alpha \bar{p} + \dfrac{\partial \bar{p}}{\partial n} & = & f_y'(\bar{y}, \bar{u})\bar{p} + \ell_y'(\bar{y}, \bar{u}) + g_y'(\bar{x}, \bar{y}(x))\bar{\lambda}_b \ on \ \Gamma, \end{cases} \tag{4}$$

$$\begin{cases} \bar{\lambda} \geq 0, \quad g(\bar{x}, y(\bar{x})) \leq M, \\ \displaystyle\int_\Omega (g(x, \bar{y}(x) - M)d\bar{\lambda}(x) = 0, \end{cases} \tag{5}$$

and

$$H(\bar{y}(x), \bar{u}(x), \bar{p}(x)) = \min_{v \in K_d} H(\bar{y}(x), v, \bar{p}(x)), \quad a.e. \ x \in \Omega, \tag{6}$$

$$h(\bar{y}(\sigma), \bar{u}(\sigma), \bar{p}(\sigma)) = \min_{v \in K_b} h(\bar{y}(\sigma), v, \bar{p}(\sigma)), \quad a.e. \ \sigma \in \Gamma. \tag{7}$$

We compare this result to the litterature. Some first-order conditions for the control of state-constrained elliptic systems can be found in Mackenroth [12] in the convex case and the authors [2] [3], the last reference dealing with ill-posed systems. A derivation of Pontryagin's principle for problems without state constraints was obtained in Bonnans-Casas [4] ; Bonnans-Tiba [6] extended this approach to the control of elliptic variational inequalities (V.I.). The result for state constrained problem was obtained in Bonnans [1] and extended by the authors [5]. However in the previous references it is assumed that the state satisfies an homogeneous Dirichlet condition at the boundary and so only the distributed Hamiltonian appears. The novelty here lies in the boundary Hamiltonian, satisfying a natural extension of Pontryagin's principle.

The paper is as follows. In section 2 we connect strong stability to exact penalization and show how to regularize the problem with exact penalty. In section 3, using an Hamiltonian formulation of the variation of the cost, spike perturbations, and Ekeland's principle, we derive some optimality conditions for the approximate problem. Finally in section 4 we show how to pass to the limit in order to get the main result.

2 Exact penalization and regularization

Here we give easy extensions of some results of Bonnans [1], Bonnans-Casas [5]. To $(P)_M$ we associate the exact penalty function

$$J_r(u) := J(y_u, u) + r\|(g(., y_u) - M)^+\|_\infty$$

and the associated optimization problem

$$\min J_r(u) \; ; \; u \in U. \tag{$Q)_r$}$$

We endow U with the metric

$$d(u, v) := mes\,\{x \in \Omega \; ; \; u_d(x) \neq v_d(x)\} + mes\,\{\sigma \in \Gamma \; ; \; u_b(\sigma) \neq v_b(\sigma)\}.$$

The following can be checked as in [4].

Proposition 2 (i) *The mapping $u \to y_u$ is compact $(U, d) \to Y$.*

(ii) *If $(P)_M$ is strongly stable then there exists $r > 0$ such that any solution of $(P)_M$ is a local solution of $(Q)_r$ in the space (U, d).*

We now fix that for some $r > 0$, $d > 0$, $\bar{u}$ is solution of

$$\min J_r(u) \; ; \; u \in U, \; d(u, \bar{u}) \leq \delta. \tag{$Q)_{r,\delta}$}$$

This problem has no more state constraints. However its cost is non-differentiable, and we will regularize it. The first idea is to approximate the L^∞-norm by the L^q-norm, q being a "large" number. The L^q norm being not differentiable at 0 we will define a regularized cost as follows :

$$J_{r,q}(u) := J(y_u, u) + r[q^{-q} + \int_\Omega [(g(x, y_u(x)) - M)^+]^q dx]^{1/q},$$

and we consider the regularized problem

$$\min J_{r,q}(u) \; ; \; u \in U, \; d(u, \bar{u}) \leq \delta. \tag{$Q)_{r,\delta,q}$}$$

Proposition 3 *The following identity holds*

$$\inf(Q)_{r,\delta} = \lim_{q \to \infty} \inf(Q)_{r,\delta,q}.$$

Proof.   Define $U_\delta := \{u \in U \; ; \; d(u, \bar{u}) \leq \delta\}$. Let $u \in U_\delta$ be given. From the relations

$$\|(g(\cdot, y_u) - M)^+\|_q \;\; \leq \;\; [q^{-q} + \int_\Omega [(g(x, y_u(x)) - M)^+]^q dx]^{1/q}$$

$$\leq \;\; \frac{1}{q} + \|(g, \cdot, y_u) - M)^+\|_q,$$

and the convergence of $\|z\|_q \to \|z\|_\infty$, for all $z \in L^\infty(\Omega)$, we deduce that $J_{r,q}(u) \to J_r(u)$, hence

$$(*) \qquad\qquad \overline{\lim}\{\inf(Q)_{r,\delta,q}\} \leq \min(Q)_{r,\delta}.$$

Let us prove the converse inequality. let $\{u_q\}$ be a sequence in U_δ (and y_q the associated states) such that $J_{r,q}(u_q) \leq \inf(Q)_{r,\delta,q} + 1/q$. Then, using $(*)$:

$$\overline{\lim} J_{r,q}(u_q) = \overline{\lim} \inf(Q)_{r,\delta,q} \leq \min(Q)_{r,\delta} \leq \underline{\lim} J_r(u_q).$$

We end the proof by checking that $\underline{\lim} J_{r,q}(u_q) = \overline{\lim} J_r(u_q)$. The sequence $\{y_q\}$ is indeed compact in $C(\bar{\Omega})$. Let $\hat{y}$ be a limit-point in $C(\bar{\Omega})$. Assume that $\hat{z} := (g(\cdot, \hat{y}) - M)^+$ is such that $|\hat{y}|$ attains its maximum at $x_0 \in \bar{\Omega}$. For all $\varepsilon > 0$ there exists $\eta > 0$ and k_0 such that for a given subsequence

$$|(g(x, y_q(x)) - M)^+| \geq \|\hat{z}\|_\infty - \varepsilon, \quad \forall x \in \bar{\Omega}, \quad \|x - x_0\| \leq \eta, \quad \forall q > k_0.$$

From this we deduce that $\|(g(\cdot, y_q) - M)^+\|_q \to \|(g(\cdot, \hat{y}) - M)^+\|_\infty$ and the result follows. $\qquad\qquad\qquad\qquad\qquad\qquad\qquad\qquad\qquad\qquad\qquad\qquad\qquad\qquad\qquad\quad\square$

3   Hamiltonian formulation of the variation of the cost and spike perturbations

We first consider a problem with differentiable cost and without state constraints, that generalizes $(Q)_{r,\delta,q}$. We show how to write the difference of costs as integral of difference of Hamiltonians. Then we will make use of spike perturbations. We consider the criterion

$$c(u) := J(y_u, u) + \Phi[\int_\Omega b(x, y_u(x)) dx].$$

Here $\Phi : \mathbb{R} \to \mathbb{R}$ is C^1 and $b : \mathbb{R}^2 \to \mathbb{R}$ is continuous, and has a continuous derivative w.r.t. y. Given two controls u, v and the associated states y_u, y_v we define, using the mean value theorem, some interpolated states y^i, $i = 1$ to 5, such that

$$y^i(x) \in [y_u(x), y_v(x)], \quad \forall x \in \Omega,$$

and

$$\begin{aligned}
F(y_v, v) &= F(y_u, v) + F'_y(y^1, v)(y_v - y_u), \text{ a.e. in } \Omega, \\
L(y_v, v) &= L(y_u, v) + L'_y(y^2, v)(y_v - y_u), \text{ a.e. in } \Omega, \\
f(y_v, v) &= f(y_u, v) + f'_y(y^3, v)(y_v - y_u), \text{ a.e. on } \Gamma, \\
\ell(y_v, v) &= \ell(y_u, v) + \ell'_y(y^4, v)(y_v - y_u), \text{ a.e. on } \Gamma, \\
\Phi[\int_\Omega b(\cdot, y_v)] &= \Phi[\int_\Omega b(\cdot, y_u)] + \Phi'[\int_\Omega b(\cdot, y^5)] \int_\Omega b'_y(\cdot, y^5)(y_v - y_u),
\end{aligned}$$

and the interpolated costate $p_{u,v}$ solution of

$$\begin{cases} -\Delta p_{u,v} = F'_y(y^1,v)p_{u,v} + L'_y(y^2,v) + \Phi'[\int_\Omega b(.,y^5)]b'_y(.,y^5) \text{ in } \Omega, \\[2mm] \alpha p_{u,v} + \dfrac{\partial p_{u,v}}{\partial n} = f'_y(y^3,v)p_{u,v} + \ell'_y(y^4,v) \text{ on } \Gamma. \end{cases} \tag{8}$$

Indeed the above equation has a unique solution in Y.

Proposition 4 (Hamiltonian formulation of the variation of the cost). *For any two control u,v in U, $p_{u,v}$ being solution of (8), the following holds :*

$$c(v) = c(u) + \int_\Omega [H(y_u,v,p_{u,v}) - H(y_u,u,p_{u,v})] + \int_\Gamma [h(y_u,v,p_{u,v}) - h(y_u,u,p_{u,v})]. \tag{9}$$

Proof. Put

$$\theta := \int_\Omega [L(y_v,v) - L(y_u,v)] + \int_\Gamma [\ell(y_v,v) - \ell(y_u,v)] + \Phi[\int_\Omega b(.,y_v)] - \Phi[\int_\Omega b(.,y_u)].$$

Then it is easily checked that

$$c(v) = c(u) + \int_\Omega [L(y_u,v) - L(y_u,u)] + \int_\Gamma [\ell(y_u,v) - \ell(y_u,u)] + \theta. \tag{10}$$

Using $\{y^i\}$, and (8) we find

$$\begin{aligned} \theta &= \int_\Omega L'_y(y^2,v)(y_v - y_u) + \int_\Gamma [\ell'(y^4,v)(y_v - y_u) \\ &\quad + \Phi'[\int_\Omega b(.,y^5)]\int_\Omega b'(.,y^5)(y_v - y_u) \\ &= \int_\Omega [-\Delta p_{u,v} - F'_y(y^1,v)p_{u,v}](y_v - y_u)dx \\ &\quad + \int_\Gamma [\alpha p_{u,v} + \frac{\partial p_{u,v}}{\partial n} - f'_y(y^3,v)p_{u,v}](y_v - y_u)d\sigma. \end{aligned}$$

Now by Green's formula

$$\int_\Omega -\Delta p_{u,v}(y_v - y_u) = -\int_\Gamma \frac{\partial p_{u,v}}{\partial n}(y_v - y_u) + \int_\Gamma p_{u,v}\frac{\partial(y_v - y_u)}{\partial n} - \int_\Omega p_{u,v}\Delta(y_v - y_u)$$

hence, after simplification

$$\begin{aligned} \theta &= \int_\Omega [-\Delta(y_v - y_u) - F'_y(y^1,v)(y_v - y_u)]p_{u,v} \\ &\quad + \int_\Gamma [\alpha(y_v - y_u) + \frac{\partial(y_v - y_u)}{\partial n} - f'_y(y^3,v)]p_{u,v}. \end{aligned}$$

Using the state equations for y_u and y_v we find

$$\theta = \int_\Omega [F(y_u,v) - F(y_u,u)]p_{u,v} + \int_\Gamma [f(y_u,v) - f(y_u,u)]p_{u,v}.$$

This and (10) give the conclusion. $\qquad\square$

We define

$$\omega_k(x_0) := \{x \in \Omega \; ; \; \|x - x_0\| \le 1/k\},$$
$$\gamma_k(\sigma_0) := \{\sigma \in \Gamma \; ; \; \|\sigma - \sigma_0\| \le 1/k\}.$$

We say that a sequence $v^k = (v_d^k, v_b^k)$ in U is a distributed (resp. boundary) spike perturbation of $u \in U$ around $x_0 \in \Omega$ (resp. $\sigma_0 \in \Gamma$) if for some $v \in K_d$ (resp. $v \in K_b$)

$$v_b^k = u_b \text{ and } v_d^k(x) = \begin{cases} v \text{ if } x \in \omega_k(\sigma_0), \\ u_d(x) \text{ if not;} \end{cases}$$

respectively :

$$v_d^k = u_b \text{ and } v_b^k(\sigma) = \begin{cases} v \text{ if } \sigma \in \gamma_k(\sigma_0), \\ u_b(\sigma) \text{ if not.} \end{cases}$$

By p_u we mean $p_{u,u}$.

Proposition 5 *Let $u \in U$ be given. Then*

(i) *For almost all $x_0 \in \Omega$ the following holds : if v^k is a distributed spike perturbation of u around x_0, and y_k is the associated state,*

$$\lim_{k \to \infty} \frac{1}{mes(\omega_k(x_0))}[c(v^k) - c(u)] = H(y_u(x_0), v, p_u(x_0)) - H(y_u(x_0), u(x_0), p_u(x_0)).$$

(ii) *For almost $\sigma_0 \in \Gamma$ the following holds : if v^k is a boundary spike perturbation of u around σ_0, and y_k is the associated state, then*

$$\lim_{k \to \infty} \frac{c(v^k) - c(u)}{mes(\gamma_k(\sigma_0))} = h(\sigma_0, y_u(\sigma_0), v, p_u(\sigma_0)) - h(\sigma_0, y_u(\sigma_0), u(\sigma_0), p_u(\sigma_0)).$$

Proof. We give only the proof of case (ii) ; case (i) can be dealt with in a similar way. We have from Proposition 4

$$c(v^k) - c(u) = \int_{\gamma_k(\sigma_0)}[h(y_u, v, p_k) - h(y_u, u, p_k)],$$

where p_k is the interpolated costate $p_{v^k,u}$. From the (easily established) uniform convergence of $p_k \to p_u$ in $C(\bar{\Omega})$, and the boundedness of the data we deduce that (if the limit exists)

$$\lim_{k \to \infty} \frac{c(v^k) - c(u)}{mes(\gamma_k(\sigma_0))} = \lim_{k \to \infty} mes(\gamma_k(\sigma_0))^{-1} \int_{\gamma_k(\sigma_0)}[h(y_u, v, p_u) - h(y_u, u, p_u)].$$

Now by continuity we always have

$$mes(\gamma_k(\sigma_0))^{-1} \int_{\gamma_k(\sigma_0)} h(y_u, v, p_u) \to h(y_u(\sigma_0), v, p_u(\sigma_0)),$$

hence the formula holds if σ_0 is a Lebesgue point of $\sigma \to h(y_u(\sigma), u(\sigma), p_u(\sigma))$; the set of Lebesgue point being of full measure, we get the result. $\qquad\square$

From the above results we deduce an optimality system for the minimization of $c(u)$ over U.

Theorem 2 *If $\hat{u}$ minimizes $c(u)$ over U then, denoting $\hat{y} := y_{\hat{u}}$ and $\hat{p} := p_{\hat{u}}$, the following holds :*

$$H(\hat{y}(x), \hat{u}_d(x), \hat{p}(x)) = \min_{v \in K_d} H(\hat{y}(x), v, \hat{p}(x)), \quad a.e. \ in \ \Omega,$$

$$h(\hat{y}(x), \hat{u}_b(x), \hat{p}(x)) = \min_{v \in K_b} h(\hat{y}(x), v, \hat{p}(x)), \quad a.e. \ on \ \Gamma.$$

4 Back to state constrained problems

We now end the proof of the main result, as follows. Assuming $(P)_M$ strongly stable it follows that $\bar{u}$ (solution of $(P)_M$) is also solution of $(Q)_{r,\delta}$ for r large enough and δ small enough (Prop. 2) ; by Prop. 3, $\bar{u}$ is an $\varepsilon(q)$-solution of $Q_{r,\delta,q}$, with

$$\varepsilon(q) = J_{r,q}(\bar{u}) - \inf(Q)_{r,\delta,q}$$

and $\varepsilon(q) \searrow 0$ as $q \nearrow \infty$. Now $u \to J_{r,q}(\bar{u})$ is continuous from (U_δ, d) into $\mathbb{R}$ and (U_δ, d) is a complete metric space. Applying Ekeland's principle [10], we deduce that there exists $u_q \in U_\delta$ such that

$$d(\bar{u}, u_q) \leq \sqrt{\varepsilon(q)}, \tag{11}$$

$$J_{r,q}(u_q) \leq J_{r,q}(u) + \sqrt{\varepsilon(q)}d(u, u_q), \quad \forall u \in U. \tag{12}$$

Now, if v^k is, say, a boundary spike perturbation of u_q around σ_0, if follows that

$$d(v^k, u_q) \leq mes\gamma_k(\sigma_0),$$

hence with (12)

$$0 \leq \lim_{k \to \infty} \frac{J_{r,q}(v^k) - J_{r,q}(u_q)}{mes(\gamma_k(\sigma_0))} + \sqrt{\varepsilon(q)}.$$

We may now apply Prop. 5 ; we obtain the following

Theorem 3 *For any $q > 1$ there exists $u_q \in U$ satisfying (11) and such that denoting $y_q := y_{u_q}$, there exists $p_q \in Y$, $\lambda_q \in L^1(\Omega)$ such that*

$$\begin{cases} -\Delta p_q = F'_y(y_q, u_q)p_q + L'_y(y_q, u_q) + g'_y(., y)\lambda_q & in \ \Omega, \\[2ex] \alpha p_q + \dfrac{\partial p_q}{\partial n} = f'_y(y_q, u_q)p_q + \ell'_y(y_q, u_q) & on \ \Gamma, \end{cases} \tag{13}$$

$$\lambda_q = r[q^{-q} + \int_\Omega [(g(x, y_q(x)) - M)^+]^q dx]^{1/q-1}[(g(., y_q) - M)^+]^{q-1} \tag{14}$$

and

$$H(y_q, u_q, p_q) \;\leq\; \min_{v \in K_d} H(y_q, v, p_q) + \sqrt{\varepsilon(q)} \quad a.e. \ in \ \Omega, \tag{15}$$

$$h(y_q, u_q, p_q) \;\leq\; \min_{v \in K_b} h(y_q, v, p_q) + \sqrt{\varepsilon(q)} \quad a.e. \ on \ \Gamma. \tag{16}$$

5 From the approximate to the original control problem

We end the proof of Thm. 1 by passing to the limit in the optimality system stated in Theorem 3. Note that, as $d(u_q, \bar{u}) \leq \sqrt{\varepsilon(q)}$, and $u \to y_u$ is continuous in (U, d), we have that $y_q \to \bar{y}$ in Y. It remains to pass to the limit in (13)-(16). First let us note that an estimate of λ_q in $L^1(\Omega)$ can be obtained as in [1], hence a subsequence of λ_q $*$-converges to $\bar{\lambda} \in M(\bar{\Omega})$. That $\lambda_q \geq 0$ implies $\bar{\lambda} \geq 0$. Obviously $\bar{\lambda}$ has support where $g(x, \bar{y}(x)) = M$. As $(g(x, \bar{y}(x)) - M)^+ = 0$ it follows that $\int_{\bar{\Omega}} g(x, \bar{y}(x)) - M)^+ d\bar{\lambda}(x) = 0$; (5) follows. Now let us pass to the limit in the costate equation (13). From a given subsequence, from $\lambda_q \xrightarrow{*} \bar{\lambda}$ in $M(\bar{\Omega})$, it follows that $g_y'(., y_q)\lambda_q \xrightarrow{*} g_y'(., \bar{y})\bar{\lambda}$ in $M(\bar{\Omega})$. We see that (4) will follows from the study of the abstract problem

$$\begin{cases} -\Delta p + ap = \lambda_d & in \quad \Omega, \\ \alpha p + \dfrac{\partial p}{\partial n} = \lambda_b & on \quad \Gamma, \end{cases}$$

with $a \in L^\infty(\Omega), a \geq 0, \lambda_d, \lambda_b$ traces on Ω and Γ of $\lambda \in M(\bar{\Omega})$. We study this equation by the method of transposition. To $(f, g) \in L^s(\Omega) \times W^{1,s}(\Gamma)$ we associate z solution of

$$\begin{cases} -\Delta z + az = f & in \quad \Omega, \\ \alpha z + \dfrac{\partial z}{\partial n} = g & on \quad \Gamma. \end{cases}$$

It is known that $z \in C(\bar{\Omega})$ and that the mapping $(f, g) \to z$ is dense in $M(\bar{\Omega})$. Integrating (formally) by parts we obtain, q being the trace of p :

$$\int_\Omega fp + \int_\Gamma gq = \int_{\bar{\Omega}} z d\lambda.$$

This equation in (p, q) has a unique solution in the dual space $L^{s'}(\Omega) \times W^{-1,s'}(\Gamma)$ with $1/s' + 1/s = 1$. Now it can be proved as in Casas [9] that $p \in W^{1,\sigma}(\Omega)$ for all $\sigma < n/(n-1)$ and the q can be interpreted as the trace of p. Also the estimate of p in $W^{1,\sigma}(\Omega)$ can be obtained independently of $a \geq 0$. Coming back to our problem, it follows that p_q is bounded in $W^{1,\sigma}(\Omega)$. We can extract a subsequence such that $\lambda_q \xrightarrow{*} \bar{\lambda}$ in $M(\bar{\Omega})$ and $p_q \to \bar{p}$ in $W^{1,\sigma}(\Omega)$ for all $\sigma < n/(n-1)$. From this follows the strong convergence of p_q in $L^1(\Omega)$. Passing to the limit we obtain (4). This allows to pass to the limit in (14),(15),(16) ; we deduce that (5), (6), (7) hold. $\square$

References

[1] J.F. Bonnans. Pontryagin's principle for the optimal control of semilinear elliptic systems with state constraints. In *30th IEEE Conference on Control and Decision*, volume 2, pages 1976–1979, Brighton, England, 1991.

[2] J.F. Bonnans and E. Casas. Contrôle de systèmes elliptiques semilinéaires comportant des contraintes sur l'état. In H. Brezis and J.L. Lions, editors, *Nonlinear Partial Differential Equations and Their Applications. Collège de France Seminar*, volume 8, pages 69–86. Longman Scientific & Technical, New York, 1988.

[3] J.F. Bonnans and E. Casas. Optimal control of semilinear multistate systems with state constraints. *SIAM J. on Control & Optim.*, 27(2):446–455, 1989.

[4] J.F. Bonnans and E. Casas. Un principe de Pontryagine pour le contrôle des systèmes elliptiques. *J. Differential Equations*, 90(2):288–303, 1991.

[5] J.F. Bonnans and E. Casas. An extension of Pontryagin's principle for state-constrained optimal control of semilinear elliptic equation and variational inequalities. To appear.

[6] J.F. Bonnans and D. Tiba. Pontryagin's principle in the control of semilinear elliptic variational equations. *J. Appl. Math. and Optim.*, 23:299–312, 1991.

[7] J.V. Burke. Calmness and exact penalization. *SIAM J. on Control & Optim.*, 29(2):493-497, 1991.

[8] E. Casas. Control of an elliptic problem with pointwise state constraints. *SIAM J. on Control & Optim.*, 24(6):1309–1318, 1986.

[9] E. Casas. Boundary control of semilinear elliptic equations with pointwise state constraints. SIAM J. on Control and Optim., to appear.

[10] I. Ekeland. Nonconvex minimization problems. *Bull. Amer. Math. Soc.*, 1(3):76–91, 1979.

[11] D. Gilbarg and N.S. Trudinger. *Elliptic Partial Differential Equations of Second Order.* Springer-Verlag, Berlin-Heidelberg-New York, 1977.

[12] U. Mackenroth. On some elliptic optimal control problems with state constraints. *Optimization*, 17(1986),595-607.

[13] G. Stampacchia. Le problème de Dirichlet pour les équations elliptiques du second ordre à coefficients discontinus. *Ann. Inst. Fourier (Grenoble)*, 15:189–258, 1965.

Author's addresses

Frédéric Bonnans

Eduardo Casas

INRIA, Domaine de Voluceau,
BP 105, 78153 Rocquencourt, France

Dpto Mat. Apl. Ciencias Comp., ETSI de Caminos
C. y P. , Avda los Castros s/n, Santander, España

bonnans@seti.inria.fr

casas@ccuvx.unican.es

International Series of Numerical Mathematics, Vol. 107, © 1992 Birkhäuser Verlag Basel   251

Propriétés de controlabilité pour les systèmes elliptiques, la méthode des domaines fictifs et problèmes de design optimal

Controllability properties for elliptic systems, the fictitious domain method and optimal shape design problems

Dan TIBA*

Résumé

La méthode des domaines fictifs est utilisée dans la résolution numérique d'équations aux dérivées partielles dans un domaine extérieur ou pour appliquer les différences finies aux systèmes définis dans des domaines avec une géométrie compliquée. Le principe de cette approche est très utile dans des problèmes avec des domaines inconnus au variables.

En tenant compte de certaines propriétés de controlabilité exacte pour les équations elliptiques linéaires ou nonlinéaires, on peut développer une méthode d'approximation pour les problèmes de design optimal par des problèmes de contrôle dans un domaine fixé.

Abstract

The fictitious domain method is used in the numerical solution of partial differential equations in exterior domains or in order to apply finite differences to systems defined in domains with a complicated geometry. The principle of this approach is very useful in problems with unknown or variable domains.

Taking into account certain exact controllability properties for linear or nonlinear elliptic equations, we develop an approximation method for optimal design problems by distributed control problems in a fixed domain.

Mots clés

Domaines fictifs. Equations elliptiques. Controlabilité exacte. Design optimal.

Key words

Fictitious domains. Elliptic equations. Exact controllability. Optimal design.

---

*Institute of Mathematics, Romanian Academy of Sciences, P.O. Box 1-764, RO-70700 Bucharest, Romania and INRIA, Domaine de Voluceau, BP 105, 78153 Rocquencourt, France

1 Introduction

The fictitious domain approach is well known in the solution of partial differential equations
by finite differences method in domains with a complicated geometry or in exterior domains,
G.P. Astrakmantsev [1], W. Proskurowski, O. Windlund [15], R. Glowinski et al. [7]. One
variant associates to the given equation a distributed control problem governed by the extension of the equation to a more advantageous larger domain. See P. Joly and C. Atamian
[9], C. Atamian [2] for an analysis along these lines of the exterior Helmholtz problem.

A similar embedding of domains idea, with boundary control, was used independently
by J. Blum [6] in free boundary problems related to the physics of plasma.

Naturally, the main numerical disadvantage of this approach of working "larger problems", disappears in applications concerning unknown or variable domain problems. In the
first setting the Blum's work enters, while for the second we quote the work of Hofmann,
Kocvara and Haslinger [8] devoted to the study of optimal shape design problems by a
geometric-distributed control procedure. Moreover, a controllability-type argument for elliptic systems shows that the control problem may generate in an implicit manner a large
class of variable domains which are considered in optimal design problems. In this way the
geometric optimization problems may be discussed by a purely control approach, in a fixed
domain. This was studied, in the case of boundary control and with a more limited range
of applications, in the papers of Barbu and Tiba [5], Tiba [17], Neitaanmaki, Makinen and
Tiba [13]. See also Tiba [18], ch. III.5 for a specific example along these lines.

One important advantage, by comparison with the standard boundary variation technique, is to avoid the new mesh generation at each step of the algorithm, which is extremely
time consuming.

The aim of this paper is two fold : first, we establish some exact controllability properties
for nonlinear elliptic systems and, next, we use them in general linear or nonlinear optimal
design problems Pironneau [14], Barbu and Friedman [3], Barbu and Stojanovic [4].

Finally, we mention that a first approximate controllability-type result for elliptic equations may be found in the classical monograph of J.L. Lions [10], p. 85.

2 Exact distributed controllability properties

Assume that D is a subdomain of the domain $\Omega \subset R^2$, such that ∂D and $\partial \Omega$ are regular.
We define the controlled variational inequality, in $\Omega - \bar{D}$:

$$-\Delta z + \beta(z) \ni u \qquad \Omega - \bar{D}, \qquad (2.1)$$
$$z = 1 \qquad \partial \Omega, \qquad (2.2)$$
$$z = 1 \qquad \partial D, \qquad (2.3)$$

where $\beta \subset R \times R$ is a maximal monotone graph given by $\beta(r) = 0$, $r > 0$, $\beta(0) =]-\infty, 0]$,
$\beta(r) = \emptyset$, $r < 0$. The boundary conditions are motivated by the subsequent applications in
optimal design and don't play an essential role in the argument. The controllability problem

we study is to find $u \in L^2(\Omega - D)$ such that

$$\frac{\partial z}{\partial n} = \varphi \text{ in } \partial D, \tag{2.4}$$

where φ is given in $H^{1/2}(\partial D)$.

Theorem 2.1 *The problem (2.4) has at least one exact solution $u \in L^2(\Omega - D)$.*

Proof Let $\tilde{z} \in H^2(\Omega - D) \cap C^1(\Omega - \bar{D})$ (not unique) be given by the trace theorem, such that

$$\frac{\partial \tilde{z}}{\partial n} = \varphi, \quad \tilde{z} = 1 \qquad \partial D, \tag{2.5}$$

$$\tilde{z} = 1 \qquad \partial\Omega. \tag{2.6}$$

The interior regularity of $\tilde{z}$ may be inferred by assuming it as the solution of a linear fourth order elliptic problem in $\Omega - \bar{D}$, with appropriate boundary conditions.

In a neighbourhood of ∂D, we have $\tilde{z} > 0$. We fix a regular curve Γ in this neighbourhood, surrounding ∂D, at some distance $c > 0$ from ∂D. To choose such a constant, we use the continuity of $\tilde{z}$ and a compactness argument. We define

$$\hat{z}(x) = \begin{cases} \tilde{z}(x) \text{ between } \partial D \text{ and } \Gamma, \\ 1 \text{ between } \Gamma \text{ and } \partial\Omega. \end{cases} \tag{2.7}$$

By a local regularization of $\hat{z}$ around Γ (see the next lemma), we construct z_ε associated to a regularization parameter $\varepsilon > 0$, which satisfies (2.2), (2.3), (2.4). We may take, obviously, $\beta(z_\varepsilon) = 0$ in $\Omega - D$ and compute u by (2.1).

Lemma 2.1 *Let $\bar{D} \hookrightarrow \Omega \subset R^n$ be regular domains and $\varphi \in C(\bar{\Omega}) \cap H^2(\Omega)$ be such that $\varphi \mid_{\Omega-D} \in C^1(\Omega - \bar{D})$, $\varphi \mid_D \in C^1(D)$. There exists $\varphi_\varepsilon \in C^1(\Omega) \cap H^2(\Omega)$ such that*

$$\varphi_\varepsilon \to \varphi \quad in \quad C(\bar{\Omega}), \tag{2.8}$$

$$\varphi_\varepsilon > \varphi \quad in \quad \Omega, \tag{2.9}$$

$$\varphi_\varepsilon \mid_{\partial\Omega} = \varphi \mid_{\partial\Omega}, \qquad \nabla\varphi_\varepsilon \mid_{\partial\Omega} = \nabla\varphi \mid_{\partial\Omega}. \tag{2.10}$$

Proof Let Γ_ε^1, Γ_ε^2 be two regular surfaces surrounding ∂D, at distance less than ε from ∂D and such that $\Gamma_\varepsilon^1 \subset D$, $\Gamma_\varepsilon^2 \subset \Omega - \bar{D}$. We define, between Γ_ε^1 and Γ_ε^2, the mapping

$$\tilde{\varphi}_\varepsilon(x) = \int_{[0,1]^n} \varphi(x - \varepsilon \ dist^2(x, \Gamma_\varepsilon^1).dist^2(x, \Gamma_\varepsilon^2)\tau)\rho(\tau)d\tau, \tag{2.11}$$

where ρ is a Friedrichs mollifier, that is $\rho \in C_0^\infty(R^n)$, $\rho \geq 0$, supp $\rho \subset [0,1]^n$ and $\int_{[0,1]^n} \rho(\tau)d\tau = 1$.

If $x \to x_0 \in \Gamma_\varepsilon^i$, $i = 1, 2$, then $dist(x, \Gamma_\varepsilon^i) \to 0$ and, by the uniform continuity of φ, we get $\tilde{\varphi}_\varepsilon(x) \to \varphi(x_0)$, therefore $\tilde{\varphi}_\varepsilon \mid_{\Gamma_\varepsilon^i} = \varphi \mid_{\Gamma_\varepsilon^i}$.

By (2.11), for $x \notin \Gamma_\varepsilon^i$, we have :

$$\frac{\partial}{\partial x_i}\tilde{\varphi}_\varepsilon(x) = \int_{[0,1]^n} \frac{\partial\varphi}{\partial x_i}(\cdots)(1 - \varepsilon\chi(x)\tau_i)\rho(\tau)d\tau, \tag{2.12}$$

where $\lim_{x \to x_0} \chi(x) = 0$, $\forall x_0 \in \Gamma^i_\varepsilon$. Then (2.12) also gives $\lim_{x \to x_0} \dfrac{\partial}{\partial x_i} \tilde{\varphi}_\varepsilon(x) = \dfrac{\partial}{\partial x_i} \varphi(x_0)$, $x_0 \in \Gamma^i_\varepsilon$, that is $\nabla \tilde{\varphi}_\varepsilon \mid_{\Gamma^i_\varepsilon} = \nabla \varphi \mid_{\Gamma^i_\varepsilon}$.

We can define

$$\hat{\varphi}_\varepsilon(x) = \begin{cases} \tilde{\varphi}_\varepsilon(x), & x \text{ between } \Gamma^1_\varepsilon, \Gamma^2_\varepsilon, \\ \varphi(x) & \text{otherwise.} \end{cases} \tag{2.13}$$

Obviously, $\hat{\varphi}_\varepsilon \in C^1(\Omega)$ and (2.10) is satisfied. Moreover, again by the uniform continuity of φ, (2.8) is true as $\varepsilon \to 0$. However, (2.9) may be violated. In order to overcome this difficulty, we denote $K_\varepsilon = |\hat{\varphi}_\varepsilon - \varphi|_{C(\bar{\Omega})}$, $K_\varepsilon \to 0$, and the mapping

$$\eta(x) = \; dist \, (x, \partial\Omega)^2. \tag{2.14}$$

As the closure of the set $\{\hat{\varphi}_\varepsilon(x) \neq \varphi(x)\}$ is a compact in Ω, then

$$\eta(x) \geq K > 0 \tag{2.15}$$

on it. By (2.8), (2.14), (2.15), we see that

$$\varphi_\varepsilon(x) = \hat{\varphi}_\varepsilon(x) + \frac{2K_\varepsilon}{K}\eta(x) \tag{2.16}$$

will check (2.8), (2.9). It also satisfies (2.10) since $\eta \in H^2_0(\Omega)$.

Remark   Higher order regularity, boundary conditions or convergence may be obtained working with higher order powers of the distance function.

Remark   The local regularization, preserving inequalities and traces has a clear intuitive geometric contents. The above lemma has the advantage to provide an explicit construction in quite a general situation with respect to the singularities of φ.

As another application of this lemma, we give an exact distributed controllability result for the coincidence set, in the obstacle problem. This may be compared with the work of Barbu and Tiba [5], where the case of boundary control and approximate controllability is discussed.

Corollary 2.1 *Let $\bar{D} \hookrightarrow \Omega$ be regular domains. There is $u \in L^2(\Omega)$ (not unique) such that the variational inequality :*

$$-\Delta y + \beta(y) \ni u \qquad \Omega \tag{2.17}$$

$$y = 1 \qquad \partial\Omega \tag{2.18}$$

has the coincidence set $E_y = \{x \in \Omega \; ; \; y(x) = 0\}$ equal to $\bar{D}$.

Proof   Consider in $\Omega - \bar{D}$ a regular surface Γ, at distance greater than $c > 0$ from ∂D and $\partial\Omega$, with c a suitable constant. Define the piecewise regular mapping ψ in $\Omega - D$, as follows :

$$\psi(x) = \begin{cases} dist^2(x, \partial D), & x \text{ between } \Gamma \text{ and } \partial D, \\ 1, & x \text{ between } \Gamma \text{ and } \partial\Omega. \end{cases}$$

Applying the regularization technique described in *lemma 2.2*, in the proof, we obtain a function $\psi_\varepsilon > 0$, satisfying the same boundary conditions. It remains only to define

$$y(x) = \begin{cases} \psi_\varepsilon & x \in \Omega - D \\ 0 & x \in \bar{D} \end{cases}$$

and $\beta(y) = 0$ in D (for instance). Then u is obtained by (2.17).

Remark The above result also ensures that $\beta(y)$ (the unknown reaction of the obstacle) may be taken 0 in Ω.

Remark Theorem 2.1 is also valid in the linear case $(\beta(y) = 0)$. A stronger statement may be obtained, by the use of the HUM method Lions [11], allowing the control u to have the support in a neighbourhood of ∂D.

3 Optimal design in electrochemical machining process

We consider the regular domains $C \subset E \subset D \subset \Omega$, D variable. In $D - C$, we solve the obstacle problem

$$-\Delta y + \beta(y) \ni f \qquad D - C, \tag{3.1}$$
$$y = 0 \qquad \partial C, \tag{3.2}$$
$$y = 1 \qquad \partial D, \tag{3.3}$$

where β is the maximal monotone graph defined in section 2 and $f \in L^2(\Omega\backslash C)$. This models the electrochemical processing in industry : ∂C and ∂D are the electrodes and the boundary condition $y = 1$ on ∂D signifies that some constant tension is applied, Ockendon and Elliott [16]. The coincidence set

$$E_y = \{x \in D\backslash C \; ; \; y(x) = 0\} \tag{3.4}$$

gives the final shape of the metal piece obtained around C.

The optimal design problem discussed by Barbu and Friedman [3], Barbu and Stojanovic [5], is to find the domain D, $E \subset D \subset \Omega$, such that

$$E_y \supset E\backslash C. \tag{3.5}$$

Taking into account the definition of E_y, an obvious approach is

$$(P) \qquad\qquad \min_D \frac{1}{2} \int_{E\backslash C} y^2 dx$$

subject to (3.1)-(3.3).

We associate to (P) a sequence of optimal control problems and we prove, by the controllablity results established in section 2, that a strong approximation relationship is valid :

$$(P_n) \qquad\qquad \min_u \left\{ \frac{1}{2} \int_{E\backslash C} y^2 dx + n \int_{\Omega_y} (u - f)^2 dx \right\}$$

subject to

$$-\Delta y + \beta(y) \ni u \qquad \Omega - C, \tag{3.6}$$
$$y = 1 \qquad \partial\Omega, \tag{3.7}$$
$$y = 0 \qquad \partial C, \tag{3.8}$$

where $E \subset \Omega_y \subset \Omega$ is the smallest Lipschitzian domain with the property that $y = 1$ on $\partial\Omega_y$. The existence of Ω_y is ensured by (3.7).

Neither (P), nor (P_n) have ensured existence of optimal pairs since no compactness assumption is made.

If we assume the existence of a Lipschitz domain D, $E \subset D \subset \Omega$, which "solves" (3.1)-(3.5), then Theorem 2.1 allows to extend the variational inequality (3.1) to the whole domain $\Omega \backslash C$ and to find a control $u^* \in L^2(\Omega \backslash C)$ such that $u^* |_D \equiv f$ and the cost associated to it is null, therefore optimal.

The following result refines this remark

Theorem 3.1 *For any $n \in N$, the problem (P) is embedded in (P_n) and*

$$\inf(P) \geq \inf(P_n) \tag{3.9}$$

conversely, if $\delta_n > 0$ is small and $[y_n, u_n]$ is a δ_n optimal pair for (P_n), then Ω_{y_n} is an ε_n optimal subdomain of (P) with $\varepsilon_n > 0$ small, depending on δ_n.

Proof Let D be a Lipschitz subdomain of Ω, $E \subset \bar{D} \hookrightarrow \Omega$. By Theorem 2.1, there is $u_D \in L^2(\Omega \backslash D)$ such that the solution y_D of

$$-\Delta y_D + \beta(y_D) \ni u_D \qquad \Omega - \bar{D} \tag{3.10}$$

$$y_D = 1 \qquad \partial\Omega \cup \partial D \tag{3.11}$$

satisfies $\dfrac{\partial y_D}{\partial n}\big|_{\partial D} = -\dfrac{\partial \tilde{y}}{\partial \nu}\big|_{\partial D}$, where $\tilde{y}$ is the solution of (3.1)-(3.3) and n, ν are the normal vectors to ∂D in both directions.

Then the pair $[\bar{y}, \bar{u}]$

$$\bar{u}(x) \;=\; \begin{cases} u_D & \Omega - \bar{D}, \\ f & D - C, \end{cases} \tag{3.12}$$

$$\bar{y}(x) \;=\; \begin{cases} y_D & \Omega - \bar{D}, \\ \tilde{y} & D - C, \end{cases} \tag{3.13}$$

is admissible for (P_n) and we may take $\Omega_{\bar{y}} \subseteq D$, so $J_n(\bar{u}) = J(D)$.

By a variant of a result of Pironneau [14], Ch. III, p. 32, we may extend these considerations to all $E \subset D \subset \Omega$, which ends the proof of (3.9).

Conversely, by (3.9) we get

$$\int_{\Omega_{y_n}} (u_n - f)^2 dx \leq \frac{C + \delta_n}{n}, \quad C = \inf(P). \tag{3.14}$$

Let $\tilde{y}_n$ denote the solution of

$$-\Delta\tilde{y}_n + \beta(\tilde{y}_n) \ni f \qquad \Omega_{y_n} - C, \tag{3.15}$$

$$\tilde{y}_n = 1 \qquad \partial\Omega_{y_n}, \tag{3.16}$$

$$\tilde{y}_n = 0 \qquad \partial C. \tag{3.17}$$

Since, obviously, y_n satisfies :

$$-\Delta y_n + \beta(y_n) \ni u_n \quad \Omega_{y_n} \tag{3.18}$$

and (3.16), (3.17), then (3.14)-(3.18) and the continuous dependence on the right-hand side in variational inequalities shows that

$$|\tilde{y}_n - y_n|^2_{H^1(\Omega_{y_n})} \leq \varepsilon_n,$$

which ends the proof.

Remark We notice the generality of the argument developped in sections 2,3 which may be, similarly, applied to other types of equations or boundary conditions.

Remark Compared with the standard mapping method, Murat and Simon [12], we underline the simplicity of the above technique, the differential operator being maintained unchanged and the penalization term in the cost functional being very simple.

4   The algorithm

We detail in this section the algorithm suggested by the previous results, on a typical linear optimal design problem, Pironneau [14] :

$$(S) \qquad \min_{D} \left\{ \int_{E} |\nabla y - y_d|^2 dx \right\}$$

$$-\Delta y = f \qquad D, \tag{4.1}$$
$$y = 0 \qquad \partial D. \tag{4.2}$$

Above, we consider Lipschitzian domains $E \subset D \subset \Omega \subset R^N$, D variable, and $f \in L^2(\Omega)$, $y_d \in L^2(E)^N$.

The associated approximating control problems are

$$(S_n) \qquad \min_{u} \left\{ \int_{E} |\nabla y - y_d|^2 dx + n \int_{\Omega_y} (u - f)^2 dx \right\}$$

$$-\Delta y = u \qquad \Omega, \tag{4.3}$$
$$y = 0 \qquad \partial\Omega, \tag{4.4}$$

with $\Omega_y \subset \Omega$ being the smallest Lipschitzian domain such that $E \subset \Omega_y$ and $y = 0$ on $\partial\Omega_y$.

Remark If $u \geq 0$ in Ω_y, then the maximum principle shows that Ω_y is a connected component of $supp(y_+)$, y_+ being the positive part of y.

In the absence of specific assumptions nor the problem (S) neither the problem (S_n) have ensured the existence of optimal solutions. However, a result similar to Theorem 3.1 may be obtained, starting with the linear variant of Theorem 2.1

Theorem 4.1 *(S) is embedded in (S_n) and*

$$\inf(S_n) \leq \inf(S), \quad \forall n. \tag{4.5}$$

Conversely, let $u_n \in L^2(\Omega)$ be an δ_n-optimal control for (S_n), $\delta_n > 0$, a small parameter. Then Ω_{y_n} is an ε_n-optimal domain for (S) with $\varepsilon_n > 0$ small, depending on δ_n.

Let $[u, y]$ be an admissible pair for (S_n) (for any n). From now on, we assume that $\nabla y \neq 0$ in a neighbourhood of $\partial\Omega_y \subset \Omega \subset R^2$.

Remark If $\partial\Omega_y$ is regular, then $\nabla y \neq 0$ is equivalent with $\dfrac{\partial y}{\partial n} \neq 0$ on $\partial\Omega_y$ as the tangential component is null.

If $f \geq 0$ in Ω, applying the maximum principle and the Hopf lemma to the solution of

$$-\Delta\tilde{y} = f \qquad \Omega_y, \tag{4.6}$$

$$\tilde{y} = 0 \qquad \partial\Omega_y, \tag{4.7}$$

we get $\dfrac{\partial\tilde{y}}{\partial n} < 0$ in $\partial\Omega_y$. Taking into account the proof of Theorem 3.1, we see that the δ_n-optimal pairs for (S_n) are "close" for $\tilde{y}$ on Ω_y, thus justifying our assumption.

We take $u, v \in H^1(\Omega) + PC(\Omega)$ (piecewise continuous and bounded in Ω) and we compute the gradient of the cost functional for (S_n) :

$$I = \lim_{\lambda \to 0} \frac{1}{\lambda} \left\{ \frac{1}{2} \int_E \{ |\nabla y_\lambda - y_d|^2 - |\nabla y - y_d|^2 \} + y \int_{\Omega_{y_\lambda}} (u + \lambda v - f)^2 - n \int_{\Omega_y} (u - f)^2 \right\}$$

where y_λ is the solution associated to $u + \lambda v$ via (4.3), (4.4). Obviously $y_\lambda = y + \lambda r$ with

$$-\Delta r = v \qquad \Omega, \tag{4.8}$$

$$r = 0 \qquad \partial\Omega. \tag{4.9}$$

Proposition 4.2 *Under the above conditions*

$$I = \int_E (\nabla y - y_d, \nabla r)dx + 2n \int_{\Omega_y} (u - f)v - n \int_{\partial\Omega_y} \frac{(u - f)^2 r}{\frac{\partial y}{\partial n}} d\sigma. \tag{4.10}$$

Proof It is enough to study the limit

$$I_1 = \lim_{\lambda \to 0} \frac{1}{\lambda} \left\{ \int_{\Omega_{y_\lambda} \cap w} (u - f)^2 dx - \int_{\Omega_y \cap w} (u - f)^2 dx \right\}, \tag{4.11}$$

and w is a neighbourhood of some arbitrary point $M \in \partial\Omega_y$. By the convergence properties of y_λ, we have $\nabla y_\lambda \neq 0$ in M, for λ small and the implicit function theorem will define C^1 curves $\Gamma_\lambda = \partial\Omega_{y_\lambda} \cap w$, $\Gamma = \partial\Omega_y \cap w$, for any M, starting with the equation $y_\lambda = 0$, respectively $y = 0$.

Choosing a new local system of coordinates in M, with $\vec{x}_1$ normal to $\partial\Omega_y$ and $\vec{x}_2$ tangent, then we can express

$$\Gamma \quad : \quad x_1 = \alpha(x_2) = \beta(x_2, 0), \tag{4.12}$$

$$\Gamma_\lambda \quad = \quad x_1 = \alpha_\lambda(x_2) = \beta(x_2, \lambda) \tag{4.13}$$

with some C^1 mappings $\alpha, \beta, \alpha_\lambda$.

Assume that $\Omega_{y_\lambda} \cap w \supset \Omega_y \cap w$ and (by further restricting it) $w = [a, b] \times [c, d]$ with a, b, c, d some constants. In the new coordinates, we have :

$$I_1 = \lim_{\lambda \to 0} \int_{(\Omega_{y_\lambda} - \Omega_y) \cap w} (u - f)^2 dx = \lim_{\lambda \to 0} \frac{1}{\lambda} \int_c^d dx_2 \int_{\alpha(x_2)}^{\alpha_\lambda(x_2)} (u - f)^2 dx_1 \tag{4.14}$$

Above, we suppose that $y > 0$, $r > 0$, $\lambda > 0$, in $\Omega_y \cap w$, the other cases being possible to discuss similarly.

Moreover, again by the implicit function theorem, we get

$$\lim_{\lambda \to 0} \frac{\alpha_\lambda(x_2) - \alpha(x_2)}{\lambda} = -\frac{r(\alpha(x_2), x_2)}{y_{x_1}(\alpha(x_2), x_2)}. \tag{4.15}$$

Then (4.14), (4.15) and a Leibniz-Newton type formula in the interior integral give :

$$\begin{aligned}
I_1 &= -\int_c^d (u - f)^2(\alpha(x_2), x_2)\frac{r(\alpha(x_2), x_2)}{y_{x_1}(\alpha(x_2), x_2)}dx_2 \\
&= \int_c^d (u - f)^2(\alpha(x_2), x_2)\frac{r(\alpha(x_2), x_2))}{|y_{x_1}(\alpha(x_2), x_2)|}\frac{\sqrt{1 + \alpha'(x_2)^2}}{\sqrt{1 + \frac{y_{x_2}^2(\alpha(x_2), x_2)}{y_{x_1}^2(\alpha(x_2), x_2)}}}dx_2 \\
&= \int_{\partial\Omega_y \cap w} (u - f)^2\frac{r}{|\nabla y|}d\sigma = -\int_{(\partial\Omega_y) \cap w} \frac{(u - f)^2 r}{\frac{\partial y}{\partial n}}d\sigma.
\end{aligned} \tag{4.16}$$

The last form is independent of the system of axes and remains valid in all the possible configurations, thus ending the proof.

We apply the adjoint system method in order to obtain an explicit representation of (4.10). We define the adjoint equation, in variational formulation :

$$\int_\Omega \nabla z.\nabla g - \int_E (\nabla y - y_d).\nabla g + n\int_{\partial\Omega_y} \frac{(u - f)^2}{\frac{\partial y}{\partial n}}g d\sigma = 0 \tag{4.17}$$
$$\forall g \in H_0^1(\Omega), \quad z \in H_0^1(\Omega).$$

The existence of a unique solution $z \in H_0^1(\Omega)$ for (4.17) follows from the minimization on $H_0^1(\Omega)$ of the functional

$$\frac{1}{2}\int_\Omega |\nabla z|^2 dx - \int_E (\nabla y - y_d).\nabla z dx + n\int_{\partial\Omega_y} \frac{(u - f)^2}{\frac{\partial y}{\partial n}}z d\sigma,$$

which is convex, lower semicontinuous, proper and coercive.

Taking $g = r$ in (4.17), we infer

$$\int_E (\nabla y - y_d)\nabla r = \int_\Omega \nabla z\nabla r + n\int_{\partial\Omega_y} \frac{(n - f)^2}{\frac{\partial y}{\partial n}}t d\sigma,$$

that is

$$I = \int_\Omega \nabla z\nabla r + 2n\int_{\Omega_y} (u - f)v = \int_\Omega zv + 2n\int_{\Omega_y} (u - f)v.$$

Finally, the gradient of the cost functional J_n, associated to (S_n) is

$$\nabla J_n(u) = z + 2n(u - f)\chi_{\Omega_y}, \tag{4.18}$$

where χ_{Ω_y} is the characteristic function of Ω_y.

Remark In a gradient algorithm, we notice that (4.18) will preserve the regularity of the control, required for the first iteration.

Remark Under regularity assumptions, the adjoint equation (4.17) may be formally interpreted as a transmission problem. If $p = z \, |_{\Omega - \bar{\Omega}_y}$ and $q = z \, |_{\Omega_y}$, then :

$$
\begin{aligned}
-\Delta p &= 0 & & \Omega - \bar{\Omega}_y, \\
-\Delta q &= -div[\chi_E(\nabla y - y_d)] & & \Omega_y, \\
p &= q & & \partial\Omega_y, \\
p &= 0 & & \partial\Omega, \\
\frac{\partial p}{\partial n} + \frac{\partial q}{\partial r} &= -\frac{n(u-f)^2}{\frac{\partial y}{\partial n}} & & \partial\Omega_y.
\end{aligned}
$$

References

[1] G.P. Astrakmantsev – Methods of fictitious domains of a second order elliptic equation with natural boundary conditions, U.S.S.R. Comp. Math. and Math. Phys., 18 (1978).

[2] C. Atamian – Thèse de l'Université de Paris VI (1991).

[3] V. Barbu, A. Friedman – Optimal design of domains with free-boundary problems, SIAM J. Control and Optimiz., 29 (2) (1991).

[4] V. Barbu, D. Tiba – Boundary controllability of the coincidence set in the obstacle problem, SIAM J. Control and Optimiz., 29 (5) (1991).

[5] V. Barbu, S. Stojanovic – Controlling the free boundary of elliptic variational inequalities on a variable domain (to appear).

[6] J. Blum – Sur quelques problèmes d'analyse numérique et de contrôle optimal en physique des plasmas. Thèse de l'Université de Paris VI (1985).

[7] R. Glowinski, C. Atamian, Q.V. Dinh, J. He, P. Périaux, H. Steve – Control approach to fictitious domain methods. Application to fluid dynamics and electromagnetism, Conférence INRIA (juin 1990).

[8] J. Haslinger, K.H. Hoffmann, M. Kocvara – Control/fictitious domain method in solving optimal shape design problems (to appear).

[9] P. Joly, C. Atamian – Une analyse de la méthode des domaines fictifs pour le problème de Helmholtz extérieur, Rapport INRIA 1378 (1991).

[10] J.L. Lions – Contrôle optimal de systèmes gouvernés par des équations aux dérivées partielles, Dunod, Paris (1968).

[11] J.L. Lions – Controllabilité exacte, perturbations et stabilization de systèmes distribués, Masson, Paris (1988).

[12] F. Murat, J. Simon – Etude de problèmes d'optimal design, in "Optimization techniques, Modelling and optimization in the service of man", J. Cea (ed.), LNCS 41, Springer Verlag, Berlin (1976).

[13] P. Neittaanmaki, R. Makinen, D. Tiba – Controllability-type propeties for elliptic systems and applications, in "Control and estimation of distributed parameter systems", F. Kappel and K. Kunisch (eds.), Birkauser-Verlag, Base (1991).

[14] O. Pironneau – Optimal shape design for elliptic systems, Springer Verlag, Berlin, (1984).

[15] W. Proskurowski, O. Windlund – On the numerical solution of Helmholtz equation by the capacitance matrix method, Math. Comp. 30 (1979).

[16] J.R. Ockendon, C.M. Elliott – Weak and variational methods for moving boundary problems, Research Notes in Mathematics 59, Pitman, London (1982).

[17] D. Tiba – Une approche par controllabilité frontière dans les problèmes de design optimal, CRAS Paris, 310, Série I (1990).

[18] D. Tiba – Optimal control of nonsmooth distributed parameter systems, LNM 1459, Springer Verlag, Berlin (1990).

International Series of Numerical Mathematics, Vol. 107, © 1992 Birkhäuser Verlag Basel

Optimal control for elliptic equation and applications

R. Tahraoui

Ceremade, University Paris-Dauphine
and I.V.F.M. Rouen
France

I. Motivations.

In this paper we investigate some questions of control arising from problems like

   i) the optimal shape for maximal torsional rigidity of a barre.
   ii) problems of control by domains [3],[4],
   iii) existence result in hydrodynamic [5],[6].

To study these problems, we propose an approach based on classical optimal control problem governed by an elliptic equation. But here, the dependancy of the control variable is non convex [7],[8].

II. Setting of the problem.

We would like to optimize a criterion of the following kind

$$(1) \qquad J(v) = J(v,\omega) = \int_\Omega l(x,v,\omega)\,dx$$

where ω is the state function satisfying the following equations :

$$\begin{cases} -\Delta\omega = b(x,v,\omega) & \text{on } \Omega, \\ \omega|_{\partial\Omega} = 0 \,; \end{cases}$$

the control variable v belongs to the control set

$$\mathcal{U}_1 = \left\{ v \in L^\infty(\Omega) \,/\, 0 < \alpha \le v \le \beta \,,\ \int_\Omega g(v)\,dx = \gamma \right\},$$

or

$$\mathcal{U}_2 = \left\{ v \in L^\infty(\Omega) \,/\, 0 < \alpha \leq v \leq \beta \,,\, v_* = u_0 \right\},$$

where v_* represents the unidimensional increasing rearrangement of v and v^* its unidimensional decreasing rearrangement. We note

$$v_0(s) = u_0(|\Omega| - s) \qquad \forall\, s \in [0, |\Omega|]$$

where $|\Omega|$ is the Lebesgue measure of Ω. In the sequel we consider the two following problems :

$$(\mathcal{P}_i) \qquad\qquad \sup \text{ or } \inf \left\{ J(v) \,/\, v \in \mathcal{U}_i \right\} \qquad i = 1,2 .$$

REMARK: The dependancy of $J(\cdot,\omega)$ and $b(x,\cdot,\omega)$ with respect to the variable control v is generally non convex ; and $\mathcal{U}_i$ is non convex too $(i = 1,2)$.

What are the difficulties ?

The direct method of the variational calculus consists to take a minimizing sequence of control v_n, and to pass to the limit as n goes to infinity.

To conclude, some convexity arguments are necessary to obtain weak sequential lower semi-continuity and weak closure in (1)-(3) respectively in (4). These properties are not true when there is no convexity [10],[11].

To simplify, we propose our results through out some simples examples, but which are very significants. For more precise details we refer to [8],[7] and [9].

III. Examples.

1) $b(x,v,\omega) = a(v) \cdot \lambda_1(v) \cdot \omega \,,\quad g(v) = v.$

$$(\mathcal{P}_1) \qquad\qquad \sup \left\{ J(v) \,/\, v \in \mathcal{U}_1 \right\} \,,\quad J(v) = \lambda_1(v)$$

where $\lambda_1(v)$ is the first eigenvalue of Laplace operator Δ on Ω.

RESULTS: There is an unique optimal solution $(\overline{u},\overline{\omega})$ of our problem. In addition this solution has the following properties :

i) if $a(v) = v$, the optimal control $\overline{u}$ is bang-bang ; more precisely there is a real $\overline{t}$ such that

$$\overline{u}(x) = \begin{cases} \alpha & \text{a.e. on } \quad E = \{x \in \Omega \,/\, \overline{\omega}(x) > \overline{t}\} \\ \beta & \text{a.e. on } \quad \Omega \setminus E \,. \end{cases}$$

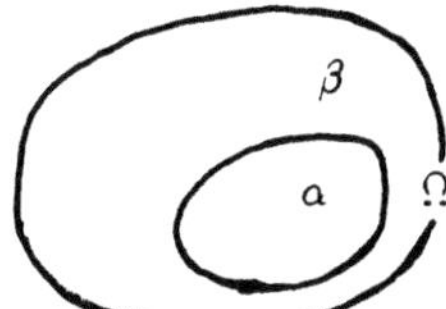

Fig. 1. Structure of the optimal control.

ii) if $a(v) = a(x) \cdot v$, there is some measurable subset of Ω where $\overline{u}$ is not bang-bang i.e. there exists $\overline{t}$ such that

$$\overline{u}(x) = \begin{cases} \alpha & \text{if } \quad a(x) \cdot \overline{\omega}^2(x) > \overline{t} \,. \\ \in]\alpha, \beta[& \text{if } \quad x \in E = \{x \in \Omega \,/\, a(x) \cdot \overline{\omega}^2(x) = \overline{t}\} \,, \\ \beta & \text{if } \quad a(x) \cdot \overline{\omega}^2(x) < \overline{t} \,; \end{cases}$$

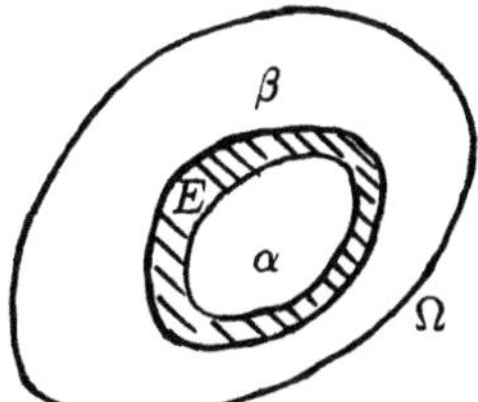

Fig. 2. Structure of the optimal control.

2) $b(x,v,\omega) = v$, $J(v) = \displaystyle\int_\Omega v \cdot \omega \, dx = \int_\Omega |\nabla \omega|^2 \, dx,$

$(\mathcal{P}_2)$ $\inf \left\{ J(v) \,/\, v \in \mathcal{U}_2 \right\} \,.$

RESULTS: There exists an optimal solution $(\overline{u},\overline{\omega})$ of the considered problem. This solution has the following properties.

i) $\overline{u}$ and $\overline{\omega}$ have the "same level sets", more precisely there exists a decreasing function φ such that

$$\begin{cases} \overline{u} = \varphi(\overline{\omega}) \\ \varphi(t) = u_0(\overline{\omega}^*)^{-1}(t) \end{cases}$$

where $\overline{\omega}^*$ is the unidimensional decreasing rearrangement of $\overline{\omega}$.

IV. Applications.

1) Problem of control by domains.

Let us consider the following set of admissible controls :

$$\mathcal{U} = \left\{ D = (D_1, D_2, D_3) \quad \text{measurable set} \quad \subset \Omega^3 \right.$$

$$\text{s.t.} \quad |D_i| = \gamma_i \, , \quad \bigcup_{i=1}^{3} D_i = \Omega \quad \text{a.e.}$$

$$\left. D_i \cap D_j = \emptyset \quad \text{if} \quad i \neq j \quad \text{and} \quad \sum_i \gamma_i = |\Omega| \right\} .$$

The problem is to find $\overline{D} \in \mathcal{U}$ satisfying

$$(5) \qquad J(\overline{D}) \leq J(D) = \int_\Omega |\nabla \omega_D|^2 \, dx \qquad \forall\, D \in \mathcal{U} \, ,$$

where the state function ω_D verifies the equation

$$\begin{cases} -\Delta\omega_D = v_D \quad \text{on} \quad \Omega \, , \\ \omega_D|_{\partial\Omega} = 0 \, ; \end{cases}$$

and v_D defining by :

$$v_D(x) = \alpha_i \quad \text{if} \quad x \in D_i \, , \; i = 1, 2, 3$$

with $\alpha_1 > \alpha_2 > \alpha_3 > 0$, three given constants. This problem is similar to $(\mathcal{P}_2)$; and then there exists an optimal solution $(\omega_{\overline{D}}, \overline{D})$. In addition there exist two reals $0 < t_1 < t_2$ such that

$$\begin{aligned} \overline{D}_1 &= \left\{ x \in \Omega \,/\, \omega_{\overline{D}}(x) < t_1 \right\} , \\ \overline{D}_2 &= \left\{ x \in \Omega \,/\, t_1 < \omega_{\overline{D}}(x) < t_2 \right\} , \\ \overline{D}_3 &= \left\{ x \in \Omega \,/\, \omega_{\overline{D}}(x) > t_2 \right\} . \end{aligned}$$

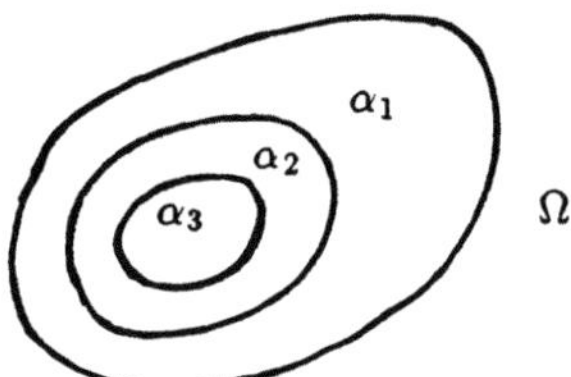

Fig. 3. Shape of the optimal domain.

QUESTION: What happens if $J(\cdot)$ is such that $(\mathcal{P}_2)$ has no solution ?

In this situation we relax the problem $(\mathcal{P}_2)$ i.e. we consider the following new problem $(\tilde{\mathcal{P}}_2)$: to minimize

$$J(v) = \int_\Omega v \cdot l(x, \omega(v)) \, dx$$

on the admissible control set

$$\tilde{\mathcal{U}}_2 = \left\{ v \in L^\infty(\Omega) \,/\, 0 < \alpha \leq v \leq \beta \,/\, \int_0^t v_*(s) \, ds \geq \int_0^t u_0(s) \, ds \right.$$
$$\left. \text{and} \quad \int_0^{|\Omega|} v_* \, ds = \int_0^{|\Omega|} u_0 \, ds \right\},$$

where $l(x, t)$ is a regular function, increasing with respect t such that $l(x, 0) = 0$, i.e. we want to solve

$$(\tilde{\mathcal{P}}_2) \qquad\qquad\qquad \inf \left\{ J(v) \,/\, v \in \tilde{\mathcal{U}}_2 \right\} .$$

This problem has an optimal solution $(\tilde{u}, \tilde{\omega})$ verifying the following properties : [9]

i)
$$\int_\Omega S(x) \, \tilde{u}(x) \, dx = \int_0^{|\Omega|} S^*(s) \tilde{u}^*(s) \, ds$$
$$= \int_0^{|\Omega|} S^*(s) \, v_0(s) \, ds$$

where $S(x) = l(x, \tilde{\omega}) + p$, p being the adjoint state associated to $\tilde{\omega} = \tilde{\omega}(\tilde{u})$.

ii) $\tilde{u}^*(s) = v_0(s)$ a.e. $s \in I \subset [0, |\Omega|]$, for any measurable set I such that
$$\left| \{ x \in S^{-1}(S^*(I)) \,/\, S(x) = t \} \right| = 0 \qquad \forall\, t \in \mathbf{R} .$$

iii) for any $t_0 > 0$ such that $\sigma_0 = |\{x \in \Omega \; / \; S(x) = t_0\}| > 0$ we have

$$\int_\gamma^t \tilde{u}^*(s)\,ds \;\leq\; \int_\gamma^t v_0(s)\,ds \qquad \forall\, t \in [\gamma, \gamma + \sigma_0] \,,$$

$$\int_\gamma^{\gamma+\sigma_0} \tilde{u}^*(s)\,ds \;=\; \int_\gamma^{\gamma+\sigma_0} v_0(s)\,ds$$

where
$$\gamma \;=\; \left|\{x \in \Omega \; / \; S(x) > t_0\}\right| \,.$$

These relations help. some time, to give some sufficient conditions to prove the existence of solution for $(\mathcal{P}_2)$.

2) Torsional maximal rigidity.

Let us consider a hollow beam P. Let Ω_0 be a fixed open set denoting the hole cross section and Σ the cross section of P with a given fixed area equal to β. The problem is to find a contour of Σ such that the rigidity of P is maximal when submitted to torsional strains. It is a typical shape optimization problem. We will show that we can apply some ideas used in [7],[8] for optimal control problems. This problem has been considered by many authors - (see for instance [2],[3],[1]) - . More recently R.B. Gonzales de Paz [12] studied this question by a different approach. Meanwhile it seems that some of the results in [12] are not complete. In this paper we show an existence theorem and give some qualitative informations not contained in [12]

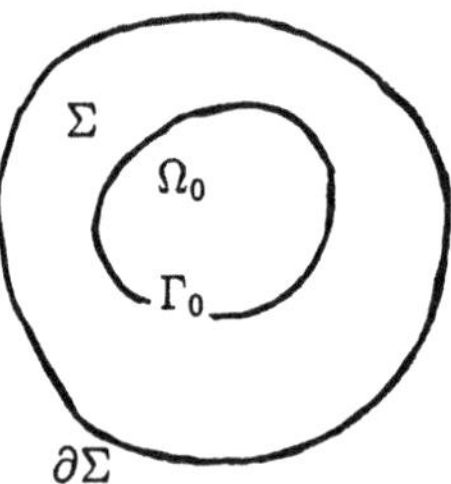

Fig. 4. Cross section of the beam.

We suppose that the beam is an homogeneous and isotropic material. We refer to [2] and [1] for a description of the mechanical model. The constraint function ω satisfy the following equations :

$$(1) \qquad \begin{cases} -\Delta\omega = 2 \quad \text{on} \quad \Sigma \\ \omega|_{\partial\Omega} = 0 \quad , \quad \omega|_{\Gamma_0} = c_\Sigma \\ \displaystyle\int_{\Gamma_0} \frac{\partial\omega}{\partial\eta}\, d\sigma = 2\beta_0 \end{cases}$$

where c_Σ is a unknown constant, $\beta_0 = |\Omega_0|$ is the area of the hole, $|\Sigma| = 3$ the area of the cross section of the material and η is the normal vector to Γ_0 oriented externally.

The torsional rigidity of P, with a cross section Σ, is

$$K(\Sigma) = \int_\Sigma |\nabla\omega|^2\, dx = 2\int_\Sigma \omega\, dx + 2c\beta_0 \; ;$$

the function $\omega = \omega_\Sigma$ is solution of the equations (1). Our goal is to find an open set Σ s.t.

$$K(\Sigma) \geq K(\Sigma')$$

for any admissible open set Σ' i.e. with the previous caracteristics. Due to physical arguments, a reasonable solution should be a connected cross section Σ with "surround" the hole Ω_0 (see fig. 4).

IDEA: Roughly speaking we give a new formulation of the initial problem. Let us consider a ball $B = B(0,R) \supset \Omega_0$ such that $|B \setminus \Omega_0| >> \beta$, and the following admissible control set :

$$\mathcal{U}(B) = \mathcal{U} = \Big\{ v \in L^\infty(B) \,/\, 0 \leq v \leq 1,$$

$$v = 1 \quad \text{a.e.} \quad x \in \Omega_0 \,, \quad \int_{B\setminus\Omega_0} v\, dx = \beta \Big\} .$$

The state equations are :

$$(2) \qquad \begin{cases} -\Delta\omega_v = 2v \qquad \text{on} \quad B \setminus \overline{\Omega}_0 \,, \\ \omega_v/\Gamma_0 = c_v \quad , \quad \text{unknown constant,} \\ \displaystyle\int_{\Gamma_0} \frac{\partial\omega_v}{\partial\eta}\, d\sigma = 2\beta_0 \quad , \quad \omega_v/\partial B = 0 \,. \end{cases}$$

The cost function to minimize is defined by :

$$J(v) \ = \ \int_{B \setminus \Omega_0} 2v \cdot \omega_v \, dx + 2c_v \cdot \beta_0 \ .$$

From the preceding results there exists an optimal solution $(\overline{u}, \overline{\omega})$ characterized by

$$\overline{u}(x) \ = \ \begin{cases} 1 & \text{a.e. on} \quad E = \{x \in B \setminus \Omega_0 \ / \ \overline{\omega}(x) > \overline{t}\} \ , \\ 0 & \text{a.e. on} \quad B \setminus (\Omega_0 \cup E) \ . \end{cases}$$

QUESTION: How to recover the solution of the initial problem ?

It is clear that we have

$$\begin{cases} -\Delta(\overline{\omega} - \overline{t}) \ = \ 2 & \text{on} \quad E \quad \text{which is an open set,} \\ (\overline{\omega} - \overline{t})|_{\Gamma_0} \ = \ \overline{c} - \overline{t} \ , \\ \displaystyle \int_{\Gamma_0} \frac{\partial}{\partial \eta}(\overline{\omega} - \overline{t}) \, d\sigma \ = \ 2\beta_0 \ , \\ (\overline{\omega} - \overline{t})/\Gamma \ = \ 0 & \text{where} \quad \Gamma = \partial E \setminus \Gamma_0 \ . \end{cases}$$

And then we see that, intituively, the optimal pair $(\overline{\omega} - \overline{t}, E)$ is a good candidate to solve our initial problem. To conclude this, the essential difficulties are :

i) the proof of the 2-connectedness of E which guarantees the existence of physical solution.

ii) the proof of the inequality

$$\int_E |\nabla(\overline{\omega} - \overline{t})|^2 \, dx \ \geq \ \int_\theta |\nabla \omega_\theta|^2 \, dx$$

for any θ bounded open set belonging to a specific class of admissible sets.

To conclude positively we need two hypothesis :

a) The possible oscillations of any maximizing sequence of domains stay in a fixed ball $B_0 = B(0, R_0)$ i.e. they do not go to infinity. This assumption seems to be reasonnable from the physicist point of view.

b) Some technical assumption which seems to be related to our method [17].

Finally we obtain the following result :

Theorem.

The optimal pair $(\overline{\omega} - \overline{\iota}, E)$ is the optimal shape for our problem.

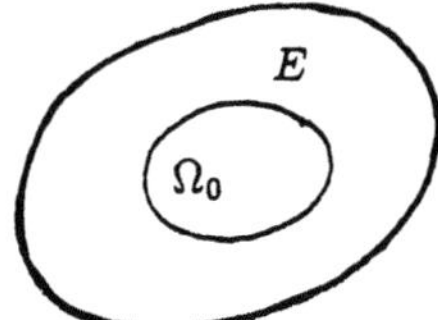

Fig. 5. Solution of our problem.

3. Ideal incompressible bidimensionnal fluid.

This question has been studied also by many authors [5],[6],[13],[14]. We consider an ideal incompressible fluid occupying a bounded connected open set of $\mathbf{R}^2$. We are interesting by solutions of Euler equations ; $u = (u_1, u_2)$ is the velocity of a particle of the fluid in a position (x, y). We have

$$\operatorname{div} u = 0 .$$

From V. Arnold [5], there exist variational principles for stationnary Euler equations :

$$
\begin{cases}
- u_1 \dfrac{\partial u_1}{\partial x} - u_2 \dfrac{\partial u_1}{\partial y} = \dfrac{\partial p}{\partial x} , \\[2mm]
- u_1 \dfrac{\partial u_2}{\partial x} - u_2 \dfrac{\partial u_2}{\partial y} = \dfrac{\partial p}{\partial y} , \\[2mm]
\vec{u} \cdot \vec{n} = 0 \quad , \quad \Gamma = \partial\Omega ,
\end{cases}
$$

where p is the pressure of the fluid. Also from [5], these solutions are maximizers or minimizers of the kinetic energy

$$E = E(u) = \frac{1}{2} \int_\Omega |u|^2 \, dx .$$

It remains to precise the function class where the maximizers or minimizers functions belong. From V. Arnold [5], in the non stationary case, between two times t_1 and t_2 the fluid vorticity $\omega = \operatorname{rot} u$ has a fixed rearrangement. This permits to define the new constraint i.e. we consider the problem

$$\inf \left\{ E(u) \,/\, u \quad \text{s.t.} \quad \operatorname{div} u = 0 \quad \text{and} \right.$$

$$\left. \operatorname{rearrang}^t = (\operatorname{rot} u) = v_0 \, , \, \text{given} \right\} .$$

As Ω is an open subset of $\mathbf{R}^2$, there exists a function φ satisfying :

$$\frac{\partial \varphi}{\partial y} = u_1 \quad , \quad \frac{\partial \varphi}{\partial x} = -u_2 \ ;$$

and consequently φ verifies :

$$\begin{cases} -\Delta\varphi = \omega \quad \text{on} \quad \Omega , \\ \varphi/\Gamma = 0 \ ; \end{cases}$$

the kinetic energy becomes then

$$E = \tilde{E}(\omega) = \frac{1}{2}\int_\Omega |u|^2 \, dx = \frac{1}{2}\int_\Omega |\nabla\varphi|^2 \, dx \ .$$

Now we would like to solve the following problem :

$$(\mathcal{P}) \qquad\qquad \inf\left\{\tilde{E}(\omega) \ / \ \omega \in \mathcal{U}\right\}$$

where

$$\mathcal{U} = \left\{v \in L^\infty(\Omega) \ / \ 0 < \alpha \le v \le \beta \, , \ v_* = v_0\right\} .$$

REMARK: We can consider

$$\mathcal{U} = \left\{v \in L^p(\Omega) \ / \ v_* = v_0\right\} \quad , \quad p > 1 \ ;$$

but in this case, we have some regularity difficulties to recover our initial problem. However the result remains valid.

RESULTS: Here our problem is similar to $(\mathcal{P}_2)$; then there exists an optimal pair $(\overline{\varphi}, \overline{v})$ for $(\mathcal{P})$, and there exists a decreasing function f satisfying
$$\overline{v} = f(\overline{\varphi}) \ .$$

QUESTION: How to recover the initial problem ?

It is classical to put

$$\overline{u}_1 = \frac{\partial\overline{\varphi}}{\partial y} \quad , \quad \overline{u}_2 = -\frac{\partial\overline{\varphi}}{\partial x} \quad ,$$

$$p_1 = \overline{u}_1 \frac{\partial\overline{u}_1}{\partial x} + \overline{u}_2 \frac{\partial\overline{u}_1}{\partial y} \ ,$$

$$p_2 \;=\; \bar{u}_1 \frac{\partial \bar{u}_2}{\partial x} + \bar{u}_2 \frac{\partial \bar{u}_2}{\partial y}\;.$$

We have

$$(1) \qquad p_1 \;=\; \frac{1}{2}\frac{\partial}{\partial x}\,[|\nabla\overline{\varphi}|^2] + \frac{\partial}{\partial x}\,F(\overline{\varphi})$$

where

$$F(t) \;=\; \int_0^t f(s)\,ds\;.$$

In the same way, we obtain that

$$(2) \qquad p_2 \;=\; \frac{1}{2}\frac{\partial}{\partial y}[|\nabla\overline{\varphi}|]^2 + \frac{\partial}{\partial y}\,F(\overline{\varphi})\;.$$

Relations (1) and (2) prove that (p_1, p_2) is a gradient vector i.e. there is a function p such that

$$p_1 \;=\; \frac{\partial p}{\partial x}\;,\quad p_2 \;=\; \frac{\partial p}{\partial y}\;.$$

In addition, this pressure function is defined by

$$p(x,y) \;=\; -\frac{1}{2}|\nabla\overline{\varphi}|^2 - F(\overline{\varphi})\;.$$

As $\varphi \in C^{1,\alpha}(\overline{\Omega}) \cap H^2(\Omega)$, $0 < \alpha < 1$, the preceding calculation is justified. In the same way justifying computation as in [13] and [15], we obtain the boundary condition

$$\vec{u}\cdot\vec{n}|_\Gamma \;=\; 0\;.$$

Finally $\bar{u}$ is the desired solution.

BIBLIOGRAPHIE

[1] N.V. Banichuk, *Problems and methods of optimal structural design*. Plenum Press, London (1983).

[2] H. Lanchon, *Torsion élastoplastique d'un arbre*. J. de mécanique, 13 (1974), p. 267-318.

[3] J. Cea, *Problems of shape optimal design*. In "Optimization of distributed parameter structures, NATO Proceedings" (Eds. Hang, E.J. and Cea, J.). Sÿthoff and Noordhoff (1981).

[4] O. Pironneau, *Optimal shape design for elliptic systems*. Springer-Verlag, New York (1984).

[5] V.I. Arnold, *Méthodes mathématiques de la mécanique classique*. Collection MIR, 1976, Moscou.

[6] T.B. Benjamin., *The alliance of practical and analytical insights into the nonlinear problems of fluid mechanics : applications of methods of functional analysis to problems in mechanics*, Lecture Notes in Math., vol. 503, Springer-Verlag, 1976, p. 8-29.

[7] R. Tahraoui, *Quelques remarques sur le contrôle des valeurs propres*, Séminaire du Prof. J.L. Lions. Collège de France, Paris, 1985, vol. 8, Pitman.

[8] R. Tahraoui, *Contrôle optimal dans les équations elliptiques*, Siam J. Control and Optimization, vol. 30, n.1, January 1992.

[9] R. Tahraoui, *Contrôle optimal à réarrangement fixé*, C.R. Acad. Sc. Paris, t. 303, Série I, n.19, 1986.

[10] G.B. Morrey, *Multiple integrals in the calculus of variations*. Springer, Berlin, 1966.

[11] B. Dacorogna, *Direct methods in the calculus of variations*. Springer-Verlag, n. 78.

[12] R.B. Gonzales de Paz, *On the optimal design of elastic shafts*, p. 61, Maths. Modelling and Num. Anal., vol. 23-4-1989, p. 615-625.

[13] D. Serre, *Sur la formulation variationnelle de l'écoulement des fluides parfaits*, Equipe d'Analyse Numérique de Lyon-Saint-Etienne, 1986.

[14] G.R. Burton, *Steady symmetric vortex pairs and rearrangements*, Proc. Roy. Soc. Edinburgh Serie A, 108, (1988), 269-290.

[15] R. Temam, *Navier-Stokes Equations*. Theory and Numerical Analysis, North-Holland, Amsterdam, 1979.

[16] B. Turkington, *On steady vortex flow in two dimension, I, II*, Comm. Partial Diff. Equat. 8, (1983), p. 999-1071.

[17] R. Tahraoui, *Rigidité maximale à la torsion : quelques remarques qualitatives*. Publication du Ceremade Paris IX, n. 9109.

Inverse Problems for Variational Inequalities

Dr. Vyacheslav Maksimov

Abstract. The problems of dynamical identification of the controls in parabolic   variational   inequalities   are   considered.   A   numerical solution methods are suggested.

1. Introduction.

The problems of dynamical identification of distributed controls and coefficients in parabolic systems described by variational inequalities are investigated. The suggested constructions are based on the discrepancy method from the theory of ill-posed problems (see Tikhonov and Arsenin 1977) and the method of positional control with a model (Krasovski 1985, Kryazhimski and Osipov 1983, Osipov et al. 1991). For other algorithms for reconstruction of controls and coefficients of an elliptic operator in parabolic inequalities based on the feedback principle (see Osipov 1989 and Maksimov 1988, 1990, 1991a,b). Note that in recent years, the control problems for parabolic variational inequalities were studied intensely by Barbu (1984), Tiba (1985) et al. However the inverse problems for these systems were considered in few papers (see Hoffmann and Sprekels 1985).

2. Problem Formulation

Let a system

$$(\dot{x}(t) - Bu(t) - f(t), x(t) - z) + \langle Ax(t), x(t) - z \rangle$$

(1)

$$+ \phi(x(t)) - \phi(z) \leq 0 \quad \text{a.e. } t \in T = [t_0, \vartheta] \quad \forall z \in V, x(t_0) = x_0$$

be given. Here H and V are real Hilbert spaces with norms $|\cdot|_H$ and $|\cdot|_V$ resp.,

$H = H^*, V \subset H$, V is densely and continuously imbedded in H, $(\cdot, \cdot)$ is the scalar product in H, $\langle \cdot, \cdot \rangle$ is the duality between V and V^*, $f(\cdot) \in L_2(T;H)$ is a given function, $A : V \to V^*$, is a linear, continuous and symmetric operator satysfying the condition $\langle Ay, y \rangle + a|y|_H^2 \geq \bar{\omega}|y|_V^2 \ \forall \ y \in V$ for certain $\bar{\omega} > 0$ and a, $(U, |\cdot|_U)$ is a uniformly by convex Banach space, $B \in \mathcal{L}(U;H)$ is a linear continuous operator. Let ϕ be the indicator function of a convex closed set $K \subset H$, i.e. $\phi(x) = 0$ if $x \in K$, $\phi(x) = +\infty$ in the opposite case. Let $x_0 \in K$ and the conditions of Corollary 4.2 (see Barbu 1984, p. 132) be fulfilled : there exists an $h \in H$ such that $(I + \lambda A_a)^{-1}(y + \lambda h) \in K$ for all $\lambda > 0$ and $y \in K$, where $A_a y = ay + Ay$ for $y \in D(A_H) = \{ v \in V : Av \in H \}$. Then for any input $u(\cdot) \in L_2(T;U)$ there is the single solution of the system (1), i.e. the function $x(\cdot) = x(\cdot; t_0, x_0, u(\cdot))$ satisfying (1) and such that $x(\cdot) \in W^{1,2}(T;H) \cap C(T;V) \cap L_2(T;D(A_H))$,

$$\dot{x}(t) = (f(t) - Ax(t) - \partial\phi(x(t)))^0 \text{ a.e. } t \in T.$$

Here $W^{1,2}(T;H) = \{ y(\cdot) \in L_2(T;H) : \dot{y}(\cdot) \in L_2(T;H) \}$, $\partial\phi$ is the subdifferential of function ϕ, $(f(t) - Ax(t) - \partial\phi(x(t)))^0$ is the element of minimal norm in $f(t) - Ax(t) - \partial\phi(x(t))$. Further we assume that $u(\cdot) \in P(\cdot) = \{ u(\cdot) \in L_2(T;H): u(t) \in P \text{ a.e. } t \in T \}$, P is a closed, bounded and convex set. The problem in question can be explained as follows. The motion $x_r(\cdot) = x(\cdot; t_0, x_0, u_r(\cdot))$ of the system (1) depending on a time-varying unknown input variable $u_r(t) \in P$ proceeds at the time interval T. The interval T is put into parts by intervals $[\tau_i, \tau_{i+1})$, $i \in [0:m-1]$, $\tau_{i+1} = \tau_i + \delta$, $\delta > 0$, $\tau_0 = t_0$, $\tau_m = \vartheta$. At time instants $\tau_i \in \Delta = \{\tau_i\}_{i=0}^m$ the phase coordinates of the system (1) are measured approximately, i.e. the elements $\psi_{\varepsilon,i} \in X$ close to $x_r(\tau_i)$ in the following sense : $| Cx_r(\tau_i) - \psi_{\varepsilon,i} |_X \leq \varepsilon$ are found. Here $C \in \mathcal{L}(H;X)$, $(X, |\cdot|_X)$ is a real Hilbert space, ε is the level of the informational noise. A motion $x_r(\cdot)$ is unknown. Let U_* be the set of all controls with values in $P(\cdot)$, generating the $\zeta(t) = Cx_r(t)$: $U_* = \{ u(\cdot) \in P(\cdot) : Cx(t; t_0, x_0, u(\cdot)) = \zeta(t) \text{ for } t \in T \}$. The problem is to calculate an approximation to a certain element $u^*(\cdot)$ from U_* synchronically with the process, basing on nonaccurate measurements of $x_r(\tau_i)$.

3. The dynamical discrepancy method.

Let us indicate an algorithm for approximating an input $u^*(\cdot) \in U_*$ based on a dynamical modification of the discrepancy method. Let $X = V$, $C = I$ (the identity operator), $f(\cdot) \in W^{1,2}(T;H)$ and at instants τ_i the elements $\psi_{\varepsilon,i} \in V \cap K$ approximating the values of the states $x_r(\tau_i)$ be calculated :

$$(2) \qquad\qquad |\psi_{\varepsilon,i} - x_r(\tau_{\varepsilon,i})|_V \leq \varepsilon.$$

For an $\varepsilon \in (0,1)$ denote by Δ_ε a partition of the interval T with diameter $\delta(\varepsilon)$: $\Delta_\varepsilon = \{\tau_{\varepsilon,i}\}_{i=0}^{m_\varepsilon}$, $m_\varepsilon = m(\delta(\varepsilon))$, $\delta(\varepsilon) = \tau_{\varepsilon,i+1} - \tau_{\varepsilon,i}$, $\tau_{\varepsilon,0} = t_0$, $\tau_{\varepsilon,m_\varepsilon} = \vartheta$. Let $\omega_x(\cdot)$ be a modulo of continuity of the function $t \to x_r(t) \in C(T;V)$, $\Phi_x(\cdot) : R^+ \to R^+$ is a function with the properties : $\Phi_x(\delta) \to 0$ as $\delta \to 0$, $\Phi_x(\delta) \geq \omega_x(\delta) + \sup \{ \int_t^{t+\delta} |\dot{x}_r(\tau)|_H^2 \, d\tau : t, t+\delta \in T \}$. Introduce the sets

$$(3) \qquad V_{b_1,b_2,i}^{\varepsilon,\delta}(\psi_{\varepsilon,i},\psi_{\varepsilon,i-1}) = \{ u \in P : \sup_{z \in S(\psi_{\varepsilon,i-1})} \{ F_{i,\delta}(u,\psi_{\varepsilon,i},\psi_{\varepsilon,i-1},$$

$$\psi_{\varepsilon,i-1} - z)\} \leq \nu(\varepsilon,\delta(\varepsilon);b_1,b_2,x_r) \}, \qquad i \in [1:m_\varepsilon],$$

where

$$S(a) = \{ x \in K : |x - a|_V \leq 1 \}, \quad \nu(\varepsilon,\delta;b_1,b_2,x_r) = b_1\varepsilon\delta^{-1} + b_2(\delta^{1/2} + \Phi_x(\delta)),$$

$$F_{i,\delta}(u,w,v,z) = ((w-v)/\delta - f(\tau_{\varepsilon,i}) - Bu,z) + \langle Av,z \rangle.$$

Let us describe the algorithm, i.e. the sequence of actions forming an approximation to $u^*(\cdot)$. First, a family Δ_ε of partitions of the interval T with diameters $\delta(\varepsilon)$, $\delta(\varepsilon) \to 0$, $\varepsilon\delta^{-1}(\varepsilon) \to 0$ as $\varepsilon \to 0$, are chosen. Before the initial time t_0 of the process, values ε, b_1, b_2 and a partition $\Delta = \Delta_\varepsilon = \{\tau_{\varepsilon,i}\}_{i=0}^{m_\varepsilon}$ are fixed. The work of the algorithm starting at time t_0 is decomposed into $m_\varepsilon - 1$ steps. At the i-th step carried out during the time interval $\delta_{\varepsilon,i} = [\tau_{\varepsilon,i},\tau_{\varepsilon,i+1})$ the element $v_i^\varepsilon = \arg\min \{|u|_U : u \in V_{b_1,b_2,i}^{\varepsilon,\delta}(\psi_{\varepsilon,i},\psi_{\varepsilon,i-1})\}$, if $V_{b_1,b_2,i}^{\varepsilon,\delta}(\psi_{\varepsilon,i},\psi_{\varepsilon,i-1}) \neq \varnothing$, $v_i^\varepsilon = 0$, in the opposite case, depending on $\psi_{\varepsilon,i},\psi_{\varepsilon,i-1} \in K$ satisfying (2) for $\tau_i = \tau_{\varepsilon,i}$ is calculated, and $v^\varepsilon(t) = v_i^\varepsilon$, $t \in \delta_{\varepsilon,i}$ is assumed, $v_0^\varepsilon = \arg\min \{|u|_U : u \in P\}$. The procedure stops at time ϑ. If values ε, $\delta(\varepsilon)$ and $\varepsilon\delta^{-1}(\varepsilon)$ are "sufficiently small", then $v^\varepsilon(\cdot)$ is a "good" approximation to some $u^*(\cdot)$ from U_*. This follows from the

Theorem 1. There exist values $b_1^*>0$ and $b_2^*>0$ such that for $b_1 \geq b_1^*$ and $b_2 \geq b_2^*$ the convergence $\|v^{\varepsilon}(\cdot) - u_*(\cdot;x_r(\cdot))\|_{L_2(T;U)} \to 0$ as $\varepsilon \to 0$ is true.

Here $u_*(\cdot;x(\cdot))$ is the element of the set U_* whoose $L_2(T;U)$-norm is minimal. The constants b_1^* and b_2^* are written out explicitly. If $\dot{x}_r(\cdot)$, $\dot{f}(\cdot) \in L_\infty(T;H)$ then in (3) it is better to put $v(\varepsilon,\delta;b_1,b_2,x_r) = b_1\varepsilon\delta^{-1} + b_2(\delta + \omega_x(\delta))$. The theorem follows from two lemmas. Let

$$U_{\Delta,v}(\psi(\cdot)) = \{\, u(\cdot) \in P(\cdot) : u(t) \in V_{b_1,b_2,i}^{\varepsilon,\delta}(\psi_{\varepsilon,i},\psi_{\varepsilon,i-1}) \text{ a.e. } t \in [\tau_{i-1},\tau_i)\, \},$$

$$U_{\Delta}(x_r(\cdot)) = \{\, u(\cdot)\in P(\cdot) : \int_{\tau_i}^{\tau_{i+1}} (Bu(t),z-x_r(t))dt \le \Psi^*(\tau_i,\tau_{i+1},z,x_r(\cdot)) \;\forall\; z\in K,$$

$$\tau_i,\tau_{i+1} \in \Delta\, \}, \quad \Psi^*(\tau_i,\tau_{i+1},z,x_r(\cdot)) = \int_{\tau_i}^{\tau_{i+1}} \langle \dot{x}_r(t)+Ax_r(t) -f(t),z-x_r(t)\rangle dt,$$

$$\Psi_2^*(\tau_i,\tau_{i+1},z,\psi(\cdot)) = (\psi_{\varepsilon,i+1}- \psi_{\varepsilon,i}- \delta f(\tau_i),z - \psi_{\varepsilon,i}) + \delta\langle A\psi_{\varepsilon,i},z - \psi_{\varepsilon,i}\rangle.$$

Lemma 1. Let $v(\cdot) \in U_{\Delta}(x_r(\cdot))$, $v_*(t) = \delta^{-1}\int_{\tau_i}^{\tau_{i+1}} v(t)dt$ a.e. $t \in [\tau_i,\tau_{i+1})$, $i \in [0:m-1]$. Then there exist values $b_1^*>0$ and $b_2^*>0$ such that $v_*(\cdot) \in U_{\Delta,v}(\psi(\cdot))$, $v(\varepsilon,\delta;\, b_1^*,b_2^*,x_r)$.

Proof. Using inclusions $x_r(\cdot) \in W^{1,2}(T;H)\cap C(T;V)$, $f(\cdot) \in W^{1,2}(T;H)$, continuity and density of imbedding V into H and correlations (2), one can easily deduce the inequality ($\tau_i = \tau_{\varepsilon,i}$, $x_i = x_r(\tau_i)$),

$$| \Psi^*(\tau_i,\tau_{i+1},z,x_r(\cdot))- \Psi_2^*(\tau_i,\tau_{i+1},z,\psi(\cdot)) | \le k_1\delta \int_{\tau_i}^{\tau_{i+1}} |\dot{x}_r(t)|_H^2 dt +$$

$$(4) \qquad k_2\delta(\varepsilon + |z - x_i|_H \int_{\tau_i}^{\tau_{i+1}} \{|\dot{x}_r(t)|_H + |f(t)|_H\}dt) +$$

$$k_3\delta(\varepsilon + \omega_x(\delta))(1 + |z - x_i|_V) + k_4\varepsilon(1 + |z - x_i|_H).$$

Further, we have

$$(5) \; | \int_{\tau_i}^{\tau_{i+1}} \{(Bv(t),z - x(t)) - (Bv(t),z - \psi_{\varepsilon,i})\}dt | \le k_5\delta(\varepsilon + \int_{\tau_i}^{\tau_{i+1}} |\dot{x}_r(t)|_H dt).$$

It follows from (4), (5) that for certain $b_p = b_p(k_j, j\in[1:5])$, $p = 1,2$ (written out explicitly) the bound

$$\delta^{-1}|(B\int_{\tau_i}^{\tau_{i+1}}v(t)dt,z - \psi_{\varepsilon,1})| \le \delta^{-1}\Psi_2^*(\tau_1,\tau_{i+1},z,\psi(\cdot)) + \nu(\varepsilon,\delta;b_1,b_2,x_r(\cdot)).$$

holds uniformly with respect to $z \in S(\psi_{\varepsilon,1-1})$. Lemma is proved.

Lemma 2. Let $\varepsilon_j \to 0$, $\delta_j \to 0$, $\varepsilon_j\delta_j^{-1} \to 0$ as $j \to 0$, $\Delta_j = \Delta_{\varepsilon_j}$, $\delta_j = \delta(\varepsilon_j)$, $u_j(\cdot) \in$ $U_{\Delta_j,\nu_j}(\psi^{(j)}(\cdot))$, $\nu_j = \nu(\varepsilon_j,\delta_j;b_1^*,b_2^*,x_r)$, $\psi^{(j)}(t) = \psi^{(j)}(\tau_{\varepsilon_j,1})$ for $t \in \delta_{\varepsilon_j,1}$, $\psi^{(j)}(\tau_{\varepsilon_j,1}) \in K$, $|\psi^{(j)}(\tau_{\varepsilon_j,1}) - x_r(\tau_{\varepsilon_j,1})|_V \le \varepsilon_j$, $u_j(\cdot) \to u_0(\cdot)$ weakly in $L_2(T;U)$. Then $u_0(\cdot) \in U_*$.

Proof. Let $u_0(\cdot) \notin U_*$. Due to the equality

$$U_* = \{u(\cdot) \in P(\cdot) \mid \int_{t_1}^{t_2} (Bu(t),v - x_r(t))dt \le \Psi^*(t_1,t_2,v,x_r(\cdot))$$

$$\forall\, t_1,t_2 \in T,\ t_1 < t_2,\ v \in K \}$$

there exists $v_* \in V\cap K$, $t_1,t_2 \in T$, $t_1 < t_2$ и $a_* > 0$ such that :

$$(6) \qquad \int_{t_1}^{t_2} (Bu_0(t),v_* - x_r(t))dt > \Psi^*(t_1,t_2,v_*,x_r(\cdot)) + a_*.$$

Let j_1 be such that for $j \ge j_1$

$$(7) \qquad \delta_j \le (t_2 - t_1)/3,$$

$$(8) \qquad \sup \{ \int_{t_*}^{t^*} |(Bu(t),v_* - x_r(t))|dt, \int_{t_*}^{t^*} |(\dot{x}_r(t) - f(t),v_* - x_r(t)) +$$

$$\langle Ax_r(t),v_* - x_r(t)\rangle|dt : u(\cdot) \in P(\cdot),\ t_*,t^* \in T,\ 0 \le t^* - t_* \le \delta_j \} \le a_*/16.$$

Denote $\tau_{1_*(j)} = \max \{\tau_1 \in \Delta_j : \tau_1 \le t_2\}$, $\tau_1^*{}_{(j)} = \min \{\tau_1 \in \Delta_j : \tau_1 \ge t_1\}$ ($\tau_1 = \tau_{\varepsilon_j,1}$). Due to (6) – (8) we have : $i_*(j) > i^*(j)$ and for $j \ge j_1$

$$(9) \qquad \int_{\tau_1^*{}_{(j)}}^{\tau_{1_*(j)}} (Bu_0(t),v_* - x_r(t))dt \ge \Psi^*(\tau_1^*{}_{(j)},\tau_{1_*(j)},v_*,x_r(\cdot)) + 3a_*/4.$$

By the definition $u_j(\cdot)$, $u_j(t) = u_j^{(l)} \in V_{b_1,b_2,l}^{\varepsilon_j,\delta_j}(\psi_{l+1}^{(j)},\psi_l^{(j)})$ for $t \in \delta_{\varepsilon_j,1}$,

$$(10) \qquad (Bu_{1j}, v - \psi_1^{(j)}) \leq \Psi_2^*(\tau_1, \tau_{1+1}, v, \psi^{(j)}(\cdot)) + \delta_j v_j \ \forall \ v \in S(\psi_1^{(j)}),$$

where $\delta_{\varepsilon_j, 1} = [\tau_1, \tau_{1+1})$, $(v_j = v \ (\varepsilon_j, \delta_j; b_1, b_2, x_r(\cdot)))$, $u_{1j} = \delta_j u_j^{(1)}$, $\psi_1^{(j)} = \psi^{(j)}(\tau_1)$. Let $\lambda_1 = 1$, $v_1 = v_*$, if $v_* \in S(\psi_1^{(j)})$, $\lambda_1 = |v_* - \psi_1^{(j)}|_v$, $v_1 = \psi_1^{(j)} + (v_* - \xi_1^{(j)})/|v_* - \psi_1^{(j)}|_v$ in the oposite case. Therefore,

$$(11) \qquad \lambda_1 \Psi_2^*(\tau_1, \tau_{1+1}, v_1, \psi^{(j)}(\cdot)) = \Psi_2^*(\tau_1, \tau_{1+1}, v_*, \psi^{(j)}(\cdot)).$$

Taking into account (4), (10), inequalities $|\psi_1^{(j)} - x_1|_v \leq \varepsilon_j$ and inclusions $u_j(\cdot) \in U_{\Delta_j, v_j}(\psi^{(j)}(\cdot))$ we conclude that

$$I_j \equiv \int_{\tau_{1^*(j)}}^{\tau_{1_*(j)}} (Bu_j(t), v_* - x_r(t))dt \leq \sum_{1=1^*(j)}^{1_*(j)-1} \lambda_1 (Bu_{1j}, v_1 - \psi_1^{(j)}) + k_0(\varepsilon_j +$$

$$\delta_j^{1/2}) \leq \sum_{1=1^*(j)}^{1_*(j)-1} \lambda_1 \{\Psi_2^*(\tau_1, \tau_{1+1}, v_1, \psi^{(j)}(\cdot)) + \delta_j v_j\} + k_0(\varepsilon_j + \delta_j^{1/2}).$$

This and (4), (11) imply

$$(12) \qquad I_j \leq \Psi^*(\tau_{1^*(j)}, \tau_{1_*(j)}, v_*, x_r(\cdot)) + f(\varepsilon_j, \delta_j),$$

$f(\varepsilon_j, \delta_j) = c_1 \varepsilon_j \delta_j^{-1} + c_2 \delta_j^{1/2} + c_3 \Phi_x(\delta_j)$. Let $j_2 \geq j_1$ be such that for $j \geq j_2$ $f(\varepsilon_j, \delta_j) \leq a_*/4$ and

$$(13) \qquad \int_{\tau_{1^*(j)}}^{\tau_{1_*(j)}} (B(u_0(t) - u_j(t)), v_* - x_r(t))dt \leq \int_{t_1}^{t_2} (B(u_0(t) - u_j(t)),$$

$$v_* - x_r(t))dt + a_*/8 \leq a_*/4.$$

Therefore, for $j \geq j_2$ it follows (12), (13) that

$$(14) \qquad \int_{\tau_{1^*(j)}}^{\tau_{1_*(j)}} (Bu_0(t), v_* - x_r(t))dt \leq \Psi^*(\tau_{1^*(j)}, \tau_{1_*(j)}, v_*, x_r(\cdot)) + a_*/2.$$

However, (14) contradicts (9). Lemma is proved.

Proof of Theorem 1. To prove the Theorem it is sufficient to show that

$$(15) \qquad v^{\varepsilon}{}_{j}(\cdot) \to u_{*}(\cdot;x_{r}(\cdot)) \text{ in } L_{2}(T;U).$$

Introduce functions $u_{j}(\cdot)$, $u_{j}(t) = v^{\varepsilon}(t+\delta(\varepsilon))$, $t \in [t_{0},\vartheta-\delta(\varepsilon))$, $u_{j}(t) = v^{\varepsilon}_{m_{\varepsilon}}$, $t \in [\vartheta-\delta(\varepsilon),\vartheta]$, $\varepsilon = \varepsilon_{j}$. Convergence (15) holds if

$$(16) \qquad u_{j}(\cdot) \to u_{*}(\cdot;x_{r}(\cdot)) \text{ in } L_{2}(T;U).$$

Prove (16). Assume that it is not true. Then there exists a subsequence of the sequence $\{u_{j}(\cdot)\}$ (denote it for simplicity by the same symbol $\{u_{j}(\cdot)\}$) such that

$$(17) \qquad u_{j}(\cdot) \to u_{0}(\cdot) \text{ weakly in } L_{2}(T;U), \quad u_{0}(\cdot) \neq u_{*}(\cdot;x_{r}(\cdot)).$$

Note that Lemma 1 and inclusion $U_{*} \subset U_{\Delta}(x_{r}(\cdot))$ imply the inclusion

$$(18) \qquad \overline{u}_{j}(\cdot;x_{r}(\cdot)) \in U_{\Delta_{j},\nu_{j}}(\zeta^{\varepsilon}{}_{j}(\cdot)),$$

where $\overline{u}_{j}(t;x_{r}(\cdot)) = \delta^{-1}\int_{\tau_{1}}^{\tau_{1+1}} u_{*}(\tau;x_{r}(\cdot))d\tau$ for a.e. $t \in [\tau_{1},\tau_{1+1})$ and Lemma 2 implies the inequality $\|u_{0}(\cdot)\|_{L_{2}(T;U)} \geq \|u_{*}(\cdot;x_{r}(\cdot))\|_{L_{2}(T;U)}$, since $u_{0}(\cdot) \in U_{*}$. Besides, due to the properts of the weak limit,

$$(19) \qquad \lim_{j\to\infty} \|u_{j}(\cdot)\|_{L_{2}(T;U)} \geq \|u_{0}(\cdot)\|_{L_{2}(T;U)} \geq \|u_{*}(\cdot;x_{r}(\cdot))\|_{L_{2}(T;U)}.$$

Thanks to (18) and the rule $u_{j}(\cdot)$ have been chosen,

$$\|u_{j}(\cdot)\|^{2}_{L_{2}(\tau_{1},\tau_{1+1};U)} \leq \|u_{*}(\cdot;x_{r}(\cdot))\|^{2}_{L_{2}(\tau_{1},\tau_{1+1};U)}, \quad i \in [0:m_{\varepsilon_{j}}-1].$$

Consequently

$$\overline{\lim_{j\to\infty}} \|u_{j}(\cdot)\|_{L_{2}(T;U)} \leq \|u_{*}(\cdot;x_{r}(\cdot))\|_{L_{2}(T;U)}.$$

Hence

$$(20) \qquad \lim_{j\to\infty} \|u_{j}(\cdot)\|_{L_{2}(T;U)} = \|u_{0}(\cdot)\|_{L_{2}(T;U)}.$$

From (17), (19) follows (16), since (17), (20) imply that in a uniformly convex

Banach space the strong convergence of $u_j(\cdot)$ to $u_0(\cdot) = u_*(\cdot;x_r(\cdot))$. Theorem is proved.

The algorithm is easily generalized for the case where ϕ is an arbitrary convex, proper and lower semicontinuous function.

4. Method of feedback control with a model.

Let us begin with an example. Let the evolution of a real physical variable $x_r(t) = x(t;t_0,x_0,u_r(\cdot))$ in the space $H = L_2(\Omega)$ be described by a free boundary problem (parabolic problem with obstacle)

$$\dot{x}_t(t,\eta)-\Delta x(t,\eta)=\omega(\eta)u(t)+f(t,\eta) \text{ a.e. in } [(t,\eta) \in T\times\Omega : x(t,\eta) > \zeta(\eta)],$$

(21)

$$\dot{x}_t(t,\eta)=\max\{\omega(\eta)u(t)+f(t,\eta)+\Delta\zeta(\eta),0\} \text{ a.e. in } [(t,\eta) \in T\times\Omega : x(t,\eta) =\zeta(\eta)],$$

$$x(t,\eta) \geq \zeta(\eta) \ \forall \ t \in T \text{ and a.e. } \eta \in \Omega, \qquad x = 0 \text{ in } \Gamma \times T,$$

$$x_0(\eta) \geq \zeta(\eta) \text{ a.e. in } \Omega, \qquad \zeta(\cdot) \in H_2(\Omega), \qquad \zeta(\eta) \leq 0 \text{ a.e. in } \Gamma.$$

which is reduced (see Brezis 1973 and Barbu 1984) to the inequality (1). Here $\Omega \subset R^n$ is a bounded open set with the smooth boundary Γ, $\omega(\eta) \in H$ is a given function. A coefficient $u = u_r(t) \in P$ is unknown. The information of the process (4) at time τ_1 is formed by the values $\psi_{\varepsilon,1} :| \ \psi_{\varepsilon,1} - Cx_r(\tau_1) | \leq \varepsilon$, $Cx_r(\tau_1) = \int_{\Omega} a(\eta)x_r(\tau_1,\eta)d\eta$, $a \in H$. This situation corresponds to observation of the average value of a physical variable. The problem consists in constructing a procedure, calculating approximately $u_r(\cdot) = u^*(\cdot) \in U_*$ synchronically with the process. These problems is included into that described in Introduction. Here $X = U = R$, $C \in \mathcal{L}(H;R)$, $Bu = \omega u$, $\omega \in H$, $u_r(t) \in R$, $V = H_1^0(\Omega)$, $K = \{ \ y \in H : y \geq \psi \text{ a.e. in } \Omega \ \}$.

To calculate approximately $u_r(\cdot)$, we apply the method of closed-loop control with a model (see Krasovski 1985, Kryazhimskii and Osipov 1983, Osipov 1989, Maksimov 1988, 1990, 1991). According to this approach, calculation of an unknown control $u_r(\cdot)$ on the basis of hindered measurement results $\psi(\cdot)$ is carried out as follows. An auxiliary system M (a model) functioning together with the real system is introduced; let $w(\cdot)$ be an output and $v^\varepsilon(\cdot)$ be a control for the model M. Then an *algorithm forming* $v^\varepsilon(\cdot)$ *by feedback* $v(\cdot) = v^\varepsilon(\cdot\,;\psi(\cdot),w(\cdot))$ is designed ; $v^\varepsilon(\cdot)$ approximates, in an appropriate sense, the unknown control $u_r(\cdot)$. Thus the problem of calculation of the unknown control is *replaced* by the problem of finding an algorithm forming a control for the model. This algorithm provides a desired approximation to the unknown control.

Let $x_0, \omega \in K$ be given functions, $\omega \in D(A_H)$, $C(\omega) \neq 0$, $P = [a,b]$, $0 < a \leq b < +\infty$, $u_r(\cdot) \in P(\cdot)$, $f(\cdot) \in L_2(T;H)$, $f(t,\eta) \geq 0$ a.e. in $T \times \Omega$. For simplicity consider $t_0 = 0$, $\zeta = 0$. The model M is given by the control system

$$\dot{w}(t) = C(\omega)v^\varepsilon(t) + \int_{t_0}^{t} v^\varepsilon(\tau)F(t,\tau)d\tau + C(\dot{x}_1(t)), \quad w(t_0) = 0, \ t \in T,$$

where $w \in R$, $x_1(\cdot) = x(\cdot\,;t_0,x_0,f)$, $F(t,\tau) = C\dot{x}_{2t}(t-\tau)$, $t \geq \tau$, $x_2(\cdot) = x(\cdot\,;0,\omega,0)$.

The general pattern of the algorithm approximating $u_r(\cdot)$ is analogous to that described in Sec.3. First a family Δ_ε is chosen, $\delta(\varepsilon) \to 0$, $\varepsilon\delta^{-1}(\varepsilon) \to 0$ as $\varepsilon \to 0$. Before the moment t_0, ε and Δ_ε are fixed. At the i-th step, during the time interval $\delta_{\varepsilon,i}$, $i \geq 1$, the following operations are carried out. At time $\tau_{\varepsilon,i}$ calculate $g_i = \{(\psi_{\varepsilon,i} - \psi_{\varepsilon,i-1}) - C(x_1(\tau_{\varepsilon,i}) - x_1(\tau_{\varepsilon,i-1}))\}\delta^{-1}(\varepsilon)$ on the basis of the collection $\psi_{\varepsilon,i}$, $\psi_{\varepsilon,i-1}$, $x_1(\tau_{\varepsilon,i})$, $x_1(\tau_{\varepsilon,i-1})$. Then, having $\psi_{\varepsilon,i-1}$, $\psi_{\varepsilon,0}$, g_i and $w(\tau_{\varepsilon,i-1})$, we determine the control $v^\varepsilon(t) = v^\varepsilon_i$ for $t \in \delta_{\varepsilon,i-1} = [\tau_{\varepsilon,i-1}, \tau_{\varepsilon,i})$, $i \in [1:m_\varepsilon]$,

$$v^\varepsilon_i = \begin{cases} c_*|g_i - p_i| \ \mathrm{sgn}\ s_{i-1}, & \text{if } s_{i-1} \neq 0 \\ 0, & \text{in the opposite case,} \end{cases}$$

$$s_{i-1} = \psi_{\varepsilon,i-1} - Cx_1(\tau_{\varepsilon,i-1}) - w(\tau_{\varepsilon,i-1}), \quad p_i = \delta^{-1}(\varepsilon)\int_{\tau_{\varepsilon,i-1}}^{\tau_{\varepsilon,i}} \int_{t_0}^{\tau_{\varepsilon,i-1}} v^\varepsilon(\tau)F(t,\tau)d\tau \ dt.$$

$c_* \in [C^{-1}(\omega),+\infty)$. After that, we transform the state $w(\tau_{\varepsilon,i-1})$ of the model into $w(\tau_{\varepsilon,i})$. The procedure stops at time ϑ.

Theorem 2. If $\varepsilon \to 0$, then $v^\varepsilon(\cdot) \to u_r(\cdot)$ weakly in $L_2(T;R)$.

The theorem follows from the two lemmas.

Lemma 3. The real input $u_r(\cdot)$ satisfied the condition

$$C(\dot{x}_r(t)) = C(\dot{x}_1(t)) + C(\omega)u_r(t) + \int_{t_0}^{t} u_r(t)F(t,\tau)d\tau \ \text{a.e. } t \in T.$$

The lemma follows from the relations $f(t,\eta) > 0$ a.e. $\eta \in \Omega$, $u(t) > 0$, $\omega \in K$,

$$\dot{x}_{2t}(t-\tau) = (-Ax_2(t-\tau) - \partial\phi(x_2(t-\tau)))^0 \text{ for a.e. } t \in [\tau,\vartheta], \; x_2(0) = \omega,$$

$$\dot{x}_1(t) = f(t) + (-Ax_1(t) - \partial\phi(x_1(t)))^0 \text{ for a.e. } t \in T,$$

$$(\int_{t_0}^{t_*} u_r(t)(Ax_j(t) + \partial\phi(x_j(t)))dt \,)^0(\eta) =$$

$$(A(\int_{t_0}^{t_*} u_r(t)x_j(t)dt) + \partial\phi(\int_{t_0}^{t_*} u_r(t)x_j(t)dt) \,)^0(\eta) \text{ a.e. in } \Omega \; \forall \, t_* \in T, \; j = 1,2.$$

Lemma 4. The bounds

$$(22) \quad \lambda(t) = |C(x_r(t)) - w(t)|^2 \le K(\delta + \varepsilon\delta^{-1}), \; t \in T, \; \|v^\varepsilon(\cdot)\|_{L_2(T;R)} \le K_1.$$

hold, uniformly with respect to all $\varepsilon \in (0,1)$, $\{\Delta_\varepsilon\}$ with diameter $\delta = \delta(\varepsilon) < \delta_0$ and $\psi_{\varepsilon,1}$, $|\psi_{\varepsilon,1} - Cx_r(\tau_1)| \le \varepsilon$.

Proof outline. Let us estimate the evolution of $\lambda(\tau_1)$ for $i \in [0:m_\varepsilon]$ ($\tau_1 = \tau_{h,1}$). We have $\lambda(\tau_{1+1}) \le \lambda(\tau_1) + \mu_1 + I_1 + \varepsilon k_1 I_1^{1/2}$. Here

$$\mu_1 = 2s_1 \int_{\tau_1}^{\tau_{1+1}} \nu_1(\tau;v^\varepsilon(\cdot))d\tau, \quad \nu_1(\tau;v^\varepsilon(\cdot)) = C(\dot{x}_r(\tau) - \dot{x}_1(\tau)) - (C(\omega) -$$

$$- \int_\tau^{\tau_{1+1}} F(\eta,\tau)v^\varepsilon(\tau)d\eta - \int_0^{\tau_1} F(\tau,\eta)v^\varepsilon(\eta)d\eta \text{ for } \tau \in [\tau_1,\tau_{1+1}], \; I_1 = |\int_{\tau_1}^{\tau_{1+1}} \nu_1(\tau;v^\varepsilon(\cdot))d\tau|^2.$$

By Theorem 4.3 (see Barbu 1984) condition $\omega \in K \cap D(A_H)$ imply $\dot{x}_2(\cdot) \in L_\infty(T;H)$. Hence, taking into account the definition of $v^\varepsilon(\cdot)$, we deduce

$$\mu_1 \le k_2\delta|s_1| \int_{\tau_1}^{\tau_{1+1}} |v^\varepsilon(\tau)|d\tau + 4\varepsilon|s_1|.$$

Note that all bounds are uniform with respect to $\varepsilon \in (0,1)$, (Δ_ε) with diameters $\delta(\varepsilon) \in (0,1)$ and measurement results $\psi_{\varepsilon,1}$, $|Cx_r(\tau_1)-\psi_{\varepsilon,1}| \le \varepsilon$. Besides

$$|s_1| \le k_3(1 + \int_0^{\tau_1} |v^\varepsilon(\tau)|^2 d\tau), \; d_1 \equiv \int_{\tau_{1-1}}^{\tau_1} |v^\varepsilon(\tau)|^2 d\tau \le a_1 + k_4\delta \sum_{j=1}^{1-1} d_j, \quad \sum_{j=1}^{m_\varepsilon-1} a_j < +\infty.$$

Thus, by the diskrete Gronuall inequality for $\delta(\varepsilon) \in (0,1)$ holds $|v^\varepsilon(\cdot)|_{L_2(T;R)}$

$\leq k_5$. Hence, $\sum\limits_{l=0}^{m_\varepsilon-1} I_l \leq k_6\delta$. Consequently, for $i \in [0:m_\varepsilon-2]$ $\lambda(\tau_{i+1}) \leq k(\delta+\varepsilon/\delta)$.

Using this inequality and the inequality $|x_r(t) - x_r(t+h)| + |w(t) - w(t+h)| \leq k_8 h^{1/2}$ ($h > 0$)we obtain easily (22). Lemma is proved.

Proof of Theorem 2. Supposing the contrary and taking into account that he set $\{v^\varepsilon(\cdot)\}$ is uniformly bounded in $L_2(T;R)$ we conclude that : there exists a sequence $h_j \to 0$ as $j \to \infty$ such that $v^{\varepsilon_j}(\cdot) \to u_*(\cdot) \neq u_r(\cdot)$ weakly in $L_2(T;R)$. Note that the integral equation

$$\Phi(t,v(\cdot)) \equiv C(\dot{x}_r(t) - \dot{x}_1(t)) - C(\omega)v(t) - \int_{t_0}^{t} v(\nu)F(t,\nu)d\nu = 0 \quad \text{for a.e. } t \in T$$

has the unique solution $v(\cdot) = u_r(\cdot)$ in $L_2(T;R)$. Therefore, the set U_* is one-element : $U_* = \{u_r(\cdot)\}$, and for a certain function $p(\cdot) \in C(T;R)$

$$(23) \qquad\qquad J(u_*) \equiv \int_{t_0}^{\vartheta} p(t)\phi(t,u_*)dt = a > 0.$$

Here $\phi(t,v) = \int_{t_0}^{t} \Phi(\tau,v(\cdot))d\tau$. Furthere, we have

$$(24) \qquad J(u_*) \leq |\int_{t_0}^{\vartheta} p(t)\phi(t,v^{\varepsilon_j})dt| + |\int_{t_0}^{\vartheta} p(t)(\phi(t,v^{\varepsilon_j})-\phi(t,u_*))dt|.$$

The first term in the right hand side of (24) goes to zero as $j \to \infty$ by Lemma, and the second one goes to zero by the weak convergence of $v^{\varepsilon_j}(\cdot)$ to $u_*(\cdot)$. Thus, we get the contradiction with (23). Theorem is proved.

Theorem 2 takes place also if $\omega \in C(\Omega')$, $x_r(\cdot)$, $x_1(\cdot)$, $x(\cdot;t_0,\omega,0) \in W^{1,\infty}(T; C(\Omega'))$, $C : C(\Omega') \to R$, $C(x(\eta)) = x(\eta_0)$, $\eta_0 \in \Omega' \subset \Omega$.

References.

Tikhonov A.N. and Arsenin V.Ya. (1977), Solution of ill-posed problems. Wiley, New York.

Krasovski N.N. (1985), Controlling of dynamical systems. Nauka, Moskow (in Russian).

Kryazhimski A.V. and Osipov Yu. S. (1983),On modelling of control in a dynamical systems. Izv. Akad. Nauk. USSR, Tech. Cybernet, Vol.2, pp.51-60 (in Russian).

Osipov Yu. S. and Kryazhimski A.V. (1983), On dynamical solution of operator equations. Dokl. Akad. Nauk USSR, Vol.269, No.3, pp. 552-556 (in Russian).

Osipov Yu.S. et al. (1991), Dynamical regularization problems for distributed parameter systems. Inst. Math. Mech. Ural Branch. Acad. Sci. USSR, 104 pp. (in Russian).

Osipov Yu.S. (1989), Inverse problems of dynamics for systems described by parabolic inequality. IIASA, Austria: WP-89-101, 11 pp..

Maksimov V.I. (1988), On dynamical modelling of unknown distrurbances in parabolic inequalities. Prikl. Mat. Mekh., Vol.52, No.5, pp.743-750 (in Russian).

Maksimov V.I. (1990), On stable solution of inverse problems for nonlinear distributed systems,*I*. Differential Equat., Vol. 26, No.12, pp.2059-2067 (in Russian).

Maksimov V.I. (1991), a) Dynamical reconstructions in nonlinear parabolic systems. 15th IFIP Conference "System Modelling and Optimization", Zurich, Switzerland, September 2-6. Abstracts, pp.256-258.

b) On modelling of controls in parabolic variational inequalities. Differential Equat., Vol. 27, No.9, pp.1603-1609 (in Russian).

Barbu V. (1984), Optimal control of variational inequalities. London, Pitman.

Tiba D. (1985). Optimality conditions for distributed control problems with nonlinear state equations. SIAM J. Control and Optim. Vol. 23, No. 1, pp.85-110.

Hoffmann K.H. and Sprekels J. (1985), On the identification of heat conductivity and latent heat in a one-phase Stefan problem. Control and Cybernet., Vol.14, No.1-3, pp. 37-51.

Brezis H. (1973), Operateurs maximaux monotones et semigroupes de contractions dans les espace de Hildert. Amsterdam-London-New York.

Author's address:

Dr. Vyacheslav Maksimov
Institute of Mathematics and Mechanics,
Ural Branch, Acad. Sci. of Russia,
S. Kovalevskoi St. 16, SU-620219,
Ekaterinburg, Russia.

The Variation of the Drag with Respect to the Domain in Navier-Stokes Flow

Juan A. Bello, Enrique Fernández-Cara and Jacques Simon

Abstract This paper deals with optimal profiles in Navier-Stokes regime. Let us introduce an initial body B and assume that an admissible variation of B is represented by vector fields $\mathbf{u}$, i.e. that a "modified" bodies is given by $B+\mathbf{u} = \{x+\mathbf{u}(x); \ x \in B\}$. We prove that the mapping $\mathbf{u} \rightarrow J(B+\mathbf{u})$, where $J(B+\mathbf{u})$ is the drag associated with $B+\mathbf{u}$, is Fréchet-differentiable at 0. We also apply some known results to the computation of the derivative.

1. The problem.

The computation of optimal profiles, i.e. those minimizing the drag, has been investigated by several authors. Frequently, the drag has been approximated by the viscous energy which is dissipated in the fluid. For instance, Pironneau (1974, 1984) computes the "derivative" of this quantity adapting Hadamard's normal variations techniques. Murat and Simon (1974) use formal calculus to deduce an expression for this derivative. More recently, the existence of a derivative of the energy has been proven rigorously in the case of a Stokes flow (Simon 1991) and also in the case of a Navier-Stokes flow (Bello et al. 1991, 1992); see Fernández Cara (1989) for some additional theoretical and numerical considerations. The goal of this paper is to prove that in Navier-Stokes regime the true drag is differentiable and, also, to compute the corresponding derivative.

Assume we are given a "large" bounded open set $D \subset \mathbf{R}^d$ (here $d = 2$ or 3; D is the fluid domain) and a "small" open set $B \subset\subset D$ (the initial shape of the body). It will be imposed to the boundaries ∂D and ∂B to be $W^{2,\infty}$ in the sense of Nečas (1967). In this paper, the family of "admissible" domains is given by

$$(1.1) \qquad \Omega_{ad} = \{D \backslash \overline{B + \mathbf{u}}; \ \mathbf{u} \in W, \ \|\mathbf{u}\|_{L^\infty} < \varepsilon_0\},$$

where $W = \{\mathbf{u}; \ \mathbf{u} \in W^{2,\infty}(\mathbf{R}^d; \mathbf{R}^d), \ \mathbf{u}\,|_{\partial D} \equiv 0\}$ and ε_0 is sufficiently small. It can be assumed that all admissible $B + \mathbf{u}$ satisfy $B + \mathbf{u} \subset \mathcal{O}$, where $\mathcal{O} \subset\subset D$ is a fixed open set. On the other hand, the admissible shapes $B + \mathbf{u}$ are assumed to be at rest and the fluid particles are assumed to travel at constant velocity $\mathbf{g}$ far from $B + \mathbf{u}$. Consequently, for every $D\backslash\overline{B + \mathbf{u}} \in \Omega_{ad}$, the following Navier-Stokes problem has to be considered :

$$(1.2) \qquad -\nu\Delta\mathbf{y} + (\mathbf{y}\cdot\nabla)\mathbf{y} + \nabla p = 0, \quad \nabla\cdot\mathbf{y} = 0 \ \text{ in } \ D\backslash\overline{B + \mathbf{u}},$$

$$(1.3) \qquad \mathbf{y} = 0 \ \text{ on } \ \partial B + \mathbf{u}, \quad \mathbf{y} = \mathbf{g} \ \text{ on } \ \partial D.$$

A solution $(\mathbf{y}(\mathbf{u}), p(\mathbf{u}))$ to (1.2)–(1.3) provides a velocity field and a pressure distribution of the fluid. In these conditions (Schlichting 1970), the drag experienced by $B + \mathbf{u}$ is given by

$$(1.4) \qquad J(B + \mathbf{u}) = -\mathbf{g}\cdot\left(\int_{\partial B+\mathbf{u}}\left(-p(\mathbf{u})\,\mathbf{Id} + \sigma(\mathbf{y}(\mathbf{u}))\right)\cdot\vec{n}\,dS\right),$$

where $\mathbf{Id}$ is the identity matrix and, for a given field $\mathbf{y}$, the components of $\sigma(\mathbf{y})$ are defined as follows :

$$\sigma_{ij}(\mathbf{y}) = \nu\left(\frac{\partial y_i}{\partial x_j} + \frac{\partial y_j}{\partial x_i}\right) \quad 1 \leq i,j \leq d.$$

In the sequel, it will be assumed that $|\mathbf{g}| < \nu\alpha$, where α is a (suitably chosen) constant only depending on D. Then, (1.2)–(1.3) possesses exactly one solution furthermore satisfying (Murat and Simon 1974)

$$(\mathbf{y}(\mathbf{u}), p(\mathbf{u})) \in H^2(D\backslash\overline{B + \mathbf{u}})^d \times H^1(D\backslash\overline{B + \mathbf{u}})/\mathbf{R}.$$

Accordingly, if $\mathbf{u}$ is given in W and its norm in $W^{2,\infty}(\mathbf{R}^d; \mathbf{R}^d)$ is small enough, the quantity $J(B + \mathbf{u})$ is correctly defined. A classical optimum design problem concerns the computation of an optimal $\mathbf{u}$, i.e. the minimization of $\mathbf{u} \to J(B + \mathbf{u})$ in a neighborhood of 0. It is thus important to know whether or not this function is differentiable and, eventually, to compute its derivative. This will be the goal of this paper.

Results of this kind have already been derived for other linear and nonlinear problems (Murat and Simon 1976, Simon 1980; for a review, see Cea and Haug 1981), by using a mapping from $\Omega = D\backslash B$ onto a fixed domain Ω_0. Here, one is faced to an additional difficulty related to the fact that the space $\{\mathbf{y} : \mathbf{y} \in H_0^1(\Omega), \ \nabla\cdot\mathbf{y} = 0\}$ is not preserved by mappings (because the null-divergence is not preserved). Thus, it is not possible to argue exactly as in the quoted papers; this difficulty will be solved here, as it was done by Bello et al. (1992), introducing for each $\mathbf{u}$ the space

$$Y(\mathbf{u}) = \{\phi \in H^1(D\backslash\bar{B}); \int_{D\backslash B}|\det\frac{\partial}{\partial x_j}(\mathbf{I} + \mathbf{u})_i|\,\phi\,dx = 0\}$$

and an appropriate isomorphism from $Y(\mathbf{u})$ onto $Y(0)$ (see below).

2. The variations of the drag.

Our goal is to obtain a formula of the kind

$$(2.1) \qquad J(B + \mathbf{u}) = J(B) + J'(B; \mathbf{u}) + \theta(\mathbf{u}),$$

which must hold for $\mathbf{u} \in W$ with $\|\mathbf{u}\|_{W^{2,\infty}}$ small enough. In (2.1), $J'(B; \cdot)$ has to be linear and continuous on W and $\theta(\mathbf{u})$ has to satisfy the following property :

$$\theta(\mathbf{u})/\|\mathbf{u}\|_{W^{2,\infty}} \to 0 \quad \text{as} \quad \|\mathbf{u}\|_{W^{2,\infty}} \to 0.$$

The main result in this paper is the following :

THEOREM 2.1 *Assume* $(\mathbf{y}, p)$ *is the unique solution of (1.2)-(1.3) for* $\mathbf{u} = 0$. *Then, one has (2.1) with*

$$(2.2) \qquad J'(B; \mathbf{u}) = -\nu \int_{\partial B} u_n \frac{\partial \mathbf{y}}{\partial \mathbf{n}} \cdot \frac{\partial \mathbf{q}}{\partial \mathbf{n}} \, dS.$$

Here, $\mathbf{n}$ *is the outward unit normal vector on* ∂B, $u_n = \mathbf{u} \cdot \mathbf{n}$ *and the pair function* $(\mathbf{q}, Q) \in H^2(D \backslash \bar{B})^d \times (H^1(D \backslash \bar{B})/\mathbf{R})$ *is the unique solution to*

$$(2.3) \qquad \begin{cases} -\nu \Delta q_i + \displaystyle\sum_{j=1}^{d} \frac{\partial y_j}{\partial x_i} q_j - \sum_{j=1}^{d} y_j \frac{\partial q_i}{\partial x_j} + \frac{\partial Q}{\partial x_i} = 0 \quad (1 \le i \le d) \quad in \ D \backslash \bar{B}, \\[2mm] \nabla \cdot \mathbf{q} = 0 \quad in \ D \backslash \bar{B}, \\[2mm] \mathbf{q} = -\mathbf{g} \quad on \ \partial B, \\[2mm] \mathbf{q} = 0 \quad on \ \partial D. \end{cases}$$

This will be demonstrated in Section 5. Notice that linear theory can be used to show that problem (2.3) possesses exactly one solution $(\mathbf{q}, Q) \in H^2(D \backslash \bar{B})^d \times (H^1(D \backslash \bar{B})/\mathbf{R})$. According to (2.2), one also observes that $J'(B; \mathbf{u})$ only depends on the behavior of u_n on ∂B.

3. Some general results concerning differentiation with respect to domains.

In order to prove Theorem 2.1, we must show that the mapping $\mathbf{u} \to (\mathbf{y}(\mathbf{u}), p(\mathbf{u}))$ is "differentiable". A relevant feature of this mapping is that, for each $\mathbf{u}$, the pair function $(\mathbf{y}(\mathbf{u}), p(\mathbf{u}))$ is defined in the open set $\Omega + \mathbf{u}$. At present, we recall some results concerning the differentiability of this kind of mappings.

Assume that Ω is a bounded open set in $\mathbf{R}^d$. For $\mathbf{u} \in W^{k,\infty}(\mathbf{R}^d; \mathbf{R}^d)$, $k \geq 1$, with norm $\|\mathbf{u}\|_{W^{k,\infty}}$ small enough, $\Omega + \mathbf{u}$ is again an open set in $\mathbf{R}^d$. Let us consider a mapping $\mathbf{u} \to z(\mathbf{u})$ such that

$$(3.1) \qquad z(\mathbf{u}) \in W^{m,r}(\Omega + \mathbf{u}) \quad \forall \mathbf{u} \in W^{k,\infty}(\mathbf{R}^d; \mathbf{R}^d) \text{ with small } \|\mathbf{u}\|_{W^{k,\infty}},$$

where $k \geq m \geq 1$ and $1 \leq r < \infty$. It is clear that such a mapping cannot be differentiated with respect to $\mathbf{u}$ in the usual form. This motivates the following two Definitions :

DEFINITION 3.1 *If the mapping* $\mathbf{u} \to z(\mathbf{u}) \circ (\mathbf{I} + \mathbf{u})$*, which is defined in a neighborhood of 0 in* $W^{k,\infty}(\mathbf{R}^d; \mathbf{R}^d)$ *and takes values in* $W^{m,r}(\Omega)$*, is differentiable at 0, we will say that* $\mathbf{u} \to z(\mathbf{u})$ *possesses a total first variation (or derivative) at 0. In such a case, the total derivative — i.e. the derivative of* $\mathbf{u} \to z(\mathbf{u}) \circ (\mathbf{I} + \mathbf{u})$ *— at 0 in the direction* $\mathbf{u}$ *will be denoted* $\dot{z}(\mathbf{u})$*.*

DEFINITION 3.2 *If, for every open set* $\omega \subset\subset \Omega$*, the mapping* $\mathbf{u} \to z_\omega(\mathbf{u}) \equiv z(\mathbf{u}) \mid_\omega$ *is differentiable at 0, we will say that* $\mathbf{u} \to z(\mathbf{u})$ *possesses a local first variation (or derivative) at 0. In such a case, the local derivative at 0 in the direction* $\mathbf{u}$ *is denoted* $z'(\mathbf{u})$ *and is well defined in the whole domain* Ω *:*

$$z'(\mathbf{u}) = \frac{d}{dt} z_\omega(t\mathbf{u}) \mid_{t=0} \qquad \text{in each } \omega \subset\subset \Omega.$$

We always consider differentiability in the sense of Fréchet.

One has the following results, which are due to Murat and Simon (1976) (see also Simon 1980, 1989) :

THEOREM 3.3 *Assume* $\mathbf{u} \to z(\mathbf{u})$ *satisfies (3.1) and possesses a total first variation at 0. Then, when it is considered a* $W^{m-1,r}$*—valued mapping,* $\mathbf{u} \to z(\mathbf{u})$ *also possesses a local first variation at 0. The local derivative* $z'(\mathbf{u})$ *is given by :*

$$z'(\mathbf{u}) = \dot{z}(\mathbf{u}) - \mathbf{u} \cdot \nabla z(0) \quad \forall \mathbf{u} \in W^{k,\infty}(\mathbf{R}^d; \mathbf{R}^d).$$

Notice that the existence of a total variation is a property which may hold or not for each particular problem. It will be demonstrated in Section 4 for the mapping $\mathbf{u} \to (\mathbf{y}(\mathbf{u}), p(\mathbf{u}))$. On the contrary, the previous result can be viewed as a reciprocal of the chain rule and provides a general criterion for the differentiability of $\mathbf{u} \to z(\mathbf{u})$.

THEOREM 3.4 *Assume that $\partial\Omega$ is of class $W^{1,\infty}$ and that $\mathbf{u} \to z(\mathbf{u})$ is as in Theorem 3.3. We also assume that, for every $\mathbf{u} \in W^{k,\infty}(\mathbf{R}^d; \mathbf{R}^d)$ of sufficiently small norm, one has :*

$$Az(\mathbf{u}) = f \quad in \quad \Omega + \mathbf{u}, \quad z(\mathbf{u}) = g \quad on \quad \partial\Omega + \mathbf{u};$$

here, $f \in \mathcal{D}'(\mathbf{R}^d)$, A is a differential operator which maps smoothly $W^{m-1,r}(\omega)$ into $\mathcal{D}'(\omega)$ for every open $\omega \subset\subset \Omega$ and $g \in W^{2,1}(\mathbf{R}^d)$. Finally, assume that $z(0) \in W^{2,1}(\Omega)$. Then, the local first variation $z'(\mathbf{u})$ satisfies :

$$(3.2) \qquad DA(z(0); z'(\mathbf{u})) = 0 \quad in \quad \Omega, \quad z'(\mathbf{u}) = -\frac{\partial}{\partial\mathbf{n}}(z(0) - g)\, u_n \quad on \quad \partial\Omega,$$

with $DA(z(0); \cdot)$ being the derivative at $z(0)$ of the mapping $z \to Az$.

THEOREM 3.5 *Assume that $\partial\Omega$ is of class $W^{1,\infty}$ and that $\mathbf{u} \to z(\mathbf{u})$ is as in Theorem 3.3. Then, the function $\mathbf{u} \to \int_{\Omega+\mathbf{u}} z(\mathbf{u})\,dx$ is differentiable at 0. Its derivative in the direction $\mathbf{u}$ is given by :*

$$(3.3) \qquad\qquad \int_{\Omega} z'(\mathbf{u})\,dx + \int_{\partial\Omega} z(0) u_n\,dS.$$

4. The existence of a total derivative of the mapping $\mathbf{u} \to (\mathbf{y}(\mathbf{u}), p(\mathbf{u}))$.

The main result in this Section is the following

THEOREM 4.1 *The mapping $(\mathbf{y}(\mathbf{u}), p(\mathbf{u})) \circ (\mathbf{I} + \mathbf{u})$, which is defined in a neighborhood of 0 in W and takes values in $H^2(D \backslash \bar{B})^d \times (H^1(D \backslash \bar{B})^d / \mathbf{R})$, is differentiable at 0. Its derivative in the direction $\mathbf{u}$ will be denoted $(\dot{\mathbf{y}}(\mathbf{u}), \dot{p}(\mathbf{u}))$.*

This result is a consequence of the Implicit Function Theorem. Here, we will only sketch the proof; for the details see Bello et al. (1992). We use the following two Lemmas :

LEMMA 4.2 *[Murat and Simon 1976, Simon 1989] $\mathbf{u} \to [M_{ij}(\mathbf{u})] \equiv {}^t[\frac{\partial}{\partial x_j}(\mathbf{I} + \mathbf{u})_i]^{-1}$ is a well defined and C^1 mapping in a neighborhood of 0 in W and takes values in $W^{1,\infty}(\mathbf{R}^d; \mathbf{R}^{d \times d})$.*

LEMMA 4.3 *[Bello et al. 1992] For each $\mathbf{u}$ in a neighborhood of 0 in W, we introduce the space*

$$Y(\mathbf{u}) = \{\phi \in H^1(D \backslash \bar{B}); \int_{D \backslash B} |\det \frac{\partial}{\partial x_j}(\mathbf{I} + \mathbf{u})_i|\, \phi\,dx = 0\}.$$

Every $Y(\mathbf{u})$ is isomorphic to $Y(0)$. More precisely, the mapping $\Lambda(\mathbf{u})$, from $Y(\mathbf{u})$ onto $Y(0)$, which is given by

$$\Lambda(\mathbf{u})\phi = \phi - \left(\int_{D\backslash B} \phi\, dx\right)\phi_0 \qquad \forall \phi \in Y(\mathbf{u}),$$

where $\phi_0 \equiv \frac{1}{c_0} = \left(\int_{D\backslash B} dx\right)^{-1}$, is an isomorphism.

In the sequel, we assume that $\mathbf{u}$ belongs to a neighborhood of 0 in W. Thus, both $M_{ij}(\mathbf{u})$ and the space $Y(\mathbf{u})$ are meaningful. As in Theorem 2.1, we use the notation $(\mathbf{y}, p) = (\mathbf{y}(0), p(0))$.

SKETCH OF THE PROOF OF THEOREM 4.1 : It is clear that $(\mathbf{y}(\mathbf{u}), p(\mathbf{u})) \circ (\mathbf{I} + \mathbf{u})$ belongs to the linear manifold $E = \tilde{G} + E_0$, where

$$E_0 = (H^2(D\backslash \bar{B})^d \cap H^1_0(D\backslash \bar{B})^d) \times (H^1(D\backslash \bar{B})/\mathbf{R}), \qquad \tilde{G} = (\mathbf{G}, 0)$$

and

$$\mathbf{G} \in H^2(D)^d, \quad \mathbf{G} \equiv 0 \text{ in } \mathcal{O}, \quad \nabla \cdot \mathbf{G} = 0 \text{ in } D, \quad \mathbf{G} = \mathbf{g} \text{ on } \partial D.$$

One has $\sum_{i,k=1}^{d} M_{ik}(\mathbf{u})\frac{\partial}{\partial x_k}(\tilde{G}_i + v_i) \in Y(\mathbf{u})$ if $\eta = (\mathbf{v}, r) \in E_0$. Now, consider the space $X = L^2(D\backslash \bar{B})^d \times Y_0$ and the function $(\mathbf{u}, \eta) \to F(\mathbf{u}, \eta)$, defined as follows :

$$\begin{cases} F_l(\mathbf{u}, \eta) = -\nu \sum_{i,j,k=1}^{d} M_{ij}(\mathbf{u})\frac{\partial}{\partial x_j}\left(M_{ik}(\mathbf{u})\frac{\partial}{\partial x_k}(\tilde{G}_l + v_l)\right) \\[2ex]
\qquad + \sum_{i=1}^{d}(\tilde{G}_i + v_i)\sum_{k=1}^{d} M_{ik}(\mathbf{u})\frac{\partial}{\partial x_k}(\tilde{G}_l + v_l) + \sum_{k=1}^{d} M_{lk}(\mathbf{u})\frac{\partial r}{\partial x_k} \qquad \text{for } 1 \le l \le d, \\[2ex]
F_{d+1}(\mathbf{u}, \eta) = \Lambda(\mathbf{u})\left(\sum_{i,k=1}^{d} M_{ik}(\mathbf{u})\frac{\partial}{\partial x_k}(\tilde{G}_i + v_i)\right). \end{cases}$$

Due to Lemma 4.2, the function F is well defined and C^1 for $\mathbf{u}$ in a neighborhood of 0 in W, and $\eta = (\mathbf{y}, p) \in E_0$. Of course, it takes values in X and it is not difficult to see that $F(\mathbf{u}, \eta) = 0$ if and only if

$$\eta = \chi(\mathbf{u}) \equiv (\mathbf{y}(\mathbf{u}), p(\mathbf{u})) \circ (\mathbf{I} + \mathbf{u}) - \tilde{G}.$$

Notice that $\frac{\partial F}{\partial \eta}(0, \chi(0))$ is an isomorphism from E_0 onto X (this stems from Lemma 4.3 and L^2 regularity for the Stokes problem; cf. Bello et al. (1992) for more details). Thus, we can apply the Implicit Function Theorem to the equation $F(\mathbf{u}, \eta) = 0$ near $(0, \chi(0))$. One easily deduces that $\mathbf{u} \to (\mathbf{y}(\mathbf{u}), p(\mathbf{u})) \circ (\mathbf{I} + \mathbf{u})$ is Fréchet-differentiable at 0.

5. The existence of a derivative of the function $u \to J(B+u)$.

This Section is devoted to the proof of Theorem 2.1. First, notice that, as a consequence of Theorem 4.1 and the results in Section 3, $\mathbf{u} \to (\mathbf{y}(\mathbf{u}), p(\mathbf{u}))$ possesses a local variation at 0 which can be characterized in terms of a linear boundary-value problem for a partial differential system. More precisely, one has :

THEOREM 5.1 *For every open set* $\omega \subset\subset D\backslash\bar{B}$, *the mapping* $\mathbf{u} \to (\mathbf{y}(\mathbf{u})|_\omega, p(\mathbf{u})|_\omega)$, *considered as a* $H^1(\omega)^d \times (L^2(\omega)/\mathbf{R})$-*valued function, is differentiable at* 0. *Hence,* $\mathbf{u} \to (\mathbf{y}(\mathbf{u}), p(\mathbf{u}))$ *is locally differentiable at* 0. *The local derivative* $(\mathbf{y}'(\mathbf{u}), p'(\mathbf{u}))$ *in the direction* $\mathbf{u}$ *satisfies :*

$$(5.1) \quad \begin{cases} -\nu\Delta\mathbf{y}'(\mathbf{u}) + (\mathbf{y}'(\mathbf{u})\cdot\nabla)\mathbf{y} + (\mathbf{y}\cdot\nabla)\mathbf{y}'(\mathbf{u}) + \nabla p'(\mathbf{u}) = 0 \ \ in \ \ D\backslash\bar{B}, \\[2ex] \nabla\cdot\mathbf{y}'(\mathbf{u}) = 0 \ \ in \ \ D\backslash\bar{B}, \\[2ex] \mathbf{y}'(\mathbf{u}) = -u_n\dfrac{\partial\mathbf{y}}{\partial\mathbf{n}} \ \ on \ \ \partial B, \ \ \ \mathbf{y}'(\mathbf{u}) = 0 \ \ on \ \ \partial D. \end{cases}$$

Again, linear theory can be used to show that problem (5.1) possesses exactly one solution. Now, consider the function (1.4), which is well defined for $\mathbf{u} \in W$ near 0. Obviously,

$$\begin{aligned} J(B+\mathbf{u}) &= \mathbf{g}\cdot\int_{\partial D}\left(-p(\mathbf{u})\,\mathbf{Id} + \sigma(\mathbf{y}(\mathbf{u}))\right)\cdot\mathbf{n}\,dS \\ &\quad - \mathbf{g}\cdot\int_{D\backslash(B+\mathbf{u})}\nabla\cdot\left(-p(\mathbf{u})\,\mathbf{Id} + \sigma(\mathbf{y}(\mathbf{u}))\right)\,dx \\ &= J_1(\mathbf{u}) - J_2(\mathbf{u}). \end{aligned}$$

In order to prove that both terms are differentiable, we will use the following result :

LEMMA 5.2 [Simon 1989] *Assume* $\mathbf{u} \to z(\mathbf{u})$ *satisfies (3.1) and possesses a total first variation at* 0. *Then, the mapping* $\mathbf{u} \to Z(\mathbf{u})\circ(\mathbf{I}+\mathbf{u}) \equiv (\frac{\partial}{\partial x_i}z(\mathbf{u}))\circ(\mathbf{I}+\mathbf{u})$, *which is defined for* $\mathbf{u}$ *near* 0 *in* $W^{k,\infty}(\mathbf{R}^d;\mathbf{R}^d)$ *and takes values in* $W^{m-1,r}(\Omega)$, *is differentiable at* 0. *Its derivative in the direction* $\mathbf{u}$ *satisfies :*

$$(5.2) \qquad \dot{Z}(\mathbf{u}) = \frac{\partial}{\partial x_i}z'(\mathbf{u}) + \mathbf{u}\cdot\nabla\left(\frac{\partial}{\partial x_i}z(0)\right).$$

As an immediate consequence, the $H^1(D\backslash\bar{B})$-valued mapping

$$\mathbf{u} \to (-p(\mathbf{u})\,\mathbf{Id} + \sigma(\mathbf{y}(\mathbf{u})))\circ(\mathbf{I}+\mathbf{u})$$

294   J. A. Bello, E. Fernandez-Cara and J. Simon

is differentiable at 0. Taking into acount that $\mathbf{u} \equiv 0$ on ∂D, one has :

$$J_1(\mathbf{u}) = \mathbf{g} \cdot \int_{\partial D} [(-p(\mathbf{u})\,\mathrm{Id} + \sigma(\mathbf{y}(\mathbf{u}))) \circ (\mathbf{I} + \mathbf{u})]\,\mathbf{n}\,dS,$$

whence $\mathbf{u} \to J_1(\mathbf{u})$ is differentiable at 0 and its derivative at 0 in the direction $\mathbf{u}$ is given by

$$(5.3) \qquad J_1'(0; \mathbf{u}) = \mathbf{g} \cdot \int_{\partial D} (-p'(\mathbf{u})\,\mathrm{Id} + \sigma(\mathbf{y}'(\mathbf{u}))) \cdot \mathbf{n}\,dS,$$

where $p'(\mathbf{u})$ and $\mathbf{y}'(\mathbf{u})$ are obtained according to Theorem 3.3 and (5.2) respectively. On the other hand, from the Navier-Stokes equations satisfied by $(\mathbf{y}(\mathbf{u}), p(\mathbf{u}))$, one has :

$$\nabla \cdot (-p(\mathbf{u})\,\mathrm{Id} + \sigma(\mathbf{y}(\mathbf{u}))) = (\mathbf{y}(\mathbf{u}) \cdot \nabla)\mathbf{y}(\mathbf{u}).$$

Using again Lemma 5.2, we find that the $W^{1,1}(D \backslash \bar{B})$-valued mapping

$$\mathbf{u} \to K(\mathbf{u}) \circ (\mathbf{I} + \mathbf{u}) \equiv [(\mathbf{y}(\mathbf{u}) \cdot \nabla)\mathbf{y}(\mathbf{u})] \circ (\mathbf{I} + \mathbf{u})$$

is differentiable at 0. According to Theorem 3.3, when it is considered a L^1-valued mapping, $\mathbf{u} \to K(\mathbf{u})$ is also locally differentiable at 0. Furthermore, it is not difficult to see that

$$K'(u) = (\mathbf{y}'(\mathbf{u}) \cdot \nabla)\mathbf{y} + (\mathbf{y}(\mathbf{u}) \cdot \nabla)\mathbf{y}'(\mathbf{u}) = \nabla \cdot (-p'(\mathbf{u})\,\mathrm{Id} + \sigma(\mathbf{y}'(\mathbf{u}))).$$

From Theorem 3.5 and the fact that $\mathbf{u}$ and $K(0)$ vanish resp. on ∂D and ∂B, we deduce that $\mathbf{u} \to J_2(\mathbf{u})$ is also differentiable at 0 and

$$(5.4) \qquad J_2'(0; \mathbf{u}) = \mathbf{g} \cdot \int_{D \backslash B} \nabla \cdot (-p'(\mathbf{u})\,\mathrm{Id} + \sigma(\mathbf{y}'(\mathbf{u})))\,dx.$$

We have thus found that $\mathbf{u} \to J(B + \mathbf{u})$ is differentiable at 0. Its derivative in the direction $\mathbf{u}$ is given by

$$(5.5) \qquad J'(B; u) = -\mathbf{g} \cdot \int_{\partial B} (-p'(\mathbf{u})\,\mathrm{Id} + \sigma(\mathbf{y}'(\mathbf{u}))) \cdot \mathbf{n}\,dS.$$

It is now easy to end the proof of Theorem 2.1. It remains only to check (2.2) with $\mathbf{q}$ being, together with Q, the unique solution to (2.3). Notice that (5.5) can also be written as follows

$$J'(B; \mathbf{u}) = \int_{\partial(D \backslash B)} [(-p'(\mathbf{u})\,\mathrm{Id} + \sigma(\mathbf{y}'(\mathbf{u}))) \cdot \mathbf{n}] \cdot \mathbf{q}\,dS$$

$$= \nu \int_{D \backslash B} \Delta \mathbf{y}'(\mathbf{u}) \cdot \mathbf{q}\,dx + \sum_{i,j} \int_{D \backslash B} \sigma_{ij}(\mathbf{y}'(\mathbf{u}))\frac{\partial q_j}{\partial x_i}\,dx$$

$$- \int_{\partial(D \backslash B)} p'(\mathbf{u})\mathbf{n} \cdot \mathbf{q}\,dS$$

$$= I_1 + I_2 - I_3.$$

We first observe that

$$I_1 = \int_{D\setminus B} [(\mathbf{y}'(\mathbf{u}) \cdot \nabla)\mathbf{y} + (\mathbf{y} \cdot \nabla)\mathbf{y}'(\mathbf{u}) + \nabla p'(\mathbf{u})] \cdot \mathbf{q}\, dx$$

$$= -\int_{D\setminus B} (\mathbf{y} \cdot \nabla)\mathbf{q} \cdot \mathbf{y}'(\mathbf{u})\, dx + \sum_{i,j} \int_{D\setminus B} \frac{\partial y_i}{\partial x_j} q_i y_j'(\mathbf{u})\, dx + I_3.$$

On the other hand,

$$I_2 = -\nu \sum_{i,j} \int_{D\setminus B} \frac{\partial^2 q_j}{\partial x_i^2} y_j'(\mathbf{u})\, dx$$

$$- \nu \int_{\partial B} u_n \left(\sum_i \frac{\partial y_i}{\partial \mathbf{n}} n_i \right) \left(\sum_j \frac{\partial q_j}{\partial \mathbf{n}} n_j \right) dS - \nu \sum_{i,j} \int_{\partial B} u_n \frac{\partial y_j}{\partial \mathbf{n}} \frac{\partial q_j}{\partial \mathbf{n}} n_i^2\, dS.$$

Hence,

$$(5.6) \qquad \begin{cases} J'(B;\mathbf{u}) = \displaystyle\sum_{i,j} \int_{D\setminus B} \left[-\nu \frac{\partial^2 q_j}{\partial x_i^2} + \frac{\partial y_i}{\partial x_j} q_i - y_i \frac{\partial q_j}{\partial x_i} \right] y_j'(\mathbf{u})\, dx \\[2ex] \qquad - \nu \displaystyle\int_{\partial B} u_n \frac{\partial \mathbf{y}}{\partial \mathbf{n}} \cdot \frac{\partial \mathbf{q}}{\partial \mathbf{n}}\, dS. \end{cases}$$

The first integral in (5.6) vanishes, since it is equal to

$$-\int_{D\setminus B} \nabla Q \cdot \mathbf{y}'(\mathbf{u})\, dx = \int_{\partial B} u_n\, Q \left(\sum_j \frac{\partial y_j}{\partial \mathbf{n}} n_j \right) dS = \int_{\partial B} u_n\, Q(\nabla \cdot \mathbf{y})\, dS.$$

Consequently, (2.2) is proven.

References

Bello, J.A., Fernández-Cara, E., Simon, J. (1991), *Variation par rapport au domaine de l'énergie visqueuse dissipée dans un fluide de Navier-Stokes.* Note C. R. Acad. Sci. Paris, t. 313, Série I, p. 447–450.

Bello, J.A., Fernández-Cara, E., Simon, J. (1992), *Optimal shape design for Navier-Stokes flow.* Proceedings of I.F.I.P. Conference in Zürich, 1991.

Cea J., Haug E.J. (1981), *Optimization of distributed parameter structures.* Sijthoff and Noordhoff.

Fernández Cara, E. (1989), *Optimal design in fluid mechanics*. In "Control of Partial Differential Equations", A. Bermúdez Ed., p. 120–131, Lecture Notes in Control and Information Sciences No. 114, Springer-Verlag.

Murat F., Simon J. (1974), *Quelques résultats sur le contrôle par un domaine géométrique*. Rapport du L.A. 189 No. 74003. Université Paris VI.

Murat F., Simon J. (1976), *Sur le contrôle par un domaine géométrique*. Rapport du L.A. 189 No. 76015. Université Paris VI.

Nečas, J. (1967), *Les méthodes directes en théorie des équations elliptiques*. Masson, Paris.

Pironneau, O. (1974), *On optimum design in fluid mechanics*. J. Fluid. Mech., Vol. 64, part. I, pp. 97-110.

Pironneau, O. (1984), *Optimal shape design for elliptic systems*. Springer-Verlag, New-York.

Schlichting, H. (1970), *Boundary-layer theory*. Academic Press, New York.

Simon, J. (1980), *Differentiation with respect to the domain in boundary value problems*. Numer. Funct. Anal. and Optimiz., 2 (7 and 8), 649–687.

Simon, J. (1989), *Diferenciación de problemas de contorno respecto del dominio*. Lectures in the University of Sevilla.

Simon, J. (1991), *Domain variation for drag in Stokes flow*. In "Control theory of distributed parameter systems and applications", X. Li and J. Yong Ed., Lectures Notes in Control and Information Sciences No 159, 28–42.

Author's address :

Juan A. Bello, Enrique Fernández-Cara
University of Sevilla
Departamento de Análisis Matemático
C/ Tarfia s/n
41012 SEVILLA, SPAIN

Jacques Simon
C.N.R.S. and Université Blaise Pascal (Clermont-Ferrand 2)
Laboratoire de Mathématiques Appliquées
63177 AUBIERE CEDEX, FRANCE

Mathematical programming and nonsmooth optimization

International Series of Numerical Mathematics, Vol. 107, © 1992 Birkhäuser Verlag Basel

SCALAR MINIMAX PROPERTIES
IN VECTORIAL OPTIMIZATION

Teodor Precupanu

Abstract. A scalar minimax characterization of the solutions of a convex vectorial optimization problem is established. As consequences, duality results are given in terms of ϵ-supernormality and semidistance on dual generated by a certain Minkowsky functional.

0. Preliminaries

Let X be a *real locally convex space* endowed with a partial ordering induced by a convex cone P (of positive elements) and let S be a nonempty subset of X. We recall some usual definitions. An element $x_0 \in S$ is called *minimal* (efficient, Pareto optimal, admissible) *element* of S if there is no $x \in S$, $x \leq x_0$, i.e.

$$S \cap (x_0 - P) = \{x_0\} \tag{0.1}$$

or equivalently

$$P \cap (x_0 - S) = \{0\}. \tag{0.1'}$$

We denote by $Eff_p(S)$ the *set of all minimal elements* of S with respect to the convex cone P. The convex cone P is called *pointed cone* if $P \cap (-P) = \{0\}$. Let us define:

$$C(A; x_0) = \cup_{\lambda \geq 0} \lambda(A - x_0) \tag{0.2}$$

the *cone of feasible directions* in A at x_0 (which coincides with the cone generated by $(A - x_0)$);

$$A_\infty = \left(\cap_{\lambda \geq 0} \lambda A \right) \cup \{0\} \tag{0.3}$$

the *asymptotic (recession) cone* of a set A;

$$A^0 = \{x^* \in X^* \mid (x^*, x) \leq 1, \forall x \in A\} \tag{0.4}$$

the *polar set* of a set A from X (in the topological dual X^*);

$$f^*(x^*) = \sup_{x \in X}\{(x^*, x) - f(x)\}, \quad x^* \in X^* \tag{0.5}$$

the *conjugate function* of the function $f : X \longrightarrow \overline{\mathbf{R}}$;

$$p_A(x) = \inf\{\lambda \mid \frac{1}{\lambda}x \in A, \ \lambda \in (0, \infty]\}, \ x \in X, \tag{0.6}$$

the *Minkowsky functional* of a set A which contains the origin.

Similar definitions hold for a set A of X^*, respectively for a function $g : X^* \longrightarrow \overline{\mathbf{R}}$.

The following dual properties of Minkowsky functional (see, e.g., Barbu and Precupanu, 1986) are known in convex analysis:

$$p_A^* = I_{A^\circ} \ , \quad I_A^* = p_{A^\circ} \tag{0.7}$$

where A is a nonempty closed convex set which contains the origin and I_A is the *indicator function* of a set A (equal zero on A and ∞ otherwise).

The following properties are obviously:

(i) $x_0 \in S$ is a minimal element of S if and only if

$$(-P) \cap C(S; x_0) = \{0\} \tag{0.1}''$$

i.e. $x_0 \in S$ is the minimal element of $x_0 + C(S; x_0)$ or zero is a minimal element of P with respect to the cone $C(S; x_0)$. Thus, if $x_0 - S$ is a convex cone then $x_0 \in S$ is a minimal element of S with respect to cone P if and only if the origin is a minimal element of P with respect to the cone $x_0 - S$.

(ii)   if $S_1 \subset S_2$, then any element of S_1 which is minimal element of S_2 is also minimal element of S_1

(iii)   if S is a closed convex set and $Eff_p(S) \neq \emptyset$, then

$$(-P) \cap S_\infty = \{0\} \tag{0.1}'''$$

(iv)   if $S = S_1 \cup S_2$, then $Eff_p(S) = Eff_p(S_1) \cup Eff_p(S_2)$.

By property (ii), for $S_1 = S$ we obtain different stronger definition of minimality taking S_2 as special sets which contains S (for instance $\overline{S}$, $conv\overline{S}$, $x_0 + \overline{C(S; x_0)}$, $P + S$). Conversely, we can obtain weaker definitions of minimality taking $S_2 = S$. Also, similar properties hold with respect to cone P (according to last part of property (i)). Particularly, any element of S which is minimal for $P + S$ is also minimal for S (see, e.g. Geofrion, 1968). Moreover, if P is a pointed cone, the converse is also true.

In this paper we establish a scalar minimax characterization of the solutions of a convex vectorial optimization problem (Theorem 1.1) as a direct consequence of the dual form by polarity of the condition $(0.1)'$ from the definition of the minimal elements. Thus it can be obtained dual characterizations of the minimality where explicity intercomes the norm of the space and the semimetric structure of dual generated by the Minkowsky functional of the polar of the translates of the closed convex set which is minimized with respect to a closed convex cone.

1. Scalar minimax properties in real linear normed spaces

Throughout this section we suppose that P is a *closed convex cone* and S is a *closed convex set* in a *real linear normed space* X. Thus, by bipolar theorem (see Barbu and Precupanu, 1986, p.97) the property (0.1) can be equivantly rewritten as follows

$$\overline{conv(P^0 \cup (x_0 - S)^0)} = X^* \tag{1.1}$$

or

$$\overline{P^0 + (x_0 - S)^0} = X^* \tag{1.2}$$

according to calculus with polars (see also Levine and Pamerol, 1972).

If $C(S; x_0)$ is a *closed convex cone with respect to* P, i.e.

$$C(S; x_0) \cap P = \overline{C(S; x_0)} \cap P \tag{1.3}$$

then the property (1.2) is also equivalent to

$$\overline{P^0 - C(S; x_0)^0} = X^*. \tag{1.2}'$$

Also, the condition $(0.1)'''$ becomes sufficient in some special cases (see Bitran and Magnati, 1979, for finite dimensional case).

The main result of this paper is the following "minimax" characterisation of minimal elements in convex case.

Theorem 1 *An element x_0 in a convex closed set S is a minimal element of S with respect to the closed convex cone P if and only if for every $\epsilon \in (0,1)$ and $x^* \in S^*(0,1) = \{x^* \in X^*; \|x^*\| \leq 1\}$ we have*

$$\inf_{p^* \in P^0} \sup_{s \in S}[(x^* - p^*, x_0 - s) - \epsilon\|x_0 - s\|] = 0. \tag{1.4}$$

Proof. Let $x_0 \in S$ be a minimal element of S with respect to P. According to (1.2), for every $x^* \in S^*(0,1)$ and $\epsilon \in (0,1)$ there exist $p_\epsilon^* \in p^0$ and $u_\epsilon^* \in (x_0 - S)^0$ such that $\|x^* - p_\epsilon^* - \epsilon u_\epsilon^*\| \leq \epsilon$. Consequently, we have

$$(x^* - p_\epsilon^* - \epsilon u_\epsilon^*, x_0 - s) \leq \epsilon, \ \forall s \in S,$$

where $v_\epsilon^* \in S^*(0,1)$. Thus, we obtain

$$\epsilon \geq (x^* - p_\epsilon^*, x_0 - s) - \epsilon(v_\epsilon^*, x_o - s) \geq (x^* - p_\epsilon^*, x_0 - s) - \epsilon\|x_0 - s\|, \ \forall s \in S,$$

i.e.

$$\epsilon \geq \inf_{p^* \in P^0} \sup_{s \in S}[(x^* - p^*, x_0 - s) - \epsilon\|x_0 - s\|].$$

Now, if $\epsilon' \in (0, \epsilon)$, we also have

$$\epsilon > \epsilon' \geq \inf_{p^* \in P^0} \sup_{s \in S}[(x^* - p^*, x_0 - s) - \epsilon'\|x_0 - s\|]$$

$$\geq \inf_{p^* \in P^0} \sup_{s \in S}[(x^* - p^*, x_0 - s) - \epsilon\|x_0 - s\|] \geq 0.$$

Taking $\epsilon' \to 0$, we obtain (1.4).

Conversely, if the minimum property (1.4) holds, then by usual minimax inequality we get

$$\sup_{s \in S} \inf_{p^* \in P^0}[(x^* - p^*, x_0 - s) - \|x_0 - s\|] \leq 0, \quad \forall x^* \in S^*(0, 1),$$

or equivalently

$$\sup_{s \in S} \sup_{\|x^*\| \leq 1} \inf_{p^* \in P^0}[(x^* - p^*, x_0 - s) - \|x_0 - s\|] \leq 0, \quad \forall x^* \in S^*(0, 1),$$

i.e.

$$0 \geq \sup_{s \in S}[(1 - \epsilon)\|x_0 - s\| - \sup_{p^* \in P^0}(p^*, x_0 - s)] =$$

$$= (1 - \epsilon)\sup_{s \in S}[\|x_0 - s\| - I^*_{P_0}(x_0 - s)] =$$

$$= (1 - \epsilon)\sup_{s \in S}[\|x_0 - s\| - I_P(x_0 - s)] =$$

$$= (1 - \epsilon)\sup_{s \in S \cap (x_0 - P)}\|x_0 - s\| \geq 0.$$

Therefore the relation (0.1) holds, i.e. x_0 is an optimal element of S.

From this result it follows that the function

$$E_{S,P}(x; \epsilon) = \sup_{\|x^*\| \leq 1} \inf_{p^* \in P^0} \sup_{s \in S}[(x^* - p^*, x - s) - \epsilon\|x - s\|] \tag{1.5}$$

defined on $S^*(0, 1)$, can give certains informations concerning the optimality property of the elements of S.

Given a subset $A \subset X$ we denote

$$\|A\|^+ = \sup_{a \in P \cap A}\|a\|. \tag{1.6}$$

From the second part of the preceding proof it is clear that we have the following inequality

$$(1 - \epsilon)\|x - S\|^+ \leq E_{S,P}(x, \epsilon), \quad \forall(x, \epsilon) \in S^*(0, 1). \tag{1.7}$$

Also, using minimax inequality we obtain

$$E_{S,P}(x, \epsilon) \leq (1 - \epsilon)\inf_{p^* \in P^0} \sup_{s \in S}[\|x - s\| - (p^*, x - s)]. \tag{1.7'}$$

Proposition 1.2. *An element x_0 in a closed convex set S is a minimal element of S with respect to a closed convex cone P if for every $\epsilon \in (0,1)$ there exists $p_\epsilon^* \in P^0$ such that*

$$\|x_0 - s\| \le p_\epsilon^*(x_0 - s) + \epsilon, \quad \forall s \in S. \tag{1.8}$$

Proof. Using $(1.7)'$ and (1.7) we obtain $\|x_0 - S\|^+ = 0$ as claimed.

Remark 1. If (1.8) is fulfilled for $\epsilon = 0$ and certain $p^* \in p^0$, then $x_0 \in S$ is s minimal element of S. In this special case it follows that $C(x,s)$ is a supernormal cone (according to Peressini, 1967). Thus, the property (1.8) can be considered as a property of ϵ–*supernormality* of the convex set $x_0 - S$. On the other hand, the property (1.8) says that the origin in an ϵ–maximum of convex function $\|\cdot\| - p^*$ on $x_0 - S$. Thus, according to Corollary 4.5 of Hiriart–Urruty (1989), in terms of η–subdifferentials we obtain the following form

$$\partial_\eta(\|\cdot\| - p^*)(0) \subset (\eta + \epsilon)(x_0 - S)^0, \quad \forall \eta \ge 0. \tag{1.8$'$}$$

Therefore, $x_0 \in Eff_p(S)$ if for every $\epsilon \in (0,1)$ there exist $p_\epsilon^* \in P^0$ such that

$$S^*(p_\epsilon^*; 1) \subset \epsilon(S - x_0)^0 \tag{1.8$''$}$$

On the other part, it is clear that $S - x_0$ can be changed by $(S - x_0)_1 = (S - x_0) \cap S(0; 1)$. By property *(i)* we can suppose that P is a ϵ–supernormal cone, i.e. P is generated by a set which have property (1.8).

Proposition 1.3. *Assuming that $C(x_0, S)$ is closed with respect to P, i.e. the regularity property (1.3) is fulfilled, then $x_0 \in S$ is a minimal element of S with respect to P if and only if for every $x^* \in S^*(0,1)$, $\epsilon \in (0,1)$, there exist $u_\epsilon^* \in C(x_0, S)^0$ such that*

$$(x^* + u_\epsilon^*, p) \le \epsilon\|p\|, \quad \forall p \in P \tag{1.9}.$$

Proof. Indeed, if (1.9) holds, the minimality of x_0 can be characterized by $(1.2)'$ and so, the replacement of P with $-C(x_0; S)$ corresponds to that of x_0 with origin. According to Theorem 1.1 it follows that x_0 is a minimal element if and only if we have

$$\inf_{u^* \in C(x_0, S)^0} \sup_{p \in P}[(x^* + u^*, p) - \epsilon\|p\|] = 0$$

for all $x^* \in S^*(0; 1)$ and $\epsilon \in (0,1)$.

Since P is a cone it is clearly that $\sup\{(x^* + u^*, p) - \epsilon\|p\| \mid p \in P\}$ takes only values 0 and ∞. Therefore, the condition (1.4) is fulfilled if and only if this "sup" becomes zero for certain element $u_\epsilon^* \in C(x_0; S)^0$, i.e. (1.9) holds.

Remark 2. Generally, in the absence of the regularity property (1.3), the condition (1.9) is only sufficient for minimality. In fact, we observe that (1.9) can be regarded as the dual form of the condition $P \cap \overline{C(x_0; S)} = \{0\}$ which characterizes the proper efficiency of element x_0 as it was defined by Geofrion (1968),(see also Pascoletti and Serafini, 1984) if P is a closed convex pointed cone.

2. Function $E_{S,P}$ in terms of Minkowsky functional

In what follows we shall emphasize that the function $E_{S,P}$ can be expressed by metric structure induced by Minkowsky functional of the polar $(x_0 - S)^0$. Let us denote

$$d^0_{x,S}(u^*, A) = \inf_{a^* \in A^0} p_{(x-S)^0}(u^* - a^*) \qquad (2.1)$$

the "distance" from $u^* \in X^*$ to the set $A \subset X^*$ with respect to semimetric structure on X^* given by $p_{(x_0-S)^0}$, when $x \in S$. Then we have the following equality

$$\begin{aligned}
E_{S,P}(x, \epsilon) &= \sup_{\|x^*\| \leq 1} d^0_{x,S}(x^*, P^0 + S^*(0; \epsilon)) = \\
&= e_{x,S}(S^*(0; 1); P^0 + S^*(0; \epsilon))
\end{aligned} \qquad (2.2)$$

where by $e_{x,S}(A_1, A_2)$ we denote the *excess* of A_1 over A_2 with respect to $p_{(x-S)^0}$. For this purpose, we consider the lower semicontinuous convex functions $f_1(u) = \epsilon\|u\|$, $f_2(u) = (p^* - x^*, u) + I_{x-S}(u)$, $u \in X$. According to (0.7), using Fenchel theorem we obtain

$$\begin{aligned}
E_{S,P}(x; \epsilon) &= -\inf_{\|x^*\| \leq 1} \sup_{p^* \in P^0} \inf_{u \in X} [f_2(u) + f_1(u)] = \\
&= \sup_{\|x^*\| \leq 1} \inf_{p^* \in P^0} \min_{u^* \in S^*(0;\epsilon)} p_{(x_0-S)^0}(x^* - p^* - u^*) = \\
&= \sup_{\|x^*\| \leq 1} \inf_{v^* \in P^0 + S^*(0;\epsilon)} p_{(x-S)^0}(x^* - v^*)
\end{aligned}$$

so that (2.2) is proved.

Proposition 2.4. *An element x_0 in a closed convex set S is a minimal element of S with respect to a closed convex cone P if and only if for every $\epsilon \in (0, 1)$ the excess of closed unit ball $S^*(0; 1)$ over $P^0 + S^*(0; \epsilon)$ is equal to zero with respect to semidistance induced on X^* by $p_{(x-S)^0}$.*

Remark 3. In fact, this dual characterization of minimality can be considered as an equivalent from the relation (1.2) in term of best approximation on X^* given by $p_{(x-S)^0}$.

Finally, we remark that the results of this paper can be extended to separated locally convex spaces.

References

Barbu V. and Precupanu T., (1986), Convexity and Optimization in Banach Spaces, Publishing House of Romanian Academy and D.Reidel Publishing Company.

Bitran G.R. and Magnati T.L., (1979), The structure of admissible points with respect to cone dominance, JOTA 29, 573–614.

Borwein J.M., (1980), A note on perfect duality and limiting lagrangians, Math. Progr. 18, 330–337.

Franchetti C. and Singer I., (1979), Deviation and farthest points in normed linear spaces, Rev. Roum. Math. Pures et Appl. 24, 373–381.

Geofrion A.M. ,(1968), Proper efficiency and theory of vector maximization, J. Math. Anal. Appl. 22, 618–630.

Henig M.I., (1982), Proper efficiency with respect to cones, JOTA 36, 387–407.

Hiriart–Urruty J.B., (1980), ϵ–subdifferential calculus, Proc. of the Colloquium "Convex Analysis and Optimization", Imperial College, Pitman (Aubin & Vinter Eds.), 43–62.

Hiriart–Urruty J.B., (1989), From convex optimization to nonconvex optimization. Necessary and sufficient conditions for global optimality, in "Nonsmooth Optimization and Related Topics", Clarke F.H., Demyanov V.F., Gianessi F., Eds., Plenum.

Jahn J., (1988 a), Mathematical Vector Optimization in Partially Ordered Linear Spaces, Verlag Peter Lang.

Jahn J., (1988 b), A generalization of theorem of Arrow, Barankin and Blackwell, SIAM J. Control and Optimization 28, 999–1005.

Levine P. et Pomerol J.-Ch., (1972), Sur une théorème de dualité, et ses applications a la programmation linéaire dans les espaces vectoriels topologiques, C.R. Acad. Sci., Paris, Sér. A, 274, 1722–1724.

Nieuwenhuis J.W., (1983),   Some minimax theorems in vector–valued functions, JOTA 40, 463–475.

Pascoletti A. and Serafini P., (1984), Scalarizing vector optimization problems, JOTA 42, 499–524.

Peressini A., Ordered Topological Vector Spaces, (1967) Harper & Row.

Pomerol J.-Ch., (1984), A note on limiting infsup theorems, Math. Progr. 30, 238–241.

Precupanu T., (1981), Duality results in convex optimization, Rev. Rouín. Math. Pures et Appl. 26, 769–780.

Precupanu T., (1984), Closedness conditions for the optimality of a family of non–convex optimization problems, Optimization. Math. Oper. und Statistik 15, 339–346.

Author's address:

Prof. Dr. Teodor Precupanu
Univ. "Al. I. Cuza"
Faculty of Mathematics
Bd. Copou 11
6600, Iaşi
Romania

International Series of Numerical Mathematics, Vol. 107, © 1992 Birkhäuser Verlag Basel

Least-Norm Regularization For Weak Two-Level Optimization Problems

Pierre Loridan and Jacqueline Morgan

Abstract. In this paper, we consider a regularization for weak two-level optimization problems by adaptation of the method presented by Solohovic (1970). Existence and approximation results are given in the case in which the constraints to the lower level problems are described by a multifunction. Convergence results for the least-norm regularization under perturbations are also presented.

1. Introduction.

Let U and V be two finite dimensional euclidian spaces, X a nonempty subset of U, K a multifunction from U to the nonempty subsets of V, f_1 and f_2 two functionals defined on $X \times V$ and valued in $\mathbf{R}$. We consider the following two-level optimization problem (with explicit constraints in the lower level problem), weak in the sense of Breton et al. (1988), corresponding to a two-player game in which the first player (leader) has the leadership whereas the second reacts by choosing an optimal response (Basar and Olsder 1982, Loridan and Morgan 1988, 1989 b, Simaan and Cruz 1973).

$$(S) \begin{cases} \text{Find } \overline{x} \in X \text{ such that} : \sup_{y \in M_2(\overline{x})} f_1(\overline{x}, y) = \min_{x \in X} \sup_{y \in M_2(x)} f_1(x, y) \\ \text{where } M_2(x) \text{ is the set of optimal solutions to the lower level problem} \\ P(x) : \inf_{y \in K(x)} f_2(x, y) \end{cases}$$

By analogy with our previous papers (Loridan and Morgan, 1988, 1989b, 1990) we shall let:

$$v_2(x) = \inf_{y \in K(x)} f_2(x, y) \qquad M_2(x) = \{y \in K(x)/f_2(x, y) = v_2(x)\}$$

$$w_1(x) = \sup_{y \in M_2(x)} f_1(x, y) \qquad v_1 = \inf_{x \in X} w_1(x)$$

Definition 1.1. Any $(\overline{x}, \overline{y}) \in X \times V$ verifying $v_1 = w_1(\overline{x})$ and $\overline{y} \in M_2(\overline{x})$ is termed a Stackelberg equilibrium pair, whereas $\overline{x}$ is also called a Stackelberg solution to (S) (see Von Stackelberg 1952 for static economic problems).

Remark 1.1. Let us note that (S) may fail to have a solution even if the decision variables x and y range over a compact set, whereas f_1 and f_2 are continuous (see Basar and Olsder 1982). This fact mainly results from the lack of lower semi-continuity for the multifunction M_2 (Loridan and Morgan 1988). Well-posedness in Stackelberg problems was studied by Morgan (1989). By another way, various notions of approximate solutions to the lower level problems and regularized Stackelberg problems were investigated by Loridan and Morgan (1989 b, 1990). In the non explicit constrained case existence results were stated by Loridan and Morgan (1992) in a sequential setting, by using semicontinuity properties of marginal functions given by Lignola and Morgan (1991a, 1992). For the explicit constrained case existence and approximation results have been given by Lignola and Morgan (1991b) in a topological setting. In this paper, by adaptation of the method given by Solohovic (1970), we shall consider a regularization, which we call the least-norm regularization, which allows to transform the two-level problem (S) in another one, (RS_ε), for which the lower level problem always has a unique solution. First results about the unexplicit constrained case and connections with a Tykhonov regularization have been given by Loridan and Morgan (1991). Here existence results for the regularized problem (RS_ε) and approximation results for (RS_ε), as $\varepsilon \to 0$, are given in the case in which the constraints to the lower level problem are described by a general multifunction. Moreover convergence results for the least-norm regularization under perturbations are presented.

2. The least-norm regularization.

2.1 Preliminaries. In the case where $M_2(x)$ is a singleton for any $x \in X$, the Stackelberg problem amounts to the following:

$$(S_1) \begin{cases} \text{Find} \quad \overline{x} \in X \text{ such that} \quad : f_1(\overline{x}, \overline{y}(\overline{x})) = \min_{x \in X} f_1(x, \overline{y}(x)) \\ \text{where } \overline{y}(x) \text{ denotes the unique solution to the lower level problem} \\ P(x): \operatorname*{Inf}_{y \in K(x)} f_2(x, y) \end{cases}$$

For the unexplicit constrained case, approximation results were presented by Loridan and Morgan (1986, 1989a) and existence results by Morgan (1989), relying on continuity properties of the function $x \to \overline{y}(x)$. So, in the case where $M_2(x)$ is not a singleton, a first idea would be to consider a selected solution $y(x)$ in $M_2(x)$ for each $x \in X$ and to insert this solution in f_1 in order to construct a new Stackelberg problem analogous with (S_1), that is with uniqueness of the solution to the lower level problem. In convex optimization problems with an infinity of solutions a well-known choice is to select the solution with the least norm. In the framework of two-level optimization, this idea would lead to the new following lower level problems associated with $P(x), x \in X$:

$$Q(x): \operatorname*{Inf}_{y \in M_2(x)} \|y\|$$

Assuming that $Q(x)$ has a unique solution $y(x)$ for each $x \in X$, the corresponding regularized Stackelberg problem in the upper level would be :

$$(RS): \operatorname*{Inf}_{x \in X} f_1(x, y(x)).$$

Unfortunately, this new problem (RS) may fail to have a solution as it is seen in the following simple example (already used by Basar and Olsder 1982, Luchetti et al. 1987, Malozzi and Morgan 1991).

Example 2.1. Let us choose $X = [0,1], K(x) = [0,1]$ for any $x \in X, f_1(x,y) = -xy, f_2(x,y) = (x - 1/2)y$. Then $M_2(x) = \{0\}$ for any $x \in]1/2,1], M_2(x) = \{1\}$ for any $x \in [0,1/2[$ and for $x = 1/2$ $M_2(x)$ is not a singleton since $M_2(x) = [0,1]$. The solution with the least norm is:

$$\begin{aligned} y(x) &= 1 \quad \text{if} \quad x \in [0,1/2[\\ &= 0 \quad \text{if} \quad x \in [1/2,1] \end{aligned}$$

Then, we get

$$\begin{aligned} f_1(x,y(x)) &= -x \quad \text{if} \quad x \in [0,1/2[\\ &= 0 \quad \text{if} \quad x \in [1/2,1] \end{aligned}$$

This function is not lower semicontinuous at $x = 1/2$ and the regularized problem (RS) has no solution since we have $v_1 = -1/2 = \inf_{x \in X} f_1(x,y(x))$.

2.2 Properties of the lower level problems. First of all we state the main assumptions.

Assumptions.
(H_1) f_2 is continuous on $X \times V$
(H_2) $K(x) \subset Y$ for any $x \in X$, where Y is a compact subset of V
(H_3) K is a closed graph and lower semicontinuous multifunction on X (for the definitions see for example Lignola and Morgan 1991 a)
(H_4) for any $x \in X$, the function $y \to f_2(x,y)$ is strictly quasiconvex on its domain D_x, that is to say (Mangarian 1969), D_x is a nonempty convex set and

$$f_2(x, \theta y + (1 - \theta)z) < max\{f_2(x,y), f_2(x,z)\}$$

for any $\theta \in]0,1[$ and any y, z in D_x verifying $f_2(x,y) \neq f_2(x,z)$
(H_5) K is a convex valued multifunction.

Remark 2.1. A strictly quasiconvex function g is not necessarily quasiconvex (let us recall that a function is quasi convex if and only if its level sets are convex). However, if g is lower semicontinuous on its domain, then g is also quasiconvex (for instance, see Mangasarian 1969). Then, from (H_1) and (H_4), we deduce that the function $y \to f_2(x,y)$ is quasi convex and lower semicontinuous on its domain D_x for any $x \in X$ (which means that, for any $\lambda \in \mathbf{R}$, the set $\{y \in V/f_2(x,y) \leq \lambda\}$ is a closed convex set in V).

Remark 2.2. $K(x)$ is compact and convex so, by using Remark 2.1, the set $M_2(x)$ of optimal solutions to the lower level problem $P(x)$ is a nonempty compact convex subset. In particular, $v_2(x) = \inf_{y \in K(x)} f_2(x,y)$ is a finite number for any $x \in X$.

Remark 2.3. From the previous remarks, for any $\varepsilon > 0$, the set $M_2(x,\varepsilon) = \{y \in K(x)/f_2(x,y) \leq v_2(x)+\varepsilon\}$ of ε-solutions to $P(x)$ is a closed convex set in V and this set is also compact.

Remark 2.4. For connections with the theory of Γ-convergence (De Giorgi and Franzoni 1975, Buttazzo and Dal Maso 1982, Attouch 1984) and variational convergence (Zolezzi 1973) see Lignola and Morgan (1991a, 1991b).

Proposition 2.1. (Lignola and Morgan 1991b, theorems 1.2.2, 3.2.1 and 3.2.3) If the assumptions $(H_1), (H_2), (H_3)$ are satisfied, then for any $x \in X$ and for any sequence $x_n \to x$ in X :

1) $\lim\limits_{n\to\infty} v_2(x_n) = v_2(x)$

2) $\limsup\limits_{n\to\infty} M_2(x_n,\varepsilon) \subset M_2(x,\varepsilon)$, for $\varepsilon \geq 0$

3) $\limsup\limits_{n\to\infty} M_2(x_n,\varepsilon_n) \subset M_2(x)$, for any sequence of real positive numbers ε_n converging to 0 whenever $n \to +\infty$. (for any sequence of subsets A_n in Y, $\limsup\limits_{n} A_n$ denotes the sequential upper limit of the sets A_n).

Proof. The first result is nothing but a sequential version of the Berge theorem (Berge 1963). (See also Lignola and Morgan 1991a, Propositions 3.2.1 and 4.2.1) Concerning the second result, for the sake of completeness, let us write the elementary proof in a sequential setting (equivalent to the topological one since U and V are finite dimensional spaces). Let $x \in X, x_n \to x$ and $y \in \limsup\limits_{n\to\infty} M_2(x_n,\varepsilon)$. There exists a subsequence $n_k \to +\infty$ and a sequence $y_{n_k} \to y$ such that: $y_{n_k} \in M_2(x_{n_k},\varepsilon)$ for n_k sufficiently large, that is to say $f_2(x_{n_k},y_{n_k}) \leq v_2(x_{n_k}) + \varepsilon$. From (H_1), we get:

$$f_2(x,y) \leq \limsup\limits_{n\to+\infty} v_2(x_{n_k}) \leq \limsup\limits_{n\to+\infty} v_2(x_n) \leq v_2(x).$$

Moreover $y_{n_k} \in K(x_{n_k})$ for n_k sufficiently large and K is closed graph, therefore $y \in K(x)$. We can conclude that $y \in M_2(x)$. The third result is nothing but an adaptation of the previous one.$\Diamond$

Proposition 2.2. (Lignola and Morgan 1991b, theorem 3.2.2) Suppose that (H_1) to (H_5) are satisfied. Then, for any $\varepsilon > 0$, for any $x \in X$ and any sequence $x_n \to x$, we have: $M_2(x,\varepsilon) \subset \liminf\limits_{n\to+\infty} M_2(x_n,\varepsilon)$ i.e. the multifunction $M_2(\cdot,\varepsilon)$ is lower semicontinuous on X. (where $\liminf\limits_{n\to+\infty} M_2(x_n,\varepsilon)$ denotes the lower limit of the sets $M_2(x_n,\varepsilon)$).

Proof. As in Proposition 2.1 we give a sequential version of the topological proof presented in Lignola and Morgan (1991b).

First, let us denote by $\tilde{M}_2(x,\varepsilon)$ the set of the strict ε-solutions to the problem $P(x)$, that is:

$$\tilde{M}_2(x,\varepsilon) = \{y \in K(x)/f_2(x,y) < v_2(x)+\varepsilon\}$$

and let us show that the multifunction $\tilde{M}_2(\cdot, \varepsilon)$ is lower semicontinuous on X. Let $x_n \to x$ and $y \in \tilde{M}_2(x, \varepsilon)$. We have $y \in K(x)$ and K lower semicontinuous then, there exists $y_n \in K(x_n)$ such that $y_n \to y$. But, from (H_1), we have:

$$\lim_{n \to +\infty} f_2(x_n, y_n) = f_2(x, y) < v_2(x) + \varepsilon = \lim_{n \to +\infty} v_2(x_n) + \varepsilon.$$

Then, for n sufficiently large :

$$f_2(x_n, y_n) < v_2(x_n) + \varepsilon \text{ and } y_n \in \tilde{M}_2(x_n, \varepsilon).$$

Now, we can prove with the arguments used in Loridan and Morgan (1989b), Proposition 6.2, that:

$$M_2(x, \varepsilon) = cl(\tilde{M}_2(x, \varepsilon)) \, (\text{where } cl(\tilde{M}_2(x, \varepsilon)) \text{ denotes the closure of the set } \tilde{M}_2(x, \varepsilon))$$

then :

$$M_2(x, \varepsilon) = cl(\tilde{M}_2(x, \varepsilon)) \subset cl(\liminf_n \tilde{M}_2(x_n, \varepsilon)) = \liminf_n \tilde{M}_2(x_n, \varepsilon)$$

$$\subset \liminf_n M_2(x_n, \varepsilon). \Diamond$$

2.3 The regularized problem (RS_ε): definition and properties. In the beginning of the section (preliminaries), we have mentioned that the regularized problem (RS) may fail to have a solution. So, for any $\varepsilon > 0$, we shall consider the following regularized problem (RS_ε):

$$(RS_\varepsilon) \begin{cases} \operatorname*{Inf}_{x \in X} f_1(x, y(x, \varepsilon)) \\ \text{where } y(x, \varepsilon) \text{ is the (unique) solution to the optimization problem } Q_\varepsilon(x) : \\ Q_\varepsilon(x) : \operatorname*{Inf}_{y \in M_2(x, \varepsilon)} \|y\| \end{cases}$$

$\| \cdot \|$ being the norm in V.

Remark 2.5. $Q_\varepsilon(x)$ is analogous with the problem introduced by Solohovic (1970). The uniqueness of $y(x, \varepsilon)$ is a classical result which we recall in the following proposition :

Proposition 2.3. Suppose that $(H_1), (H_2), (H_3), (H_4)$ and (H_5) are satisfied. For any $\varepsilon > 0$, for any $x \in X$, the problem $Q_\varepsilon(x)$ has a unique solution denoted by $y(x, \varepsilon)$. Furthermore, if $\varepsilon \to 0$ then $y(x, \varepsilon) \to y(x)$ (where $y(x)$ is the unique solution to the optimization problem $Q(x) : \operatorname*{Inf}_{y \in M_2(x)} \|y\|$).

Proof. The first part is obvious since $M_2(x, \varepsilon)$ is a closed convex set in V. The second part is a well-known result (see Solohovic 1970). Let us recall the sketch of the proof. Let $(\varepsilon_n)_{n \in \mathbb{N}}$ a sequence of real positive numbers converging to 0. Since $\|y(x, \varepsilon_n)\| \leq \|y(x)\|$ for any $n \in \mathbb{N}$, there exists a subsequence $y(x, \varepsilon_{n_k}) \to y^*$ and $y^* \in M_2(x)$ by using the third result of Proposition 2.1 (with $x_n = x$ for any $n \in \mathbb{N}$). Furthermore:

$$\|y^*\| \leq \liminf_{k \to +\infty} \|y(x, \varepsilon_{n_k})\| \leq \|y(x)\|.$$

Now the conclusion follows for the whole sequence $(y(x, \varepsilon_n))_{n \in \mathbb{N}}$ since $y^* = y(x)$. $\Diamond$

Remark 2.6. We also have $\lim_{n \to +\infty} \|y(x, \varepsilon_n) - y(x)\| = 0$, for any $x \in X$ and any sequence $\varepsilon_n \to 0^+$ but the result is no longer true when x is replaced by a sequence x_n converging to x in X. Indeed, let us return to the example given in section 2.1. Let $\varepsilon > 0$ and $x \in X = [0, 1]$. It is easy to see that:

$$\begin{aligned} y(x, \varepsilon) &= 1 - (2\varepsilon/1 - 2x) \quad &&\text{if } x \in [0, 1/2 - \varepsilon[\ (\varepsilon < 1/2) \\ &= 0 \quad &&\text{if } x \in [1/2 - \varepsilon, 1] \end{aligned}$$

Let x_n be a sequence converging to $1/2$ with $x_n \in [0, 1/2[$ for any n. Let us choose $\varepsilon_n = (1/2 - x_n)^2$. Then $y(x_n, \varepsilon_n) = 1 - (1/2 - x_n)$ and $y(x_n, \varepsilon_n) \to 1$ when $n \to +\infty$. Consequently $y(x_n, \varepsilon_n)$ does not converge to $y(1/2) = 0$.

Remark 2.7. From the discussion at the beginning of the paper, (RS) may fail to have a solution under mild assumptions. So we shall now consider the regularized problem (RS_ε). From the previous results, our aim is twofold:
1) for any fixed $\varepsilon > 0$, find sufficient conditions in order to prove that the set of solutions to (RS_ε) is nonempty
2) answer the question: what can be said when $\varepsilon \to 0^+$?
In order to achieve this purpose, we give complementary assumptions which will be considered in the rest of this section.

Additional assumptions. In order to obtain a non-trivial problem (RS_ε), it goes without saying that it is assumed the function $x \to w_1(x, \varepsilon) = \sup_{y \in M_2(x, \varepsilon)} f_1(x, y)$ has a nonempty domain (at least for ε "sufficiently small"). Furthermore the following assumptions will be eventually considered:
(H_6) f_1 is lower semicontinuous on $X \times V$
(H_7) for any $x \in X$, there exists a sequence $x_n \to x$ such that, for any $y \in V$ and any $y_n \to y$.

$$\limsup_{n \to +\infty} f_1(x_n, y_n) \leq f_1(x, y)$$

We begin with the following result for the lower level problem $Q_\varepsilon(x), x \in X$.

Proposition 2.4. Suppose that (H_1) to (H_5) are satisfied. Let $\varepsilon > 0, x \in X$. Then, for any sequence $x_n \to x$ in X we have: $y(x_n, \varepsilon) \to y(x, \varepsilon)$ when $n \to +\infty$.

Proof. Let us consider $Q_\varepsilon(x)$: $\inf_{y \in M_2(x, \varepsilon)} \|y\|$. The function $y \to \|y\|$ is obviously continuous and, from Proposition 2.2, the multifunction $M_2(\cdot, \varepsilon)$ is lower semicontinuous. Then, from Proposition 3.3.1 in Lignola and Morgan (1991a), the marginal function $x \to \|y(x, \varepsilon)\|$ associated with $Q_\varepsilon(x)$ is upper semicontinuous, that is to say:

$$\limsup_{n \to +\infty} \|y(x_n, \varepsilon)\| \leq \|y(x, \varepsilon)\|.$$

Now, for any sequence $x_n \to x$, any sequence $(y_n)_n$ verifying $y_n \in M_2(x_n, \varepsilon)$ admits obviously a convergent subsequence. Moreover, from Proposition 2.1, for any sequence $x_n \to x$ and any sequence $y_n \to y$ with $y_n \in M_2(x_n, \varepsilon)$ for any $n \in \mathbb{N}$, we have $y \in M_2(x, \varepsilon)$. Then, from Proposition 4.2.1 in Lignola and Morgan (1991a), we get: $\liminf_{n \to +\infty} \|y(x_n, \varepsilon)\| \geq \|y(x, \varepsilon)\|$. So, $\lim_{n \to +\infty} \|y(x_n, \varepsilon\| = \|y(x, \varepsilon\|$. Now, there exists a subsequence $y(x_{n_k}, \varepsilon) \to y^*$ and $y^* \in M_2(x, \varepsilon)$ from Proposition 2.2. Then, the last part of the proposition follows from the uniqueness of $y(x, \varepsilon)$.

Proposition 2.5. If the assumptions (H_1) to (H_6) are satisfied and if X is compact, then the set of solutions to (RS_ε) is nonempty.

Proof. From Proposition 2.5, $y(x_n, \varepsilon) \to y(x, \varepsilon)$ for any sequence $x_n \to x$. Consequently, with (H_6), the function $x \to f_1(x, y(x, \varepsilon))$ is l.s.c. and the result follows from the compactness of X.

Remark 2.8. Let x_ε be a solution to (RS_ε). By letting :

$$w_1(x, \varepsilon) = \sup_{y \in M_2(x, \varepsilon)} f_1(x, y) \text{ and } v_1(\varepsilon) = \inf_{x \in X} w_1(x, \varepsilon)$$

we get the following inequalities :

$$f_1(x_\varepsilon, y(x_\varepsilon, \varepsilon)) \leq f_1(x, y(x, \varepsilon)) \leq w_1(x, \varepsilon) \text{ for any } x \in X.$$

We then deduce : $f_1(x_\varepsilon, y(x_\varepsilon, \varepsilon)) = \bar{v}_1(\varepsilon) \leq v_1(\varepsilon)$. Now, we shall examine what can be said when $\varepsilon \to 0^+$.

Proposition 2.6. Suppose that (H_1) to (H_7) are satisfied and X is compact. Let $(\varepsilon_n)_n$ be a sequence of real positive numbers converging to 0 and x_{ε_n} be a solution to $(RS(\varepsilon_n))$ for any $n \in \mathbb{N}$. Then any convergent subsequence of the sequence $(x_{\varepsilon_n}, y(x_{\varepsilon_n}, \varepsilon_n))$, in $X \times Y$, has a limit (x^*, y^*) which is a lower Stackelberg equilibrium pair for the initial Stackelberg problem (S), that is to say (see Loridan and Morgan 1988) : $f_1(x^*, y^*) \leq v_1$ and $y^* \in M_2(x^*)$.

Proof. From Remark 2.8, we have $f_1(x_{\varepsilon_n}, y(x_{\varepsilon_n}, \varepsilon_n)) \leq v_1(\varepsilon_n)$. Furthermore from Theorem 1.8 in Lignola and Morgan (1991b), we have $\limsup_{n \to +\infty} v_1(\varepsilon_n) \leq v_1$. Hence, the first inequality $f_1(x^*, y^*) \leq v_1$ is obvious. Now, $y(x_{\varepsilon_n}, \varepsilon_n) \in M_2(x_{\varepsilon_n})$ and by using Proposition 2.1 we deduce that $y^* \in M_2(x^*)$. $\Diamond$

Remark 2.9. Let $\bar{v}_{1,n} = \inf_{x \in X} f_1(x, y(x, \varepsilon_n))$. Then, with the previous proof, we also get :

$$\limsup_{n \to +\infty} \bar{v}_{1,n} \leq v_1.$$

Nevertheless, let us note that in the example 2.1 we have :

$$\bar{v}_{1,n} = -1/2 - \sqrt{\varepsilon_n}/\sqrt{2}, v_1 = -1/2 \quad \text{and} \quad \lim_{n \to +\infty} \bar{v}_{1,n} = v_1.$$

Remark 2.10. From Loridan and Morgan (1991), if $K(x) = Y$ for any $x \in X$, then, in the Propositions 2.1 to 2.6, the assumption (H_1) can be substituted by: f_2 is lower semicontinuous on $X \times Y$ and for any $(x,y) \in X \times Y$, for any sequence $x_n \to x$ in X there exists a sequence y_n in Y such that :

$$\limsup_{n \to +\infty} f_2(x_n, y_n) \leq f_2(x,y).$$

3. Least-norm regularization under perturbations.

Let $f_{i,n}, i = 1,2$, be two sequences of functions from $X \times V$ to $\mathbf{R}$ and K_n a sequence of multifunctions from X to the nonempty subsets of V. We are concerned with the following sequence of perturbed problems:

$$(RS_n(\varepsilon)) \begin{cases} \underset{x \in X}{\mathrm{Inf}}\ f_{1,n}(x, y_n(x,\varepsilon)) \\[2mm] \text{where} \quad y_n(x,\varepsilon) \text{ is the (unique) solution to the optimization problem} \\[2mm] Q_n(x,\varepsilon) : \underset{y \in M_{2,n}(x,\varepsilon)}{\mathrm{Inf}} \|y\| \end{cases}$$

$$\text{where} \quad M_{2,n}(x,\varepsilon) = \{y \in K_n(x)/f_{2,n}(x,y) \leq v_{2,n}(x) + \varepsilon\}$$

$$\text{with} \quad v_{2,n}(x) = \inf_{y \in K_n} f_{2,n}(x,y)$$

Let us denote :

$$w_{1,n}(x,\varepsilon) = \sup_{y \in M_{2,n}(x,\varepsilon)} f_{1,n}(x,y) \quad v_{1,n}(\varepsilon) = \inf_{x \in X} w_{1,n}(x,\varepsilon).$$

In the following, *we suppose that the problem* $Q_n(x,\varepsilon)$ *has a unique solution* $y_n(x,\varepsilon)$. Let us note that Proposition 2.3 give sufficient conditions for existence and uniqueness of $y_n(x,\varepsilon)$. The following assumptions will be now considered.

(A_1) for any $x \in X$, any $y \in V$, any $x_n \to x$ and any $y_n \to y$ we have :

$$\lim_{n \to \infty} f_{2,n}(x_n, y_n) = f_2(x,y)$$

(A_2) $K_n(x) \subset Y$ for any $x \in X$ and any $n \in \mathbf{N}$, where Y is a compact subset of V

(A_3) K_n is lower and upper convergent to K (Lignola and Morgan 1990) that is , for any $x_n \to x$ we have :

$$\limsup_{n \to +\infty} K_n(x_n) \subset K(x) \subset \liminf_{n \to +\infty} K_n(x_n)$$

(A_4) for any $x \in X$ and any $n \in \mathbf{N}$, the function $y \to f_{2,n}(x,y)$ is strictly quasiconvex on its domain

(A_5) K_n is a convex valued multifunction for any $n \in \mathbf{N}$.

Proposition 3.1. (Lignola and Morgan 1991b, theorems 1.4.2 and 3.2.3). If the assumptions $(A_1), (A_2)$ and (A_3) are satisfied, then for any $x \in X$ and for any sequence $x_n \to x$ in X we have:

1) $\lim\limits_{n \to \infty} v_{2,n}(x_n) = v_2(x)$

2) $\limsup\limits_{n \to \infty} M_{2,n}(x_n, \varepsilon) \subset M_2(x, \varepsilon)$ for $\varepsilon \geq 0$

3) $\limsup\limits_{n \to \infty} M_{2,n}(x_n, \varepsilon_n) \subset M_2(x)$,

   for any sequence of real positive numbers ε_n converging to 0 whenever $n \to +\infty$.

Proposition 3.2. (Lignola and Morgan 1991b, theorem 3.2.4). Suppose that (A_1) to (A_5) are satisfied. then, for any $\varepsilon > 0$, for any $x \in X$ and any sequence $x_n \to x$, we have: $M_2(x, \varepsilon) \subset \liminf\limits_{n \to +\infty} M_{2,n}(x_n, \varepsilon)$ i.e. the sequence $M_{2,n}(\cdot, \varepsilon)$ is lower convergent to $M_2(\cdot, \varepsilon)$.

Our aim is now to answer the questions :

1) what can be said for the problem $(RS_n(\varepsilon))$ for $\varepsilon > 0$ and $n \to +\infty$?

2) what can be said for the problem $(RS_n(\varepsilon_n))$ when $n \to +\infty$ and $\varepsilon_n \to 0^+$?

We begin with the following result for the lower level problem $Q_n(x, \varepsilon), x \in X$.

Proposition 3.3. Suppose that (A_1) to (A_5) are satisfied. Let $\varepsilon > 0, x \in X$. Then, for any sequence $x_n \to x$ in X we have : $y_n(x_n, \varepsilon) \to y(x, \varepsilon)$ when $n \to +\infty$.

Proof. Let us consider $(Q_n(x, \varepsilon))$: $\inf\limits_{y \in M_{2,n}(x, \varepsilon)} \|y\|$. The function $y \to \|y\|$ is obviously continuous and, from Proposition 3.1 and 3.2, the multifunction $M_{2,n}(\cdot, \varepsilon)$ is lower and upper convergent to $M_2(\cdot, \varepsilon)$. Then, from Propositions 3.2.1 and 4.3.1 in Lignola and Morgan (1990) the marginal function $x \to \|y_n(x, \varepsilon)\|$ associated with the problem $(Q_n(x, \varepsilon))$ is continuously convergent to the marginal function $x \to \|y(x, \varepsilon)\|$ associated with the problem $(Q(x, \varepsilon))$, that is to say:

$$\lim\limits_{n \to +\infty} \|y_n(x_n, \varepsilon)\| = \|y(x, \varepsilon)\|.$$

Now, there exists a subsequence $y_{n_k}(x_{n_k}, \varepsilon) \to y^*$ and $y^* \in M_2(x, \varepsilon)$ from Proposition 3.1. Then the proposition follows from the uniqueness of $y(x, \varepsilon) \Diamond$

Let us denote :

$$RM(\varepsilon) = \{x_\varepsilon \in X / x_\varepsilon \text{ is a solution to} (RS(\varepsilon))\} \text{ and}$$
$$RM_n(\varepsilon) = \{x_{\varepsilon,n} \in X / x_{\varepsilon,n} \text{ is a solution to} (RS_n(\varepsilon))\}$$
$$\overline{v}_1(\varepsilon) = f_1(x_\varepsilon, y(x_\varepsilon, \varepsilon))$$
$$\overline{v}_{1,n}(\varepsilon) = f_{1,n}(x_\varepsilon, y_n(x_\varepsilon, \varepsilon))$$

Furthermore the following assumptions will be considered.

(A_6) for any $x \in X$, any $y \in Y$, any $x_n \to x$ and any $y_n \to y$ we have :

$$\liminf_{n \to +\infty} f_{1,n}(x_n, y_n) \geq f_1(x,y)$$

(A_7) for any $x \in X$, there exists a sequence $\overline{x}_n \to x$ such that, for any $y \in Y$ and any $y_n \to y$

$$\limsup_{n \to +\infty} f_{1,n}(\overline{x}_n, y_n) \leq f_1(x,y)$$

Proposition 3.4. If (A_1) to (A_7) are satisfied and X is compact then for any $\varepsilon > 0$ we have:

$$\limsup_{n \to +\infty} RM_n(\varepsilon) \subset RM(\varepsilon) \text{ and } \lim_{n \to +\infty} \overline{v}_{1,n}(\varepsilon) = \overline{v}_1(\varepsilon).$$

Proof. Let us consider the functions g and g_n defined by:

$$g(x) = f_1(x, y(x,\varepsilon)) \text{ and } g_n(x) = f_{1,n}(x, y_n(x,\varepsilon))$$

From Proposition 3.3 $y_n(x,\varepsilon) \to y(x,\varepsilon)$ for any sequence $x_n \to x$. Consequently, from (A_6) we have:

$$\liminf_{n \to +\infty} f_{1,n}(x_n, y_n(x_n,\varepsilon)) \geq f_1(x, y(x,\varepsilon)).$$

Moreover, from (A_7), there exists a sequence $\overline{x}_n \to x$ such that:

$$\limsup_{n \to +\infty} f_{1,n}(\overline{x}_n, y_n(\overline{x}_n,\varepsilon)) \geq f_1(x, y(x,\varepsilon)).$$

Then the sequence $(g_n)_n$ epiconverges to g (De Giorgi and Franzoni 1975, Butazzo and Dal Mazo 1982, Attouch 1984) and the result follows. $\Diamond$

Proposition 3.5. Suppose that (A_1) to (A_7) are satisfied and X is compact. Let $(\varepsilon_n)_n$ be a sequence of real positive numbers converging to 0 and x_n be a solution to $(RS_n(\varepsilon_n))$ for any $n \in \mathbb{N}$. Then any convergent subsequence of the sequence $(x_n, y_n(x_n, \varepsilon_n))$, in $X \times Y$, has a limit (x^*, y^*) which is a lower Stackelberg equilibrium pair for the initial Stackelberg problem (S), that is to say (from the definition given in Loridan and Morgan 1988) :

$$f_1(x^*, y^*) \leq v_1 \quad \text{and} \quad y^* \in M_2(x^*).$$

Proof. From Remark 2.8, we have $f_{1,n}(x_n, y_n(x_n, \varepsilon_n)) \leq \overline{v}_{1,n}(\varepsilon_n) \leq v_{1,n}(\varepsilon_n)$. From Theorem 1.8 in Lignola and Morgan (1991b) we have $\limsup_{n \to +\infty} v_{1,n}(\varepsilon_n) \leq v_1$. Hence, the first inequality $f_1(x^*, y^*) \leq v_1$ is obvious. Now, $y_n(x_n, \varepsilon_n) \in M_{2,n}(x_n, \varepsilon_n)$ and by using Proposition 3.1 we deduce that $y^* \in M_2(x^*).\Diamond$

Remark 3.1. If $K_n(x) = Y$ for any $x \in X$ and any $n \in \mathbb{N}$, by using the results of Loridan and Morgan (1986), in the Propositions 3.1 to 3.5 the assumption (A_1) can be substituted by : for any $x \in X$, any $y \in V$, any $x_n \to x$ and any $y_n \to y$ we have :

$$\liminf_{n \to +\infty} f_{2,n}(x_n, y_n) \geq f_2(x,y)$$

for any $x \in X$, any $y \in V$ and any $x_n \to x$ there exists a sequence $y_n \to y$ such that

$$\limsup_{n \to +\infty} f_{2,n}(x_n, y_n) \leq f_2(x,y)$$

References.

Attouch H. (1984), Variational convergence for functions and operators, Pitman, Boston.

Basar T.; Olsder G.J. (1982), Dynamic noncooperative game theory, Academic Press, New York.

Berge C. (1963), Topological spaces, Mac Millan, New York.

Breton M.,et al. (1988), Sequential Stackelberg Equilibria in Two-Person Games, Journal of Opt. Th. and Appl. , vol.59, 71-97.

Buttazzo G.; Dal Maso G. (1982), Γ-convergence and Optimal Control Problems, Journal of Opt.Th. and Appl., 17, 385-407.

De Giorgi E.; Franzoni T. (1975), Su un tipo di convergenza variazionale, Atti Acad.Naz.Lincei, Scienze Fisiche Matematiche e Naturali, 58, 842-850.

Lignola M.B.; Morgan J. Convergence of marginal functions with dependent constraints, Preprint n° 14, Dipartimento di Matematica e Applicazioni, Universitá di Napli. To appear in Optimization.

Lignola M.B.; Morgan J. (1991a), Semicontinuities for marginal functions, Preprint n° 46, Dipartimento di Matematica e Applicazioni, Universitá di Napoli. To appear in Optimization.

Lignola M.B.; Morgan J. (1991b), Topological existence and stability of Stackelberg problems, Preprint n° 3, Istituto di Matematica, Universitá di Salerno.

Lignola M.B.; Morgan J. (1992), Existence and approximation results for Min Sup problems with dependent constraints, Operation Research Proceedings of the International Conference on Operation Research 90 in Vienna, Edited by W.Buhler, G.Feichtinger, F.Hartl, F.J.Radermacher and P.Stahly, Springer Verlag, Berlin, 157-164.

Loridan P.; Morgan J. (1986), Approximation of the Stackelberg problem and applications in Control Theory, Proceedings of the Vth I.F.A.C. Workshop on "Control Applications of Nonlinear Programming and Optimization" (Capri,1985) Pergamon Press, Oxford, 121-124.

Loridan P.; Morgan J. (1988), Approximate solutions for two-level optimization problems, in: Hoffmann K.H., J.B. Hiriart-Urruty; C.Lemarechal; J.Zowe (eds.): Trends in Mathematical Optimization, International Series of Num.Math. 84, Birkhauser Verlag, Basel, 181-196.

Loridan P.; Morgan J. (1989a), A theoretical approximation scheme for Stackelberg problems, Journal of Opt.Th. and Appl., vol 61, n° 1, 95-110.

Loridan P.; Morgan J. (1989b), New results on approximate solutions in two-level optimization, Optimization, 20, 819-836.

Loridan P.; Morgan J. (1990), Quasi convex lower level problems and applications in two-level optimization, Lecture Notes in Economic and Mathematical Systems, Springer-Verlag, n° 345, 325-341.

Loridan P.; Morgan J. (1991), Regularizations for two-level optimization problems, Preprint n° 43, Dipartimento di Matematica e Applicazioni, Universitá di Napoli.

Loridan P.; Morgan J. (1992), On strict ε-solutions for a two-level optimization problem, Operations Research Proceedings of the International Conference on Operation Research 90 in Vienna, Edited by W.Buhler,G.Feichtinger, F.Hartl, F.J.Radermacher and P.Stahly, Springer Verlag, Berlin, 165-172.

Lucchetti R.et al. (1987), Existence theorems of equilibrium points in Stackelberg games with constraints, Optimization, 18, 857-866.

Mallozzi L.; Morgan J., (1991), ε-mixed strategies for static continuous Stackelberg problems, Preprint n° 6, Dipartimento di Matematica e Applicazioni, Universitá di Napoli.

Mangasarian O.L. (1969), Non linear Programming, Mc Graw-Hill, New-York.

Morgan J. (1989), Constrained well-posed two-level optimization problems, Nonsmooth Optimization and related topics, Ettore Majorana Internat. Sciences Series, Edited by F.Clarke, V.Demianov and F.Giannessi, Plenum Press, New-York, 307-326.

Simaan M.; Cruz J. (1973), On the Stackelberg strategies in non zero sum games, Journal of Opt.Th.and Appl., 11, 533-555.

Solohovic V.F. (1970), Unstable extremal problem and geometric properties of Banach spaces, Soviet Math. Dokl., 11, 1470-1472.

Von Stackelberg H. (1952), The theory of market economy, Oxford university Press, Oxford.

Tykhonov A.N. (1965), Methods for the regularization of optimal control problems, Soviet Math. Dokl., 6, 761-763.

Zolezzi T. (1989), On convergence of minima, Bollettino U.M.I., 8, 246-257.

Authors'address.

Pierre Loridan
Département de Mathématiques
Laboratoire d'Analyse Numérique
Université de Bourgogne
B.P. 138
21004 Dijon Cedex
France

Jacqueline Morgan
Dipartimento di Matematicæ Applicazioni
Universitá di Napoli
Via Cinthia
Complesso di Monte San Angelo
80122 Napolo
Italy

Continuity of the value function with respect to the set of constraints

Yves Sonntag

Abstract. For E a metric space, $f:E \to \mathbb{R}$, $\emptyset \neq C \subset E$ closed, we consider the value function $\mu(f,C) = \inf\{f(c) : c \in C\}$. One studies the continuity of $\mu(f,.)$ by using a family of set convergences recently introduced. "Conversely", one obtains results concerning these convergences, proving a posteriori the interest of the principle of classification of convergences proposed by Sonntag and Zalinescu (1991a), (1991b).

1. Semi-metric structures of type p and of class $\mathfrak{X}$ on the class of closed sets

Let E_d be a metric space and $\mathfrak{F}$ be the class of nonempty closed subsets of E_d. Let $\emptyset \neq \mathfrak{A} \subset \mathfrak{F}$ and $\emptyset \neq \mathfrak{X} \subset \mathfrak{F}$. We consider the coarser uniformity on $\mathfrak{A}$ for which the functions $\mathfrak{A} \to \mathbb{R}$, $A \to d(A,X)$, are uniformly continuous for all $X \in \mathfrak{X}$.

$$(d(U,V) = \inf\{d(u,v) : u \in U, v \in V\}, d(u,V) = d(\{u\},V))$$

This initial (weak) uniformity can be described by the family $(p_X)_{X \in \mathfrak{X}}$ of semi-metrics

$$A, B \in \mathfrak{A}, X \in \mathfrak{X}: p_X(A,B) = |d(A,X) - d(B,X)|.$$

It is known , see Sonntag and Zalinescu (1991a) and (1991b), that a number of already known convergences, but also potentially interesting convergences , correspond to topologies deduced from the above uniformity.

Note that a sufficient condition for the uniformity be separated is that $\mathfrak{b} \subset \mathfrak{X}$, where $\mathfrak{b}$

$= \{\{u\} : u \in E\}$.

For an arbitrary directed set I and for A_i, A in $\mathfrak{A}$, the convergence (i.e. the topology) $\mathfrak{X}(p)$ [$\mathfrak{A}$ being clear from the context], is given by:

$$\mathfrak{X}(p)\text{-}\lim(A_i) = A \text{ iff } \forall\ A \in \mathfrak{X} : \lim_{i \in I} d(A_i, X) = d(A, X).$$

For example, if (A_i) is increasing $(i \leq j \Rightarrow A_i \subset A_j)$, (A_i) is $\mathfrak{F}(p)$-convergent with limit $A = \text{adh}(\cup_{i \in I} A_i)$ iff $A \in \mathfrak{A}$.

Examples. Here $\mathfrak{A}$ is $\mathfrak{F}$, and if E is a normed vector space (n.v.s.), $\mathfrak{A}$ is frequently $\mathfrak{C} = \{X \in \mathfrak{F} : X \text{ convex}\}$.

The uniformity that corresponds to $\mathfrak{X} = \mathfrak{F}$ is the finest: it is the *proximal* uniformity (topology = convergence, respectively); see B. Fisher, G. Beer - A. Lechicki - S. Levi - S. Naimpally.

The uniformities for which $\mathcal{S} \subset \mathfrak{X} \subset \mathfrak{K} = \{X \in \mathfrak{F} : X \text{ compact}\}$ are all equivalent: one obtains the *Wijsman* uniformity.

The class $\mathfrak{B} = \{X \in \mathfrak{F} : X \text{ bounded}\}$ defines the *b-proximal* uniformity [P. Shunmugaraj-D. Pai, H. Attouch-D. Azé-G. Beer].

When E is a n.v.s. the uniformity defined by $\mathfrak{X} = \mathfrak{C}$ is the uniformity of G. Beer - C. Hess. The class $\mathfrak{X} = \mathfrak{V} = \{X \in \mathfrak{F} : X \text{ linear manifold}\}$ defines the same uniformity.

The class $\mathfrak{X} = \mathfrak{C} \cap \mathfrak{B}$ defines the *slice* uniformity. Introduced by Sonntag and Zalinescu (1991a), (1991b) as a possible substitute of the "celebrated" convergence of U. Mosco for *non*-reflexive n.v.s., it was studied by Beer (1991a), (1991b) (cf. also H. Attouch - G. Beer) who showed its importance.

Denoting by σ the weak topology $\sigma(E,E')$, one may take for $\mathfrak{X}$: $\mathfrak{F}_\sigma$, $\mathfrak{B}_\sigma$, $\mathfrak{K}_\sigma$, $\mathfrak{K}_\sigma \cap \mathfrak{C}$, etc... etc...

In the reflexive Banach spaces, for $I = \mathbb{N}$, $\mathfrak{C} \cap \mathfrak{B}$, $\mathfrak{K}_\sigma$, $\mathfrak{K}_\sigma \cap \mathfrak{C}$ give the convergence of U. Mosco.

The uniformity defined by $\mathfrak{K} = \{X \in \mathfrak{F} : X \text{ affine hyperplane}\}$ is the *scalar* uniformity, see Sonntag and Zalinescu (1992). The convergence of Beer and Hess is the supremum of Wijsman and scalar convergences.

We insist on the fact that the initial definitions for all these convergences were very different of those given here. One of our aims, here, is to that the new definitions generates interesting results.

2. The value function

Let $f:E \to \mathbb{R}$ be a function and $\emptyset \neq C \in \mathcal{F}$; C is the set of constraints. The *value function* , of variables f and C, is defined by

$$\mu(f,C) = \inf\{f(c) : c \in C\}.$$

From the beginning of functional/set convergences, called "variational" (cf. the fundamental paper of U. Mosco), the continuity of μ presented a particular interest. Here f will be supposed to be fixed; only C will move. A similar study for C fixed and f variable and f and C variable is envisaged by R. Lucchetti and the author.
A great number of works were dedicated to this subject: H. Attouch and R. Wets, G. Beer, R. Lucchetti, T. Zolezzi, etc... made important contributions.
In the direction of uniformities of type p, solely envisaged here, see Sonntag and Zalinescu 1991a, 1991b, Beer et al. 1990 for other types, we cite the thesis of P. Shunmugaraj, D. Pai, Sonntag and Zalinescu (1991b), Lucchetti et al. (1991).
Lucchetti et al. (1991) envisaged the continuity of $\varepsilon\text{-argmin}(f,C) = \{c \in C : f(c) \leq \mu(f,C) + \varepsilon\}$, where $\varepsilon > 0$.
For getting an idea concerning the "tendancy for continuity" of $\mu(f,.)$, let us examine the natural example furnished by the function $h_X : E \to \mathbb{R}$, $u \to d(u,X)$, where $X \in \mathcal{F}$ is given. For $C \in \mathcal{F}$ we have $\mu(h_X,C) = d(C,X)$. Consequently, the continuity of $\mu(h_X,.)$ on $\mathcal{U}$ for all X from a given class $\mathcal{X}$ is equivalent to $\mathcal{X}(p)\text{-}\lim(C_i) = C$, where C_i, C are in $\mathcal{U}$.
Therefore, examples for the discontinuity of $\mu(h_X,.)$ will be obtained by studing fineness of the convergences $\mathcal{X}(p)$. But simple enough counter-examples show that if E is not very particular (compact for instance) we have

$$\mathcal{F}(p) \supset \mathcal{B}(p) \supset \mathcal{S}(p) \text{ in metric spaces,}$$

$$\mathcal{F}_\sigma(p) \supset \mathcal{C}(p) \supset (\mathcal{C} \cap \mathcal{B})(p) \supset (\mathcal{K}_\sigma \cap \mathcal{C})(p) \text{ and}$$

$$\mathcal{B}_\sigma(p) \supset \mathcal{K}_\sigma(p) \supset (\mathcal{K}_\sigma \cap \mathcal{C})(p) \text{ in n.v.s.,}$$

where $\supset$ represents *strict* inclusion.
For example, if (e_n) is the usual Schauder basis of the space l^p ($1 \leq p < \infty$):
$\bullet$ $E = l^1$, $A_n = [e_1, 1/2e_n]$, $A = \{e_1\}$ (example of M. Baronti and P.L. Papini). $(A_n) \to A$ in the sense of Mosco, but (A_n) is not convergent in the sense of Wijsman. The sequence

$(\mu(h_X,A_n))$ is not convergent for a well chosen $X = \{x\}$.

• $E = l^2$, $A_n = [0,e_n]$, $A = \{0\}.(A_n) \to A$ for $\mathcal{B}(p)$ and for $\mathcal{F}_\sigma(p)$, but (A_n) is not

convergent for $\mathcal{F}(p)$. The sequence $(\mu(h_X,A_n))$ is not convergent for some $X \in \mathcal{F}$ (for

details see Sonntag and Zalinescu (1991b)).

It is possible to continue with other examples (see Sonntag (1988)).

Although h_X is a "good" function, from other point of view, because, for example,

$|h_X(u) - h_X(v)| \leq d(u,v)$. Remind that if E is a n.v.s. and $X \in \mathcal{C}$, h_X is convex.

Let us consider the level sets of h_X: $N_\alpha(h_X) = \{u \in E : h_X(u) \leq \alpha\}$, where $\alpha \in \mathbb{R}$; N_α

$= \emptyset$ for $\alpha < 0$, $N_0 = X$, $N_\alpha = \overline{B}(X,\alpha)$ for $\alpha > 0$. The function h_X is inf-bounded $[= \forall$

$\alpha \in \mathbb{R}: N_\alpha$ is bounded $\Leftrightarrow$ coercive, i.e. $\lim_{\|u\| \to \infty} f(u) = +\infty$ in n.v.s.] iff $X \in \mathcal{B}$. h_X

is very rarely inf-compact $[= \forall \ \alpha \in \mathbb{R}: N_\alpha$ is compact].

Another theoretical example is given by the function $g_X(u) = \sup\{d(u,x) : x \in X\}$. With

this function one obtains results concerning Chebyshev radius.

3. Other examples

See Lucchetti et al. (1991) for b), c), d).

a) $E = \mathbb{R}$, $A_n = \{1/n\}$, $A = \{0\}$: $(A_n) \to A$ in the sense of Hausdorff. The indicator

function $I_{\{0\}}$ is l.s.c., convex, proper, coercive, inf-compact. But $\mu(I_{\{0\}},A_n) = \infty$ for

all n and $\mu(I_{\{0\}},A_n) = 0$. We must impose the continuity of f.

b) $E = \mathbb{R}^2$, $A_n = \{(s,1/n) : s \in \mathbb{R}\}$, $A = \{(s,0) : s \in \mathbb{R}\}$. $(A_n) \to A$ in a very strong

sense (Hausdorff distance) which implies $\mathcal{F}(p)$. Take $f(s,t) = \max\{s^2 - s^4 t,-1\}$: $\mu(f,A_n)$

$= -1$ for all n, but $\mu(f,A) = 0$.

c) $E = l^2$, $A_n = [0,e_n]$, $A = \{0\}$. $(A_n) \to A$ for $\mathcal{B}(p)$. The function f given by $f(u) =$

$\max\{-\|u\|,\|u\| + 2\}$ is Lipschitzian and coercive, but $\mu(f,A_n) = -1$ for all n and $\mu(f,A) =$

0.

d) $E = l^2$, $A_n = \overline{B}(0,1+1/n)$, $A = \overline{B}(0,1)$. $(A_n) \to A$ in the sense of Hausdorff. The

function f given by $f(u) = \max\{-\sum_{1 \leq n < \infty}(u \mid e_n)^{2n},\|u\| - 10\}$ is continuous and coercive

but $\mu(f,A_n) \leq -8$ for all n and $\mu(f,A) = -1$.

4. A general framework for proving the continuity of $\mu(f,.)$

Let A_i, A be elements of $\mathcal{Q}$, $i \in I$ - directed set. We shall see that $\mathcal{X}$-$\lim(A_i) = A$ implies

-and sometimes is implied by- the continuity of $\mu(f,.)$, on $\mathcal{Q}$ endowed with $\mathcal{X}(p)$, for all functions f belonging to a certain class of functions, $C(E,\mathbb{R})$, that must be determined. The next result will solve half of the problem.

Lemma (see Sonntag and Zalinescu (1991b)). The following properties are equivalent:

(i) $A \subset \liminf(A_i)$, i.e. $\forall\ a \in A,\ \forall\ i \in I,\ \exists\ a_i \in A_i : \lim(a_i) = a$,

(ii) $\forall\ u \in E : \limsup d(u,A_i) \leq d(u,A)$,

(iii) $\forall\ X \in \mathfrak{F} : \limsup d(A_i,X) \leq d(A,X)$,

(iv) $\forall\ f:E \to \mathbb{R}$ u.s.c. on E $: \limsup \mu(f,A_i) \leq \mu(f,A)$.

We the aim to benefit from this lemma we shall suppose in the sequel that the uniformity $\mathfrak{X}(p)$ on $\mathcal{Q}$ satisfies $\mathfrak{X} \supset \mathcal{A}$. In this case the uniformity is separated and we have:

$$\mathfrak{X}(p)\text{-}\lim(A_i) = A \Rightarrow (i)\ (\text{easy and well known}).$$

Therefore, if we consider a function $f:E \to \mathbb{R}$ which is u.s.c. we have

$$\mathfrak{X}(p)\text{-}\lim(A_i) = A \Rightarrow \limsup\ \mu(f,A_i) \leq \mu(f,A).$$

We shall make this hypothesis (u.s.c.) and we consider in the sequel only the problem of having $\mu(f,A) \leq \liminf \mu(f,A_i)$. Suppose that we have *not* $\mu \leq \liminf(\mu_i)$ [where $\mu = \mu(f,A)$, $\mu_i = \mu(f,A_i)$], i.e. there exits $\lambda \in \mathbb{R}$ such that $\liminf(\mu_i) < \lambda < \mu$. This implies that $J = \{i \in I : \mu_i < \lambda\}$ is cofinal in I $[= \forall\ i \in I, \exists\ j \in J : j \geq i]$. By the definition of μ_i, for every $i \in J$ there exists $a_i \in A_i$ such that $f(a_i) < \lambda$. Let us consider the set $X = \{u \in E: f(u) \leq \lambda\}$; as $a_i \in X$ for every $i \in J$, we have $X \neq \emptyset$. In order to assure that $X \in \mathfrak{F}$ we shall suppose that f is l.s.c. on E; because f was already supposed to be u.s.c., we shall suppose that f is *continuous on E* in the sequel.
Note that $\mu \leq f(v) \leq \lambda < \mu$ if $v \in A \cap X$, which is impossible, whence $A \cap X = \emptyset$.
Note also that $a_i \in A_i \cap X$ for $i \in J$, and therefore $d(A_i,X) = 0$ for $i \in J$. Consequently, if we introduce hypotheses which assure that $X \in \mathfrak{X}$, since $\mathfrak{X}(p)\text{-}\lim(A_i) = A$, we shall have

$$\lim_{i \in J} d(A_i,X) = 0 = d(A,X).$$

Using what we have till now, we can obtain two types of contradiction and thus the wanted conclusion: $\mu = \lim(\mu_i)$.

(1) Without any supplementary hypothesis on f besides those assuring that $X \in \mathfrak{X}$, if

$d(A,X) = 0$ implies $A \cap X \neq \emptyset$ we get a contradiction.

(2) $d(A,X) = 0$ implies the existence of two sequences $(a_n) \subset A$ and $(x_n) \subset X$ such that $\lim_{n \to \infty} d(a_n, x_n) = 0$, and so

$$\mu \leq f(a_n) = f(a_n) - f(x_n) + f(x_n) \leq f(a_n) - f(x_n) + \lambda < \mu - \lambda + \lambda = \mu,$$

whence a contradiction if we can find an n such that $f(a_n) - f(x_n) < \mu - \lambda$.

This contradiction is assured if $\lim(f(a_n) - f(x_n)) = 0$.

(2a) If A and X are "arbitrary" we would have this result if f is uniformly continuous on E.

(2b) If one of the sequences (a_n) or (x_n) is bounded, the other one is too; if we consider a bounded set B containing (a_n) and (x_n), we shall obtain the desired conclusion if f is uniformly continuous on every bounded subset of E.

Remark 1. Note that if f is inf-bounded then X is bounded and (2b) applies.

Remark 2. The implication $\mathfrak{X}\text{-}\lim(A_i) = A \Rightarrow [\forall\ f \in C(E,\mathbb{R}) : \lim \mu(f,A_i) = \mu(f,A)]$ is reversible and provide an *equivalence* if the function h_X defined by $h_X(u) = d(u,X)$, $X \in \mathfrak{X}$, verifies $h_X \in C(E,\mathbb{R})$.

Remark 3. If $f \in C(E,\mathbb{R})$ implies $-f \in C(E,\mathbb{R})$, we shall have the supplementary convergence result $\lim v(f,A_i) = v(f,A)$, where $v(f,C) = -\mu(-f,C) = \sup\{f(c) : c \in C\}$.

5. Continuity results for the value function and characterizations of convergences

Let us apply the program established in the last section for the uniformities described in Section 1.

$1°)$ E *a metric space,* $\mathfrak{A} = \mathfrak{F}$, $\mathfrak{S} \subset \mathfrak{X} \subset \mathfrak{K}$.

In order to have $X \in \mathfrak{K}$ we suppose that f is inf-compact.

The contradiction evoked in (1) comes from the well known implication

$$[A \in \mathfrak{F}, X \in \mathfrak{K}, A \cap X = \emptyset] \Rightarrow d(A,X) > 0.$$

Let $CI(E,\mathbb{R})$ be the class of *continuous* and *inf-compact* functions $f:E \to \mathbb{R}$.

Proposition 1. Wijsman-$\lim(A_i) = A$ implies $\lim \mu(f,A_i) = \mu(f,A)$ for all $f \in CI(E,\mathbb{R})$. If the closed balls of E_d are compact then the converse is true.

Remark. If $\mathcal{F}$-$\lim(A_i) = A$ with $A \in \mathcal{K}$, the same argument shows that $\lim(\mu_i) = \mu$ for all continuous function f.

$2°)$ E *is a n.v.s.,* $\mathcal{U} = \mathcal{F}_\sigma$, $\mathcal{X} = \mathcal{K}_\sigma$.
Let $CI_\sigma((E,\mathbb{R})$ be the class of *continuous* and $\sigma(E,E')$-*inf-compact* functions $f{:}E \to \mathbb{R}$.
If $f \in CI_\sigma$ then $X \in \mathcal{K}_\sigma$.
On the other hand: $d(A,X) = 0 = d(0,X - A) \Rightarrow 0 \in \overline{X - A}$; but $X - A \in \mathcal{F}_\sigma$ (see the book of Dunford and Schwartz, the lemma of p.414), and so $0 \in X - A$, i.e. $A \cap X \neq \emptyset$.
Hence we have the case (1).
The function h_X, $X \in \mathcal{K}_\sigma$, is generally not σ-inf-compact: it is so when E is a reflexive Banach space.

Proposition 2. Let E be a n.v.s., $\mathcal{U} = \mathcal{F}_\sigma$, $\mathcal{X} = \mathcal{K}_\sigma$.

$$\mathcal{K}_\sigma\text{-}\lim(A_i) = A \Rightarrow [\lim \mu(f,A_i) = \mu(f,A) \ \forall \ f \in CI_\sigma(E,\mathbb{R})].$$

If E is a reflexive Banach space the converse implication holds.

Remarks. a) The proposition is also true for $\mathcal{U} = \mathcal{F}$ if $A \in \mathcal{F}_\sigma$.
b) As in $1°)$, the result is also valid if $\mathcal{X} = \mathcal{F}$ with $A \in \mathcal{K}_\sigma$ and $X \in \mathcal{F}_\sigma$.

Another form of Proposition 2. Instead of $\mathcal{X} = \mathcal{K}_\sigma$ we take $\mathcal{X} = \mathcal{K}_\sigma \cap \mathcal{C}$. To assure that $X \in \mathcal{C}$ we must suppose that f is convex, so that we consider the class $CCI_\sigma(E,\mathbb{R})$ of *convex, continuous and* σ-*inf-compact* functions $f{:}E \to \mathbb{R}$.For A_i, $A \in \mathcal{F}_\sigma$ we have

$$(\mathcal{K}_\sigma \cap \mathcal{C})\text{-}\lim(A_i) = A \Rightarrow [\lim \mu(f,A_i) = \mu(f,A) \ \forall \ f \in CCI_\sigma(E,\mathbb{R})].$$

Remind that if $\mathcal{U} = \mathcal{C}$, $I = \mathbb{N}$, $\mathcal{X} = \mathcal{K}_\sigma$ and $\mathcal{X} = \mathcal{C} \cap \mathcal{K}_\sigma$ correspond to the convergence in the sense of Mosco when E is a reflexive Banach space (in this case Prop. 1 belongs to P. Shunmugaraj).

$3°)$ E *is a metric space,* $\mathcal{U} = \mathcal{F}$, $\mathcal{X} = \mathcal{F}$.
The continuity of f is sufficient to assure that $X \in \mathcal{X}$, and we make use of the contradiction (2a) assuming that f is uniformly continuous.
Denote by $U(E,\mathbb{R})$ the class of *uniformly continuous* functions $f{:}E \to \mathbb{R}$.

Proposition 3. We have

$$\mathcal{F}\text{-lim}(A_i) = A \text{ (proximal convergence)} \Leftrightarrow \lim \mu(f,A_i) = \mu(f,A) \quad \forall \ f \in U(E,\mathbb{R}) \Leftrightarrow$$

$$\lim v(f,A_i) = v(f,A) \quad \forall \ f \in U(E,\mathbb{R}) \Leftrightarrow \lim e(A_i,X) = e(A,X)$$

(result of Sonntag and Zalinescu (1991b) and Beer and Lucchetti (1990))

$$\Rightarrow \sup_{u \in A_i} \sup_{x \in X} d(u,x) \to \sup_{u \in A} \sup_{x \in X} d(u,x).$$

4°) E *is a metric space,* $\mathcal{U} = \mathcal{F}$, $\mathcal{X} = \mathcal{B}$.
In order to have that $X \in \mathcal{X}$ we suppose that f is inf-bounded on E and the contradiction
is obtained from (2b).
Let $I_b UB(E,\mathbb{R})$ be the class of *inf-bounded and uniformly continuous on bounded*
subsets functions f:E $\to \mathbb{R}$.

Proposition 4. We have

$$\mathcal{B}\text{-lim}(A_i) = A \text{ (b-proximal convergence)} \Leftrightarrow \lim \mu(f,A_i) = \mu(f,A) \quad \forall \ f \in I_b UB(E,\mathbb{R}).$$

5°) E *is a n.v.s.,* $\mathcal{U} = \mathcal{C}$, $\mathcal{X} = \mathcal{C}$.
To have $X \in \mathcal{X}$ we impose that f is convex; the conclusion will follow by (2a).
Let $CU(E,\mathbb{R})$ be the class of *convex and uniformly continuous* functions f:E $\to \mathbb{R}$.

Proposition 5. Let A_i, $A \in \mathcal{C}$. Then

$$\mathcal{C}\text{-lim}(A_i) = A \text{ (convergence of G. Beer and C. Hess)} \Leftrightarrow$$

$$\lim \mu(f,A_i) = \mu(f,A) \quad \forall \ f \in CU(E,\mathbb{R}).$$

5°) E *is a n.v.s.,* $\mathcal{U} = \mathcal{C}$, $\mathcal{X} = \mathcal{C} \cap \mathcal{B}$.

In order to have that $X \in \mathcal{X}$ it is sufficient to suppose that f is convex and inf-bounded
on E and the contradiction is obtained from (2b).
Let $CI_b UB(E,\mathbb{R})$ be the class of *convex, inf-bounded and uniformly continuous on*
bounded subsets functions f:E $\to \mathbb{R}$.

Proposition 6. Let A_i, $A \in \mathcal{C}$. Then

$$(\mathcal{C} \cap \mathcal{B})\text{-}\lim(A_i) = A \text{ (slice convergence)} \Leftrightarrow$$

$$\lim \mu(f,A_i) = \mu(f,A) \quad \forall \ f \in CI_b UB(E,\mathbb{R}).$$

Let $B \in \mathcal{C} \cap \mathcal{B}$ and consider $g_B(u) = \sup\{d(u,v) : v \in B\}$. It is obvious that $g_B \in CI_b UB(E,\mathbb{R})$. Therefore we have

Corollary. Let A_i, $A \in \mathcal{C}$. Then

$$(\mathcal{C} \cap \mathcal{B})\text{-}\lim(A_i) = A \ \Rightarrow \ \inf_{u \in A_i}\sup_{v \in B}d(u,v) \to \inf_{u \in A}\sup_{v \in B}d(u,v),$$

i.e. the slice convergence implies the "distal" convergence of G. Beer and D. Pai.

Nota. In 5°) and 6°) on can take $\mathcal{Q} = \mathcal{F}$ (one takes $\mathcal{Q} = \mathcal{C}$ to obtain again the original definitions).

References

Beer G. (1991a), Topologies on closed convex sets and the Effros measurability of set-valued functions, preprint.

Beer G. (1991b), The slice topology: a viable alternative to Mosco convergence in nonreflexive spaces, preprint.

Beer G. and Lucchetti R. (1990), Weak topologies for the closed subsets of a metric space, to appear in Trans. Amer. Math. Soc.

Lucchetti R., Shunmugaraj P. and Sonntag Y. (1991), Recent hypertopologies and continuity of the value function and of the constrained level sets, preprint.

Sonntag Y. (1988), Convergence des suites d'ensemble, monograph in preparation.

Sonntag Y. and Zalinescu C. (1991a), Set convergence. An attempt of classification, in "Differential Equations and Control Theory", V. Barbu ed., Pitman Research Notes in Math. Series, n° 250, 312-323.

Sonntag Y. and Zalinescu C. (1991b), Set convergence. An attempt of classification, submitted.

Sonntag Y. and Zalinescu C. (1992), Scalar convergence of convex sets, J. Math. Anal. Appl. 164, 219-241.

Author's address:

Yves Sonntag
Université de Provence, case X
U.F.R.-M.I.M.
3, Place Victor Hugo
13331-Marseille Cedex 3
France
Fax (33)91106102

ON INTEGRAL INEQUALITIES INVOLVING LOGCONCAVE FUNCTIONS

René Michel, Michel Volle

Abstract. We show that the majorization of the integral of a product is reinforced when dealing with longconcave functions. The case where one term of the product is what we call a simple kernel is detailed. As a by product we obtain an inequality between the integral convolution and the supconvolution.

Key words : logconcavity, d.c.programming, nonconvex duality.

The purpose of the talk is to emphasize noteworthy improvements due to the presence of logconcave functions in some integral inequalities. Logconcavity is understood in the following sense : a function $\varphi : \mathbf{R}^n \to [0, +\infty[$ is said to be *logconcave* if and only if $\log\varphi : \mathbf{R}^n \to \mathbf{R} \cup \{-\infty\}$, with the convention $\log 0 = -\infty$, is a concave function. In other words,

$$\varphi(tu + (1 - t)v) \geq \varphi^t(u)\varphi^{1-t}(v)$$

for all $u, v \in \mathbf{R}^n$, $\quad t \in \ ]0, 1[$.

The class of logconcave functions interferes notably in probability, heat equation theory, Laplacian's eigenfunctions study (H.-J.Brascamp and E.-H.Lieb 1976, G.Deslauriers and S.Dubuc 1979, S.Dubuc 1978, R.Michel and M.Volle 1991,...).

One says that a logconcave function $g : \mathbf{R}^n \to [0, +\infty[$ tends strongly to zero at infinity if

$$\forall c > 0 , \quad \lim_{||x|| \to +\infty} g(x)e^{c||x||} = 0 .$$

Now we consider the following general problem : given a logconcave function g over $\mathbf{R}^n$, tending strongly to zero at infinity, we search out the better constant γ such that

$$(1) \qquad \int_{\mathbf{R}^n} g(x)\varphi(x)dx \leq \gamma \sup_{\mathbf{R}^n}(g\varphi)$$

for any logconcave function φ on $\mathbf{R}^n$.

We first prove that the underlying maximization problem can be reduced to a finite dimensional one ; more precisely we show that, in the computation of γ, we can limit ourselves to the consideration of functions φ loglinear ($\varphi(x) = e^{<x,y>}, \quad y \in \mathbf{R}^n$).

THEOREM 1

$$\gamma = \sup_{y \in \mathbf{R}^n} \left\{ \frac{\int_{\mathbf{R}^n} g(x)e^{<x,y>}dx}{\sup_{x \in \mathbf{R}^n} g(x)e^{<x,y>}} \right\}$$

The proof ot Theorem 1 involves classical variational principles and uses some subdifferential calculus rules from convex analysis (see R.Michel and M.Volle, 1991).

The maximization problem stated in Theorem 1 admits a dual version. In fact the numerator is nothing but the Laplace transform $\hat{g}$ of g, which is a logconvex function, while the denominator can be written

$$\sup_{x \in \mathbf{R}^n} g(x)e^{<x,y>} = \exp(-\log g)^*(y)$$

where $(-\log g)^*$ is set for the Fenchel conjugate of the convex function $(-\log g)$ (see e.g. R.-T.Rockafellar, 1970).

Hence we have

$$\log \gamma = \sup_{y \in \mathbb{R}^n} (\log \hat{g})(y) - (-\log g)^*(y) \, ,$$

and we deal with the maximization of the difference of two convex functions (see e.g. J.-B.Hiriart-Urruty, 1985).

By introducing the Cramer's transform (see e.g. R.Azencott, 1980)

$$g^c = (\log \hat{g})^*$$

of g, the Toland-Singer's duality formula (I.Singer 1979, J.-F.Toland 1979) gives the following

PROPOSITION 1 $-\log \gamma = \inf\{g^c(x) + \log g(x) : g(x) > 0\} \, .$

An interesting situation occurs where the supremum in Theorem 1 is attained for $y = 0$. We have then

$$\gamma = \frac{\int_{\mathbb{R}^n} g(x) \, dx}{\sup_{\mathbb{R}^n} g}$$

or, equivalently,

$$(\sup_{\mathbb{R}^n} g) \times \int_{\mathbb{R}^n} g(x)\, \varphi(x)\, dx \;\le\; \sup_{\mathbb{R}^n}(g\varphi) \int_{\mathbb{R}^n} g(x)\, dx$$

for any function φ logconcave on $\mathbb{R}^n$.

In such a case we say that

$$g \text{ is a } simple\ kernel\,.$$

Let us give some examples of simple kernels. To this end we introduce an *euclidean* norm $\|\ \|$ on $\mathbb{R}^n$.

Example 1   For each $\alpha > 0$ the function

$$g(x) = \begin{cases} (1 - \|x\|^2)^\alpha & \text{if } \|x\| \le 1 \\ 0 & \text{if } \|x\| \ge 1 \end{cases}$$

is a simple kernel. This fact, proved in (R.Michel, 1989), generalizes (P.Erdös and T.Grünwald 1939, K.S.K. Iyengar 1941). Here we have

$$\gamma = \frac{\Gamma(\frac{n}{2} + 1)\Gamma(\alpha + 1)}{\Gamma(\frac{n}{2} + \alpha + 1)} \text{ vol } B_n$$

where B_n is the euclidian unit ball.

Example 2   For each $\alpha > 0$

$$g(x) = e^{-\alpha\|x\|^2}$$

is a simple kernel (R.Michel, 1989) and we have

$$\gamma = (\frac{\pi}{\alpha})^{\frac{n}{2}}\,.$$

A mere translation enables one to obtain a majorization of the fundamental solution of heat equation when the initial temperature is given by a logconcave function φ_0 ; for any $(t, y) \in \mathbf{R}_+^* \times \mathbf{R}^n$ we have then

$$(4\pi t)^{-\frac{n}{2}} \int_{\mathbf{R}^n} e^{-\frac{\|x-y\|^2}{4t}} \varphi_0(x) dx \leq \sup_{x \in \mathbf{R}^n} e^{-\frac{\|x-y\|^2}{4t}} \varphi_0(x) \ .$$

Example 3 For $\alpha \geq 1$, the logconcave function g defined by

$$g(x) = \begin{cases} 1 - \|x\|^\alpha & \text{if} \quad x \in B_n \\ 0 & \text{if} \quad x \notin B_n \end{cases}$$

is a simple kernel if and only if $\alpha \geq 2$ (R.Michel and M.Volle, 1991).

By using Proposition 1 a simple kernel can be characterized in terms of a relation between the kernel itself and its Cramer's transform :

PROPOSITION 2 *g is a simple kernel if and only if*

$$- \log g(x) + \sup_{\mathbf{R}^n} \log g \leq g^c(x) + \log \int_{\mathbf{R}^n} g(u) du \qquad \text{for any } x \in \mathbf{R}^n \ .$$

Moreover, in the case when g is *even*, it turns out that g is a simple kernel if and only if zero is an optimal solution of the dual problem formulated in Proposition 1 :

PROPOSITION 3   *Assume that g is even ; then g is a simple kernel if and only if*

$$\log \hat{g}(y) - \log \hat{g}(0) \leq (-\log g)^*(y) - \log g(0) \quad for\ any \quad y \in \mathbf{R}^n \ .$$

Theorem 2 below furnishes a large class of simple kernels which contains the one presented in examples 1,2,3. The proof of Theorem 2 is done by using Theorem 1 ; it requires also Jensen's inequality and some spherical integral calculus (see R.Michel and M.Volle, 1991).

THEOREM 2   *Assume that*

$$h :]-R, R[\to \mathbf{R} \qquad (R \in \,]0, +\infty])$$

is an even concave differentiable function such that

$$\lim_{|t| \to R} h(t) = -\infty \qquad if \quad R < +\infty$$

$$\lim_{|t| \to +\infty} \frac{h(t)}{t} = -\infty \qquad if \quad R = +\infty \ .$$

Let us suppose that

$$h' :]-R, R[\to \mathbf{R}$$

is one to one and onto, and that the function $t \longmapsto |h'(t)|$ is convex.

Then, $\| \ \|$ being an euclidean norm, the logconcave function

$$g(x) = \begin{cases} e^{h(||x||)} & \text{if} \quad ||x|| < R \\ 0 & \text{if} \quad ||x|| \geq R \end{cases}$$

is a simple kernel.

Remarks

1) Various examples of simple kernels verifying the assumptions of theorem 2 can be drawn from (I.-I.Hirschmann and D.V.Widder, 1955).

2) We don't know if Theorem 2 remains valid for a non euclidean norm.

When the simple kernel is a probability density one has interest in taking φ as a characteristic function of a convex set. The probability of a measurable set $A \subset \mathbf{R}^n$ being defined by

$$P(A) = \int_A g(x) \, dx \ ,$$

we obtain :

PROPOSITION 4   *Assume that the probability density g is a simple kernel; then, for any convex subset A of $\mathbf{R}^n$, we have*

$$P(A) \leq \sup_A g \ / \ \sup_{\mathbf{R}^n} g \ .$$

The simple kernel notion serves also to establish a relation between the classical integral convolution

$$(g * \varphi)(x) = \int_{\mathbb{R}^n} g(x - u)\varphi(u)du$$

and the supconvolution

$$(g \,\square\, \varphi)(x) = \sup_{u \in \mathbb{R}^n} g(x - u))\varphi(u) \, .$$

PROPOSITION 5 *Let g be a simple kernel. For any logconcave function φ over $\mathbf{R}^n$ we have*

$$\frac{g}{\|g\|_1} * \varphi \leq \frac{g}{\|g\|_\infty} \,\square\, \varphi \ \ .$$

References

Azencott R. (1980), Grandes déviations et applications. Lecture Notes in Mathematics 774, 1-176.

Brascamp H.-J. and Lieb E.-H. (1976), On extensions of the Brunn-Minkowski and Prékopa-Leindler Theorems, including inequalities for Logconcave Functions, and with an Application to the Diffusion Equation. J. of Funct.Anal.22, 366-389.

Deslauriers G. and Dubuc S. (1979), Logconcavity of the cooling of a convex body. Proc.Amer.Math.Soc., vol.74, n°2, 291-294.

Dubuc S. (1978), Problèmes d'optimisation en calcul de probabilités. Séminaire de Math.Supérieures. Les Presses de l'Université de Montréal.

Erdös P. and Grünwald T. (1939), On polynomial with only real roots. Annals of Math., vol.40, n° 3, 537-548.

Hiriart-Urruty J.-B. (1985), Generalized differentiability, duality and optimization for problems dealing with differences of convex functions. Lecture Notes in Economics and Mathematical Systems, 256, 37-69, Springer Verlag, Berlin.

Hirschmann I.-I. and Widder D.-V. (1955), The convolution transform. Princeton.

Iyengar K.S.K. (1941), A property of integral functions of order less than two with real roots. Annals of Math., vol.42 n°4, 823-828.

Michel R. (1989), Inégalités intégrales et logconcavité : des applications liées à l'équation de la chaleur. Preprint, Université d'Avignon.

Michel R. and Volle M. (1991), Logconcavité et inégalités intégrales. Preprint, Université d'Avignon.

Prekopa A. (1972), On logarithmic concave measures and functions. Acta Sci.Math., vol.34, 335-343.

Rockafellar R.-T. (1970), Convex analysis. Princeton.

Singer I. (1979), A Fenchel-Rockafellar type duality theorem for maximization. Bull.Aust. Math.Soc., vol.20, 193-198.

Toland J.-F. (1979), Duality in nonconvex optimization. J.Math.Anal. Appl., vol.71, 41-61.

Author's address :

Professors René Michel and Michel Volle
Université d'Avignon
Faculté des Sciences
Département de Mathématiques
33, rue Louis Pasteur
84000 Avignon - France

International Series of Numerical Mathematics, Vol. 107, © 1992 Birkhäuser Verlag Basel

NUMERICAL SOLUTION OF FREE BOUNDARY PROBLEMS IN SOLIDS MECHANICS

LABORDE Patrick

Abstract. The present study is concerned with a variational inequality of evolution modelizing some irreversible phenomena of damage in Solids Mechanics. By approximating the associated boundary-value problem we obtain a discrete equation $F(u) = 0$ of the same form that in theory of Plasticity. In order to solve this Euler equation, a generalized Newton algorithm is formulated by using the directional differentiability of F. This iterative method extends the so-called "tangent stiffness algorithm" in Computational Mechanics.

1. Introduction

The behavior of some materials as metals or composites presents *two aspects* very different. Physically, the first one corresponds to (reversible) elastic strains. On the other hand, the mechanical properties are modified of an *irreversible* manner (production of permanent or plastic strains , weakening of the elastic stiffness, ...).Within these media, the elastic range and the plastic or damaged range are delimited by a *free boundary* depending on the (unknown) stress field. A class of such *hysteresis* behaviors is modelized thanks to some variational inequality of evolution (defined by relationship (1) below). This model of *damage* can be considered as a generalization of the constitutive equations studied in classical *Plasticity*.

The associated partial derivative problem is approximated by means of an implicit time discretization together with a standard finite element method. In this way, we are led to consider some Euler equation $F(u) = 0$ where u is the finite element displacement field at the current discrete time . The nonlinearity of this problem comes from the presence of an operator of *projection* onto the convex set of admissible stresses.

In the context of Plasticity, engineers often use the so-called *tangent stiffness algorithm*. This computationally efficient procedure is usually formulated in the spirit of the original Newton method, though the mapping F is *not* differentiable.

The aim of this present communication is to clarify this practice from a mathematical point of view. We give a general formulation of the algorithm using

the directional derivatives of the projection operator . This investigation allows
us to apply the iterative method for solving new problems in engineering, such as
the above-mentioned damage problem .

2. The model.

2.1. Constitutive equations. In order to describe some problems of *damage* (see
Lemaitre and Chaboche 1985) we consider a general enough model of behavior
inspired by the so-called *Generalized Standard Materials*.

Let α be the set of parameters which define the state of *irreversibility*
(hardening, damage,...) at the current point in the body; see Germain et al.
1983. The *internal variable* α lies in some Euclidean space E, and we denote by
$\alpha \cdot \beta$ the inner product in E. The space S of symmetric tensors $\tau = (\tau_{ij})$ of order
two is equipped with its usual inner product $\sigma \cdot \tau = \sum_{i,j} \sigma_{ij}\tau_{ij}$.

The following *constitutive equations* define the relationship between the linearized *strain* tensor ϵ on the one hand, and the *stress* tensor σ together with the
internal variable α of the other hand:

$$\varphi(\sigma, \alpha) \leq 0 \tag{1a}$$

$$\begin{cases} L(\sigma, \alpha)\dot{\sigma} \cdot (\tau - \sigma) + A(\sigma, \alpha)\dot{\alpha} \cdot (\beta - \alpha) \geq \dot{\epsilon} \cdot (\tau - \sigma) \\ \text{for all } (\tau, \beta) \text{ such that } \varphi(\tau, \beta) \leq 0 \end{cases} \tag{1b}$$

The point above a character stands for the derivation with respect to time.

The main *hypotheses* are the following. The *yield function* $\varphi : S \times E \to R$ must
be convex and $\varphi(0, 0) < 0$. The *generalized elastic compliances* L and A satisfy :

$$L(\sigma, \alpha) \in \mathcal{L}(S), \quad A(\sigma, \alpha) \in \mathcal{L}(E), \quad L(\sigma, \alpha)\tau \cdot \tau + A(\sigma, \alpha)\beta \cdot \beta \geq c(\tau \cdot \tau + \beta \cdot \beta)$$

for all $\sigma, \tau \in S$, $\alpha, \beta \in E$, where c is a strictly positive constant. Moreover, L and
A obey to a Lipschitz condition with respect to (σ, α).

The *particular case* where $L(\sigma, \alpha) = L_0$, $A(\sigma, \alpha) = A_0$ for all (σ, α) corresponds to the model of *Generalized Standard Materials* (with a quadratic free
energy) used for hardening problems; see Halphen and Nguyen 1975. In addition,
if $A_0 = 0$ and $\varphi(\sigma, \alpha) = $ for all (σ, α), then conditions (1) give the classical relations
of *perfect Plasticity* ; see Lemaitre and Chaboche 1985.

In general the dependence of L on α considered in relation (1) allows us to
take into account some damage phenomena. Moreover, a difference of behavior
between traction and compression can be described through the fact that L can be
a function of α. In practice for a concrete problem of damage, the definition of the
yield function φ may present significant differences with respect to the classical
Plasticity.

The framework of the modelization (1) was been analysed in Laborde and Michrafy 1991 . In this reference, we also give an application to a problem of damage for *Ceramic-ceramic composite materials.*

2.2. **The boundary-value problem.** The body lies in the domain $\Omega \subset R^3$. We are concerned with the problem of seeking a stress field $\sigma(t, x)$, a displacement field $u(t, x)$ and an internal variable field $\alpha(t, x)$ defined in Ω and verifying :

$$\text{The } constitutive\ relations(1) \text{ in } \Omega, \text{ where } \dot{\epsilon} = \epsilon(\dot{u}), \qquad (2a)$$

$$div\ \sigma = 0 \quad \text{in } \Omega, \qquad (2b)$$

$$\sigma(0, x) = \sigma(x), \quad u(0, x) = u(x), \quad \alpha(0, x) = \alpha(x) \qquad (2d)$$

in $(0, T)$, with the usual following notations:

$$\epsilon_{ij}(v) = \frac{1}{2}\left(\frac{\partial u_i}{\partial x_j} + \frac{\partial u_j}{\partial x_i}\right), \quad (div\sigma)_i = \sum_j \frac{\partial \sigma_{ij}}{\partial x_j},$$

n being the outward unit normal to the boundary Γ of Ω , Γ_0 and Γ_1 a partition of Γ. The *surface forces f* as also the *initial state* σ^0, u^0, α^0 are prescribed such that $\varphi(\sigma^0, \alpha^0) \leq 0$.

Contrary to the classical Plasticity, the natural inner product in the problem (2):

$$(\tau, \beta), \ (\tau^*, \beta^*) \ \rightarrow \ L(\sigma, \alpha)\tau \cdot \tau^* + A(\sigma, \alpha)\beta \cdot \beta^*$$

depends on the state (σ, α), because of the nonlinearity occuring in (1) for modelizing damage. To the knowledge of the author, there is any existence (or uniqueness) result of a solution to the boundary-value problem (2). Let us only indicate a study on hand by Laborde and Petitjean ; see also Laborde and Nguyen 1990 for a connected problem.

3. The equation to solve.

The nonlinear equation under consideration in the present study is now formulated (see relationship (6) below) by approximating the previous boundary-value problem

3.1. **Time discretization.** Let $t_0 = 0 < t_1 < t_2 \ldots < T$ be a subdivision of the interval $[0,T]$. By discretizing with respect to time the original problem (2), we are led to set the problem at step n of finding functions σ^n, u^n, α^n defined in Ω such that:

$$\begin{cases} L_n(\sigma^n - \sigma^{n-1}) \cdot (\tau - \sigma^n) + A_n(\alpha^n - \alpha^{n-1}) \cdot (\beta - \alpha^n) \geq \\ \qquad\qquad\qquad\qquad \geq (\epsilon^n - \epsilon^{n-1}) \cdot (\tau - \sigma^n) \qquad (3a) \\ \text{for all } (\tau, \beta) \in K, \quad \text{with } (\sigma^n, \alpha^n) \in K \end{cases}$$

$$div\sigma^n = 0 \tag{3b}$$

$$u^n = 0 \text{ on } \Gamma_0, \quad \sigma^n n = f^n \text{ on } \Gamma_1 \tag{3c}$$

for $n = 1, 2, \ldots$, using the notations :

$$K = \{(\tau, \beta) : (\tau, \beta) \in S \times E, \ \varphi(\tau, \beta) \le 0\}$$

$$L_n = L(\sigma^{n-1}, \alpha^{n-1}), \quad A_n = A(\sigma^{n-1}, \alpha^{n-1}).$$

The non-linearity connected with the admissiblity condition in the problem (2) (via relationship (1)) is approximated by using an *implicit method* as in Plasticity theory. Let us give an *interpretation* of this step-by-step scheme. At the step n, and for a given $x \in \Omega$, let $(P_n\tau, Q_n\tau)$ be the projection of $(\tau, 0)$ onto K_n in $S \times E$ equipped of the elastic energy norm $(\xi, \gamma) \to L_n\xi \cdot \xi + A_n\gamma \cdot \gamma$. In this definition K_n is a convex set obtained by translating the constraint set : $K_n = K - (\sigma^{n-1}, \alpha^{n-1})$. We notice that the operators P_n, Q_n depend on x, as well as the matrices L_n, A_n and the set K_n. Then, inequality (3a) can be written in the form:

$$\delta\sigma = P_n L_n^{-1}\delta\epsilon, \quad \delta\alpha = Q_n L_n^{-1}\delta\epsilon \tag{4}$$

putting $\delta\sigma = \sigma^n - \sigma^{n-1}, \ldots$.

We have before observed the dependence of the operators L and A on the unknowns (σ, α) in the problem (2) (see (1)). This non-linearity is treated by means of an *explicit* procedure in the present time discretization (see (3a)). It follows that the step-by-step problem (3) is similar to a *Plasticity problem* (precisely, a Hencky problem with strictly positive hardening). So it may be proved an *existence* result for the solution σ^n, u^n, α^n to the problem (3); moreover the pair (σ^n, α^n) is unique. About the mathematical theory of Plasticity, see e.g. Johnson 1978, Suquet 1981, Témam and Strang 1980, Nečas and Hlaváček 1981; see also Moreau 1977 for some closely aspects.

3.2. The finite element problem. Within the context of a finite element method, let resp. H, V and W be the spaces of stress, displacement and internal variable fields. In the present paper the definition of these finite element spaces is not precised by sake of simplicity. Let us only indicate that the definition of V takes into account the boundary condition (3c) on Γ_0.

Let $D \in \mathcal{L}(V, H')$ such that

$$< Dv, \tau) = \int_\Omega \epsilon(v) \cdot \tau dx.$$

We approximate the step-by-step problem (3) by seeking $\delta\sigma \in H$, $\delta u \in V$, $\delta\alpha \in W$ such that we have

$$D \, \delta u = \delta\epsilon \quad \text{with (4)} \tag{5a}$$

$$D^T \delta\sigma = \delta f \tag{5b}$$

where D^T stands for the transposed operator of D, and $\delta f \in V'$ is defined from the prescribed increment of forces at step n.

By eliminating $\delta\sigma$, the discrete problem (5) is reduced to the (finite element) *displacement problem* :

$$F(u) = D^T P L^{-1} D u - f = 0 \tag{6}$$

where the step indice n and the incremental symbol δ are omitted.

In the particular case of the *Elasticity* (say $\varphi \equiv -1$) the operator P equals the identity and (6) becomes the Euler equation for the minimization problem of the quadratic elastic energy. In the general case, the mathematical theory of Plasticity ensures the existence of a solution u to equation (6) (but without having any uniqueness result), see the aforementioned references. The formulation of the displacement problem in Plasticity given by eq.(6) can be found e.g. in Strang et al. 1980.

4. A differentiability result.

The nonlinearity of the mapping F defining eq.(6) comes from the presence of the projection P. In Section 4.1, we begin to examine the question relative to the differentiability of an abstact projection operator. Then we apply this general result for studying F in Section 4.2.

4.1. About the projection operator. Let X be a Banach space and C a nonempty closed convex subset of X. From a notational point of view : $proj_C$ strands for the operator of projection onto C, the *directional derivatives* of a function $g : X \to X$ at point y are denoted:

$$g'(y; z) = \lim_{h \to 0_+} h^{-1}(g(y + hz) - g(y))$$

and $T_C\, y$ is the *tangent cone* to C at $y \in C$, i.e. the set of the elements z such that

$$z = \lim_{h \to 0_+} r_h(y_h - y), \quad r_h > 0, \quad y_h \in C, \quad \lim_{h \to 0_+} y_h = y.$$

We have the following general result, see Zarantonello 1971, Mignot 1976 :

Proposition 1.

For every $y \in C$ the projection onto C admits directional derivatives and

$$(proj_C)'(y; z) = proj_{T_C\, y}\, z \quad \text{for all } \ z \in X.$$

Let us give an important *example* in order to illustrate this property. We consider a subset of an Hilbert space

$$C = \{y : y \in X, \ \psi(y) \le 0\}$$

where the convex function $\psi : X \to R$ is *Gateaux-differentiable*. Let us resp. denote $(.,.)$ and $\|.\|$ the inner product and the associated norm of X, and $\psi'(y)$ the gradient of ψ at y.

Proposition 2.

(i) Let $\psi(y) < 0$; then $(proj_C)'(y; z) = z$ for all $z \in X$.

(ii) Let $\psi(y) = 0$; then

$$(proj_C)'(y; z) = \begin{cases} z & \text{if } (n, z) < 0 \\ z - (n, z)n & \text{if otherwise} \end{cases} \quad where \quad n = \frac{\psi'(y)}{\|\psi'(y)\|}.$$

This property results from Proposition 1 by noting that

$$T_c y = \begin{cases} X & \text{when } \psi(y) < 0 \\ \{z : (n, z) \le 0\} & \text{when } \psi(y) = 0 \end{cases}$$

Proposition 2 can be also formulated as follows. Let $y \in C$ be *given*, then we have

$$(proj_C)'(y; z) = G(z)\, z \quad \text{for all } \ z \in X \tag{7}$$

where $G : X \to \mathcal{L}(X)$ is the piecewise constant operator such that

$$G(z) = \begin{cases} I & \text{if } z \in T_c y \\ I - n \otimes n & \text{if otherwise} \end{cases} \tag{8}$$

(The operator G depends on y).

4.2. **An application.** The definition of F involves the projection onto the translated constraint set K_n which contains the origin; see Section 3.1. Thus, Proposition 1 implies that P has directional derivatives at 0.

First suppose the yield function φ be *differentiable*. Then Proposition 2 ensures that

$$P'(0, \tau) = N(\tau)\, \tau \quad \text{for all} \ \ \tau \in S \tag{9}$$

where $N : S \to \mathcal{L}(S)$ is a piecewise constant operator defined according to the previous process.

Precisely, we deduce from (7) that operators P, Q introduced in Section 3.2 verify

$$(P'(0, \tau), Q'(0, \tau)) = G(0, \tau)\, (\tau, 0)$$

where $G : S \times E \to \mathcal{L}(S \times E)$ is defined as in (8).

This method can be used even if the yield function is *not differentiable*, for instance if

$$\varphi = \max(\varphi_1, \varphi_2) \tag{10}$$

where φ_1, φ_2 are two differentiable convex functions.

Now let us apply the previous result for studying F.

Proposition 3.

The mapping $F : V \to V'$ defined in (7) admits directional derivatives at the origin and

$$F'(0; v) = R(v)\, v \qquad \text{for all } \ v \in V. \tag{11}$$

where

$$R(v) = D^T N(L^{-1} D v) L^{-1} D. \tag{12}$$

This property is deduced from (9) since we have

$$F'(0; v) = D^T P'(0; L^{-1} D v).$$

If $\varphi(\sigma^n, \alpha^n) < 0$ in Ω, note that $N(\tau) = I$ and $R(v)$ is reduced to the usual stiffness matrix of Elasticity :

$$R(v) = D^T L^{-1} D.$$

5. The algorithm.

Now we are in a position to indicate the key aspect of an algorithm constructed from the previous study. For solving the equation $F(u) = 0$ formulated in (6) we consider the sequence $u^{(k)}$ such that

$$R_k(u^{(k+1)} - u^{(k)}) = -F(u^{(k)}) \tag{13}$$

for $k = 0, 1, \ldots$, where u^0 is given.

In this *generalized Newton method*, the matrix R_k is defined from the directional derivative (see(11)) :

$$F'(0; u^{(k)}) = R(u^{(k)})\, u^{(k)}$$

putting

$$R_k = R(u^{(k)}).$$

It follows that $R_k = D^T M_k D$ where

$$M_k = N(L^{-1} D u^{(k)}) L^{-1} \tag{14}$$

owing to (12).

This definition of M_k offers a mathematical viewpoint on the construction of the so-called *"tangent moduli"* matrix used for the popular *"tangent stiffness algorithm"* in engineering; see Bathe 1982, Owen and Hinton.

In practice, the algorithm is generally implemented for plasticity problems involving a very simple projection operator (defined by the so-called Von Mises yield function φ). We are indeed able to apply this method for treating more complicated (non differentiable) functions φ including many internal variables such as in (10).

It seems there is any theoretical result concerning the convergence of the above algorithm. However, experiments show the computationally efficiency of the method for solving the present variational inequality, see a paper in preparation by Laborde and Michrafy.

Finally, let us indicate how the definition of the matrix R_k in the algorithm takes into account the free boundary. For example, consider the finite element method where displacements are approximated by piecewise affine functions continuous on Ω, and stresses or internal variables by piecewise constant functions. Assume that the function φ is differentiable. The matrix R_k is computed by assembling the element stiffness matrices, and to construct the latter we use the matrix M_k introduced in (14). For each element T belonging to a given triangulation of Ω, the definition of M_k on T is decomposed into different cases.

(i). If $\varphi(\sigma^{n-1}, \alpha^{n-1}) < 0$ on T (elastic range), then $T_K(\sigma^{n-1})$ is the whole space and $N(.) = I$. Hence $M_k = L^{-1}$, i.e. M_k is equal to the elastic moduli matrix on T.

(ii). If otherwise $\varphi(\sigma^{n-1}, \alpha^{n-1}) = 0$ on T (anelastic range), then $T_K(\sigma^{n-1}, \alpha^{n-1})$ is the polar half-space to the gradient $(\dfrac{\partial \varphi}{\partial \sigma}, \dfrac{\partial \varphi}{\partial \alpha})$ of φ at $(\sigma^{n-1}, \alpha^{n-1})$. Therefore two cases may occur.

(ii_1) Either $(L^{-1}\epsilon^{(k)}, 0) \notin T_K(\sigma^{n-1}, \alpha^{n-1})$ on T, where $\epsilon^{(k)} = Du^{(k)}$, i.e.

$$\frac{\partial \varphi}{\partial \sigma}(\sigma^{n-1}, \alpha^{n-1}) \cdot (L^{-1}\epsilon^{(k)}) > 0.$$

Then $N(L^{-1}\epsilon^{(k)})$ is obtained by the (effective) projection of $(L^{-1}\epsilon^{(k)}, 0)$ onto $T_K(\sigma^{n-1}, \alpha^{n-1})$. Hence the corresponding matrix M_k is the so-called "*plastic moduli* " matrix, according to the terminology of the Plasticity theory.

(ii_2) Or else $(L^{-1}\epsilon^{(k)}, 0)$ is its own projection onto $T_K(\sigma^{n-1}, \alpha^{n-1})$ and $N(L^{-1}\epsilon^{(k)}) = I$. Hence M_k equals to the elastic moduli matrix on T (the so-called "*elastic unloading*" case).

REFERENCES

BATHE K.J., *Finite element procedures in engineering analysis*, Prentice-Hall, 1982.
GERMAIN P., NGUYEN Q.S. and SUQUET P., Continuum Thermodynamics, *J. Applied Mech.*, 1983, Vol. 50, p.1010-1020.

HALPHEN B. and NGUYEN Q.S., Sur les matériaux standard généralisés, *J. Mécan.*, 1975, Vol. 14, p.39-63

JOHNSON C., On plasticity with hardening, *J.Math Anal. Appli.*, 1978, Vol. 62, p.325-336.

LABORDE P. and NGUYEN Q.S., Etude de l'équation d'évolution des systèmes dissipatifs standards, *Math. Model. Num. Anal.*, 1990, Vol. 24, p.67-84.

LABORDE P. and MICHRAFY A., On general constitutive equations involving damage, *Eur. J. Mech. A/Solids*, 1991, Vol. 10, p. 215-240.

LEMAITRE J. and CHABOCHE J.L., *Mécanique des matériaux solides*, Dunod, Paris, 1985.

MIGNOT F., Controle dans les Inequations variationnelles elliptiques, *J. Funct. Anal.*, 1976, Vol. 22, p. 130-185.

MOREAU J. J., Evolution problem associated with a moving convex set in a Hilbert space, *J. Diff. Equ.* , 1977, Vol. 26, p. 347-374.

NEČAS J. and HLAVAČEK I., *Mathematical theory of elastic and elastic-plastic bodies: an introduction*, Elsevier, New York, 1981.

OWEN D.R.J. and HINTON E., *Finite elements in plasticity: theory and practice*, Pineridge Press, Swansea.

STRANG G., MATTHIES H. and TÉMAM R., Mathematical and computational methods in Plasticity, *Variational Methods in the Mechanics of Solids*, Evanston, 1978, ed. S. NEMAT-NASSER, Pergamon Press, 1980, p. 20-28.

SUQUET P., Sur les équations de la plasticité: existence et régularité des solutions, *J.Mécan.*, 1981, Vol. 21, p.3-39.

TÉMAM R. and STRANG G., Duality and relaxation in the variational problems of plasticity, *J. Mecan.*, 1980, Vol. 19, p. 493-527.

ZARANTONELLO E. Projection on convex sets, *Contribution to non linear Functional Analysis*, Symposium, Madison, 1971, p. 237-424.

Université Paul Sabatier, Laboratoire d'Analyse Numérique
118, route de Narbonne, F-31062 TOULOUSE CEDEX

If you have any concerns about our products,
you can contact us on
ProductSafety@springernature.com

In case Publisher is established outside the EU,
the EU authorized representative is:
Springer Nature Customer Service Center GmbH
Europaplatz 3, 69115 Heidelberg, Germany

Printed by Libri Plureos GmbH
in Hamburg, Germany